Learning ArcGIS Pro

Create, analyze, maintain, and share 2D and 3D maps with the powerful tools of ArcGIS Pro

Tripp Corbin, GISP

BIRMINGHAM - MUMBAI

Learning ArcGIS Pro

First published: November 2015

Production reference: 1261115

Published by Packt Publishing Ltd.
Livery Place
35 Livery Street
Birmingham B3 2PB, UK.

ISBN 978-1-78528-449-6

www.packtpub.com

Credits

Author
Tripp Corbin, GISP

Reviewer
Ian Broad

Commissioning Editor
Akram Hussain

Acquisition Editor
Meeta Rajani

Content Development Editor
Anish Dhurat

Technical Editor
Gaurav Suri

Copy Editors
Dipti Mankame
Jonathan Todd

Project Coordinator
Harshal Ved

Proofreader
Safis Editing

Indexer
Priya Sane

Production Coordinator
Manu Joseph

Cover Work
Manu Joseph

About the Author

Tripp Corbin, GISP is the CEO and a cofounder of eGIS Associates, Inc. He has over 20 years of surveying, mapping, and GIS-related experience. He is recognized as an industry expert with a variety of geospatial software packages, including Esri, Autodesk, and Trimble products. He holds multiple certifications, including Microsoft Certified Professional, Certified Floodplain Manager, Certified GIS Professional, CompTIA Certified Technical Trainer, Esri Certified Enterprise System Design Associate, and Esri Certified Desktop Professional.

During the course of his career, Tripp has assisted many private and public organizations to implement and maintain GIS solutions. Cities and counties have relied on his expertise to help them make the transition from hard copy maps into the digital world of GIS. Private companies, including engineering, surveying, planning, and forestry, often seek Tripp's experience to help them in various projects. Tripp's knowledge of GIS and surveying along with the ability to use multiple software packages has proven to be an invaluable asset.

As a GIS instructor, Tripp has taught students worldwide the power of GIS. He is a GIS instructor for the Institute for Transportation Research and Education at North Carolina State University, the Continuing Studies Center at University of North Alabama, and Davidson County Community College. He has authored many GIS classes on topics ranging from the beginner level, such as Introduction to GIS Fundamentals using ArcGIS to more advanced classes, such as Using AutoCAD in ArcGIS and Performing GIS Analysis with ArcGIS. In addition to teaching and authoring classes, Tripp has overseen the establishment of GIS training programs at these institutions.

Tripp is a very active member of the GIS professional community. He is a past president of Georgia URISA and URISA International Board Member. He currently serves on the GISP Application Review committee and as an At-Large GITA Southeast Board Member. Tripp is continually asked to present and conduct workshops at GIS-related conferences. He has conducted workshops on the GISP Certification, Deploying Mobile Solutions, GIS and Land Surveying, and Using GNSS Technologies at conferences, such as GIS Pro, Georgia Geospatial Conference, North Carolina GIS Conference, Ottawa GIS Conference, and others. Other contributions Tripp has made to the GIS Profession include helping to draft the Geospatial Technology Competency Model that was adopted by the US Department of Labor and providing help to other GIS Professionals around the world on various blogs, lists, and forums. In recognition of his contributions, he has received several awards, including the URISA Exemplary Leadership Award and the Barbara Hirsch Special Service Award.

I would like to thank several people who helped make this book a reality. First, I want to thank Polly Corbin, my wife. Her encouragement and support were invaluable. Thanks to Robert LeClair for helping me to work through software issues to determine whether they were bugs or user error. Finally, thanks to all those in the eGIS Associates' family, who picked up the slack, which allowed me to have the time to work on the book.

About the Reviewer

Ian Broad was first introduced to GIS at Fort Lewis College, his alma mater, located in Durango, CO. He now has nearly 7 years of experience in utility-based GIS, and has had the opportunity to work on many different projects, including those utilizing cutting-edge lidar technology. He's employed by Tillamook People's Utility District, a customer-owned electric company located on the beautiful Oregon Coast, where he provides a combination of GIS and IT expertise and support. Ian also operates Oregon Coast Drones an enterprise providing comprehensive aerial services.

He enjoys Python and JavaScript development, database design, open source solutions, participating on GIS StackExchange and other GIS communities, hiking, and aerial photography.

You can check out his personal website at `http://www.ianbroad.com`, and his latest project at `http://www.oregoncoastdrones.com`.

www.PacktPub.com

Support files, eBooks, discount offers, and more

For support files and downloads related to your book, please visit `www.PacktPub.com`.

Did you know that Packt offers eBook versions of every book published, with PDF and ePub files available? You can upgrade to the eBook version at `www.PacktPub.com` and as a print book customer, you are entitled to a discount on the eBook copy. Get in touch with us at `service@packtpub.com` for more details.

At `www.PacktPub.com`, you can also read a collection of free technical articles, sign up for a range of free newsletters and receive exclusive discounts and offers on Packt books and eBooks.

`https://www2.packtpub.com/books/subscription/packtlib`

Do you need instant solutions to your IT questions? PacktLib is Packt's online digital book library. Here, you can search, access, and read Packt's entire library of books.

Why subscribe?

- Fully searchable across every book published by Packt
- Copy and paste, print, and bookmark content
- On demand and accessible via a web browser

Free access for Packt account holders

If you have an account with Packt at `www.PacktPub.com`, you can use this to access PacktLib today and view 9 entirely free books. Simply use your login credentials for immediate access.

Table of Contents

Preface

Esri's ArcGIS Platform is the premiere GIS solution for those wishing to build a fully functional and scalable GIS. ArcGIS includes many integrated components, such as desktop applications, databases, image and web servers, cloud applications, and data along with mobile and web applications. ArcGIS Pro is Esri's latest addition to this already amazing GIS platform.

ArcGIS Pro is a 64-bit desktop GIS application that makes use of a modern ribbon interface. This means that it has the ability to make full use of modern computer hardware while at the same time presenting users with an easy-to-use and intuitive interface. It contains powerful tools for visualizing and analyzing 2D and 3D data.

This book will take you from software installation to performing geospatial analysis. It is packed with how-tos for a host of commonly performed tasks. You start by learning how to download and install the software, including hardware limitations and recommendations. Then, you are exposed to the new ribbon interface and how its smart design can make finding tools easier. After you are exposed to the new interface, you are taken through the steps of creating a new GIS project to provide quick access to project resources. With a project created, you learn how to construct 2D and 3D maps, including how to add layers, adjust symbology, and control labeling. Next, you learn how to access and use analysis tools to help answer real-world questions. Finally, you will learn how processes can be automated and standardized in ArcGIS Pro using tasks, models, and Python scripts.

This book will provide an invaluable resource for all those seeking to use ArcGIS Pro as their primary GIS application or for those looking to migrate from ArcMap and ArcCatalog.

What this book covers

Chapter 1, Introducing ArcGIS Pro, explains the installation process and requirements of ArcGIS Pro. You will learn how to manage software licenses and the differences between the three license levels. Finally, you will launch ArcGIS Pro and open a project to begin exploring the application.

Chapter 2, Using ArcGIS Pro – Navigating through the Interface, introduces you to ArcGIS Pro's new ribbon-based interface. You will learn how the interface works and the terminology associated with it. You will begin to explore tools that will allow you to access data, maps, and tools within an ArcGIS Pro project.

Chapter 3, Creating and Working with ArcGIS Pro Projects, explains what an ArcGIS Pro project is and how it works. You will learn how to create new projects, add items to a project, and how to use project templates.

Chapter 4, Creating 2D Maps, shows you how to create 2D maps. You will learn how to add layers, control layer display and draw order, and configure different types of layer symbology. You will also learn how to configure labeling for layers along with other layer properties.

Chapter 5, Creating 3D Maps, shows how ArcGIS Pro can be used to create 3D maps. You will examine how to combine 2D and 3D data to create amazing 3D scenes. You will learn how to configure 3D layer symbology to produces realistic-looking views and navigate within a 3D map.

Chapter 6, Creating a Layout, teaches you how to create a layout that contains 2D and/or 3D maps, dynamic legends, scales, and so on. Often the final step of any project is to present your findings and work. In GIS, this is typically done with a map. The layout provides the frame for that map.

Chapter 7, Editing Spatial and Tabular Data, explains what types of data can be edited and outlines the recommended workflows to maintain GIS data using ArcGIS Pro. ArcGIS Pro allows users to edit a wealth of GIS data including both spatial and tabular information.

Chapter 8, Geoprocessing, introduces you to many of the most commonly used tools, where they can be accessed, and what will determine which tools are available to them within ArcGIS Pro. ArcGIS Pro includes a wealth of tools to analyze and manipulate GIS data, which are referred to as geoprocessing tools.

Chapter 9, Creating and Using Tasks, shows readers how they can create tasks for common workflows within their offices to improve efficiency and standardization. ArcGIS Pro Tasks allow GIS managers to standardize processes and workflows.

Chapter 10, Automating Processes with ModelBuilder and Python, introduces readers to the basic concepts and skills needed to create simple models and Python scripts for ArcGIS Pro. ModelBuilder and Python can be used to automate and streamline analysis, conversion, and integration processes within ArcGIS Pro.

Chapter 11, Sharing Your Work, illustrates different methods within ArcGIS Pro to share maps, data, and processes with others, both on your network and off. As more and more people are becoming geospatially savvy, it is increasingly important to be able to share our GIS content with others.

Appendix A, ArcGIS Pro Glossary, provides a quick reference for the terminology associated with ArcGIS Pro.

Appendix B, Chapter Questions and Answers, provides the answers to questions contained in the chapters, so readers can verify the answers they gave for each question.

What you need for this book

- ArcGIS Pro 1.1 or higher (basic, standard, or advanced)
- Web browser
- Internet Access

Who this book is for

This book is for anyone wishing to learn how ArcGIS Pro can be used to create maps and perform geospatial analysis. It will be especially helpful for those who have used ArcMap and ArcCatalog in the past and are looking to migrate to Esri's newest desktop GIS solution. Although previous GIS experience is not required, you must have a good solid foundation using Microsoft Windows and a mouse. It is also helpful if you understand how to manage folders and files within the Microsoft Windows environment.

Conventions

In this book, you will find a number of text styles that distinguish between different kinds of information. Here are some examples of these styles and an explanation of their meaning.

Code words in text, database table names, folder names, filenames, file extensions, pathnames, dummy URLs, user input, and Twitter handles are shown as follows: "Right-click on the `Student` folder you just created and select **Paste**"

A block of code is set as follows:

```
env.workspace = "C:\\student\\IntroArcPro\\Databases\\Trippville_GIS.
gdb"
#Runs Union Geoprocessing tool on 2 Feature classes
arcpy.Union_analysis (["Parcels", "Floodplains"], "Parcels_Floodplain_
Union", "NO_FID", 0.0003)
```

Any command-line input or output is written as follows:

```
msiexec.exe /i <setup staging location>\ArcGISPro.msi ALLUSERS=1
INSTALLDIR="C:\MyArcGISPro\" /qb
```

New terms and **important words** are shown in bold. Words that you see on the screen, for example, in menus or dialog boxes, appear in the text like this: "Click on **Computer** under the **Open** pane. Then, select **Browse** under the **Computer** pane."

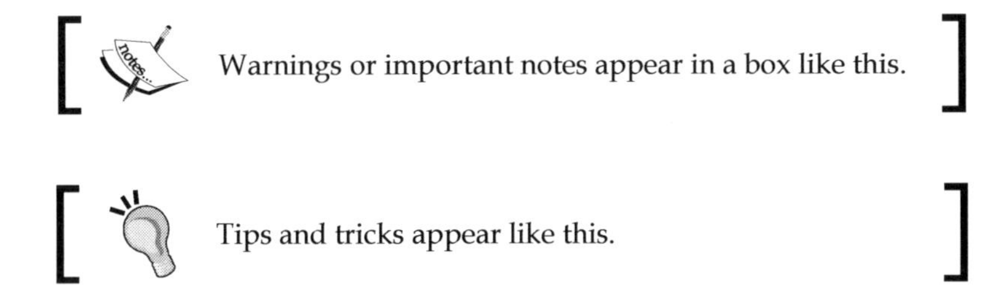

The chapters are sprinkled with questions to help you assess what you have learned and to verify if you have completed exercise tasks properly, for example:

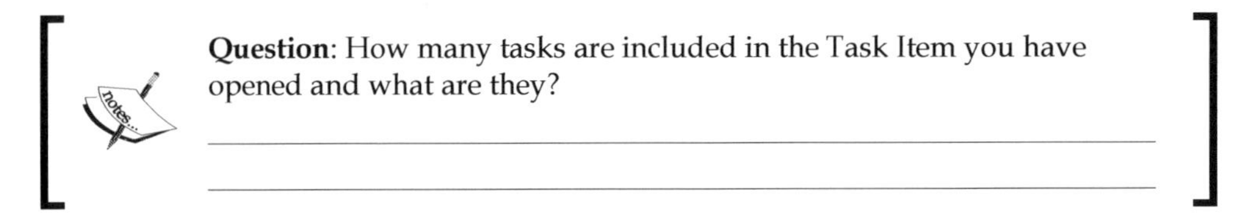

The answers to all the questions are listed chapter-wise in the Appendix towards the end of the book.

Reader feedback

Feedback from our readers is always welcome. Let us know what you think about this book—what you liked or disliked. Reader feedback is important for us as it helps us develop titles that you will really get the most out of.

To send us general feedback, simply e-mail feedback@packtpub.com, and mention the book's title in the subject of your message.

If there is a topic that you have expertise in and you are interested in either writing or contributing to a book, see our author guide at www.packtpub.com/authors.

Customer support

Now that you are the proud owner of a Packt book, we have a number of things to help you to get the most from your purchase.

Downloading the exercise data

You can download the example code files from your account at http://www.packtpub.com for all the Packt Publishing books you have purchased. If you purchased this book elsewhere, you can visit http://www.packtpub.com/support and register to have the files e-mailed directly to you.

Downloading the color images of this book

We also provide you with a PDF file that has color images of the screenshots/diagrams used in this book. The color images will help you better understand the changes in the output. You can download this file from https://www.packtpub.com/sites/default/files/downloads/4496OT_ColoredImages.pdf.

Errata

Although we have taken every care to ensure the accuracy of our content, mistakes do happen. If you find a mistake in one of our books—maybe a mistake in the text or the code—we would be grateful if you could report this to us. By doing so, you can save other readers from frustration and help us improve subsequent versions of this book. If you find any errata, please report them by visiting http://www.packtpub.com/submit-errata, selecting your book, clicking on the **Errata Submission Form** link, and entering the details of your errata. Once your errata are verified, your submission will be accepted and the errata will be uploaded to our website or added to any list of existing errata under the Errata section of that title.

To view the previously submitted errata, go to https://www.packtpub.com/books/content/support and enter the name of the book in the search field. The required information will appear under the **Errata** section.

Piracy

Piracy of copyrighted material on the Internet is an ongoing problem across all media. At Packt, we take the protection of our copyright and licenses very seriously. If you come across any illegal copies of our works in any form on the Internet, please provide us with the location address or website name immediately so that we can pursue a remedy.

Please contact us at `copyright@packtpub.com` with a link to the suspected pirated material.

We appreciate your help in protecting our authors and our ability to bring you valuable content.

Questions

If you have a problem with any aspect of this book, you can contact us at `questions@packtpub.com`, and we will do our best to address the problem.

1
Introducing ArcGIS Pro

Esri's ArcGIS platform has become well-entrenched as the primary solution for GIS professionals seeking to implement a scalable integrated solution, which can start with a single user and grow to support multiple users across various platforms. With **ArcGIS for Desktop**, **ArcGIS for Server**, **ArcGIS Online**, and **ArcGIS for Mobile**, GIS professionals can design and implement a robust GIS solution that provides a wide range of functionality to meet the needs of a growing enterprise, which includes users of various skill levels, requirements, differing platforms, and data formats. However, Esri has not been content to rest on their laurels. They continue to push the GIS envelope. With the release of ArcGIS 10.3, Esri released a new desktop application named **ArcGIS Pro**.

ArcGIS Pro is a completely new application from Esri. It is not just an update to their venerable ArcGIS for Desktop platform. ArcGIS Pro has a modern ribbon interface that has become common in most current desktop applications. It is also designed from the ground up to take full advantage of modern hyperthreaded 64-bit architecture. This greatly improves its performance compared to **ArcMap** and **ArcCatalog**. In this chapter, we will begin to take a look at this new robust and smart interface.

ArcGIS Pro also introduces a new licensing model to Esri users. It makes use of a subscription license, opposed to the standard single use or concurrent use license. Licenses of ArcGIS Pro are tied directly to the annual maintenance of your traditional ArcGIS for Desktop software and managed through ArcGIS Online. You will learn more about this later in the chapter.

ArcGIS Pro allows you to combine 2D and 3D content within a single application. In the past, you would need to utilize multiple applications, such as **ArcMap**, **ArcScene**, and **ArcGlobe**, to view 2D and 3D content. In this chapter, you will examine how you can view both 2D and 3D data within a single ArcGIS Pro project.

In this chapter, you will

- Learn the requirements to install ArcGIS Pro
- Get a general overview of ArcGIS Pro functionality
- Learn how to manage ArcGIS Pro licenses through ArcGIS Online
- Understand the relationship between ArcGIS Pro and ArcGIS for Desktop licenses

Installing ArcGIS Pro

In order to use ArcGIS Pro, you must first install it. To install ArcGIS Pro, you will need to verify that your system meets or exceeds the minimum requirements, download or have access to the install files, and finally have rights to install the software on the computer you are using. We will look at each of these steps next.

It is possible to install and run ArcGIS Pro on a computer that already has ArcGIS for Desktop installed even if ArcGIS for Desktop is an older version. The two applications may exist side by side.

It can also be installed on a computer that has no other Esri products installed.

ArcGIS Pro minimum system requirements

As mentioned earlier, ArcGIS Pro is a 64-bit application that supports hyperthreaded processing. This allows ArcGIS Pro to take full advantage of modern processors, such as the **Intel i7** and RAM larger than 4 GB. The ability of ArcGIS Pro to make use of this increased computing capability also means the computer you run it on needs to have the power to run ArcGIS Pro.

Compared with ArcGIS for Desktop, ArcGIS Pro has higher minimum computer specifications. This includes increased hardware and operating system requirements.

Operating system requirements

ArcGIS Pro requires a 64-bit **operating system** (**OS**). It will not run on a 32-bit OS. ArcGIS Pro currently supports the following operating systems:

- Windows 8.1 Pro and Enterprise
- Windows 8.0 Pro and Enterprise
- Window 7 Ultimate, Professional, and Enterprise (with SP 1)
- Windows Server 2012 R2 Standard and Datacenter

- Windows Server 2012 Standard and Datacenter
- Windows Server 2008 Standard, Enterprise, and Datacenter (with SP 1)

ArcGIS Pro versions 1.0 and 1.1 are currently not supported on Windows 10, Linux or iOS natively. Esri has said ArcGIS Pro version 1.2 will be supported on Windows 10. They have not indicated whether ArcGIS Pro will support Linux or iOS in the future. It is possible to run ArcGIS Pro on a Linux or iOS computer if you create a virtual windows machine.

Hardware requirements

ArcGIS Pro has some hefty hardware requirements. It requires modern processors and large amounts of RAM. It also requires a graphics card powerful enough to display 2D and 3D data. To ensure that ArcGIS Pro runs smoothly, you need to make sure that your computer meets or exceeds the requirements shown here:

- Hyperthreaded dual core or better processor. 2x hyperthreaded hexa-core processor is optimal such as the Intel Core i7-3930K.
- Minimum of 4 GB of RAM with 8 GB of RAM recommended. 16 GB is considered optimal by Esri.
- Minimum of 4 GB of hard disk space to install software. 6 GB or higher is recommend.
- Graphics card that supports DirectX 9 and OpenGL 2.0 with 512 MB of video RAM minimum. Esri recommends a graphics card that supports DirectX 11 and OpenGL 3.2 with 2 GB of video RAM.

Unlike ArcGIS for Desktop, which has limits on the amount of RAM and processor resources it can use, ArcGIS Pro will use all the resources you can throw at it. So, the more processing power and memory your computer has the better ArcGIS Pro will perform.

Other software requirements

ArcGIS Pro is dependent on other applications that must be installed before installation. They provide supporting services which ArcGIS Pro relies on to function. These supporting applications include the following:

- Microsoft .NET Framework 4.5
- Microsoft Internet Explorer 8 or above
- ArcGIS Online or Portal for ArcGIS organizational account

ArcGIS Pro does not require ArcGIS for Desktop in order to be installed on a computer.

Recommendations from the author

I have been working with ArcGIS Pro since it was first released in beta. I have had the opportunity to run it on various computers as it has moved through multiple beta versions to the release of version 1.0 and now 1.1. This firsthand experience, along with my experience running ArcGIS for Desktop since it was first released, has led me to several recommendations when it comes to a system to run ArcGIS Pro.

Based on my experience, I recommend the following specifications when purchasing a computer to run ArcGIS Pro:

- Make sure that you have a graphics card with its own dedicated **Graphics Processing Unit** (**GPU**). ArcGIS Pro is a graphics-intensive program. Every time you pan, zoom, or add a new layer, you will be taxing the graphics capability of your computer. A dedicated GPU will allow your computer to handle this load by processing graphic rendering requests without burdening your computer's CPU and RAM. If your computer uses integrated video, then the computer's CPU and system RAM are used to handle all processing requests including graphic rendering. ArcGIS Pro is both a processor and graphics-intensive application due to the 2D and 3D maps it creates and the analysis it performs. This can put a tremendous load on your computer's resources. Having the dedicated GPU to handle the graphics rendering load on your computer will greatly improve the performance of ArcGIS Pro and your user experience.

  To get a dedicated GPU usually requires your computer to have a separate video card or adaptor. Computers with integrated video typically do not have a dedicated GPU but instead use the CPU to handle the graphics processing. This will slow your computer down. A separate video card also has a dedicated video RAM, which will also enhance your computer's ability to run ArcGIS Pro.

- Use a solid state drive. Solid state drives are incredibly fast at accessing and storing data. They are almost as fast as RAM and are much faster than even the fastest hard disk drives. ArcGIS Pro will run at lightning speeds when installed on a solid state drive.

- There is no such thing as too much RAM. The more RAM your computer has, the better ArcGIS Pro will perform, especially if you don't have a solid state drive.
- If you are going to be doing a lot of analysis or editing and cannot get a solid state drive, try using a **raid** system to improve performance. Raid systems utilize multiple hard drives to store data. They can be configured in multiple ways. A strip set-based raid, such as **RAID 0**, **RAID 3**, or **RAID 5**, provides the best performance. By storing the data across multiple drives, the computer can access the requested data from each one at the same time. This creates a multilane highway for your data to travel along. A raid system will require at least two hard drives and often a separate controller card.
- Run two or more monitors. ArcGIS Pro, like ArcGIS for Desktop, has multiple windows. Being able to display multiple windows at one time will increase your production. It can also allow you to have multiple applications open at one time. While I have not attempted to measure the increased production with ArcGIS Pro, I do know that it increased the production of my team using ArcGIS for Desktop by 10 percent to 15 percent.

All these recommendations can also be applied for those running other applications, such as ArcGIS for Desktop, AutoCAD, MicroStation, Photoshop, and other graphics-intensive applications. I have run all these applications and can say that based on those experiences, any of these recommendations will help them run better.

Downloading the install files

In order to install ArcGIS Pro, you will need to download the install files from the **My Esri** website at `https://my.esri.com/`. Once there you will need to log in using your Esri global account. You will need to make sure that your global account is linked to your organization's Esri customer account in order to do this.

Once you are logged in to **My Esri**, you will need to click on the **My Organizations** tab and select the **Downloads** option, as shown in the following screenshot:

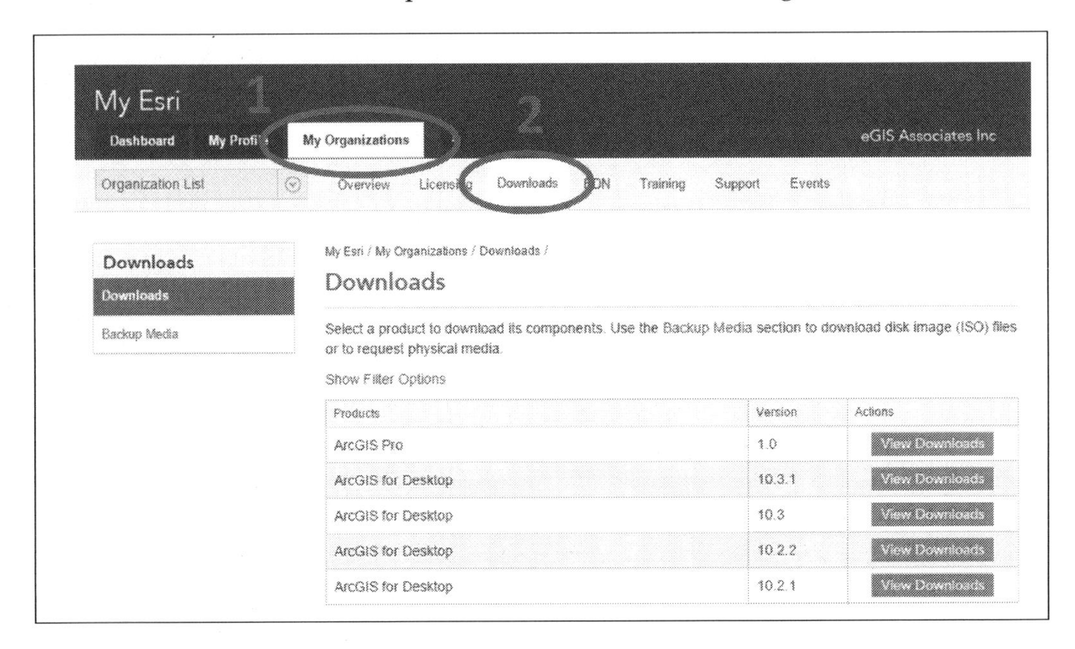

If you are not the administrator for your organizations, Esri account, you might not see the **Downloads** tab. If this is the case, you will need to request permissions from Esri to view the **Downloads** tab. To do this, you need to go to **My Organizations** and then the **Overview** tab.

Then, you need to click on **Request Permissions** and fill out the online form, as shown in the following image:

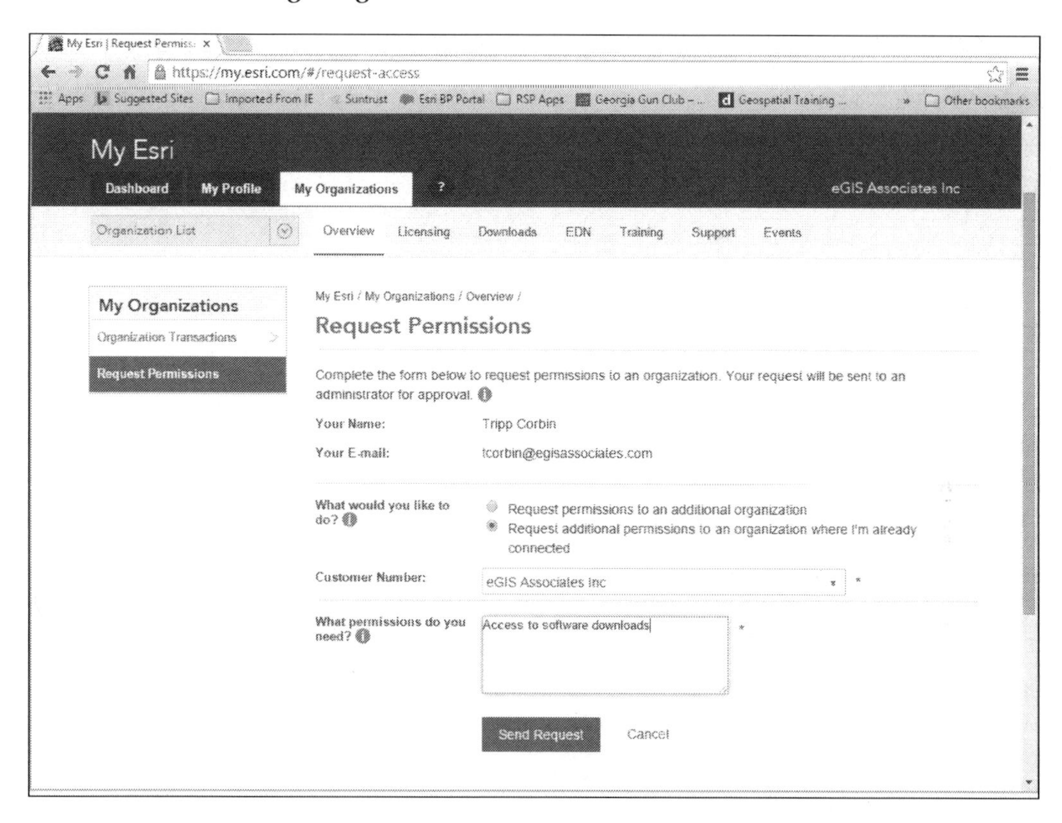

Once you get to the **Downloads** page, you will need to locate ArcGIS Pro and click on **View Downloads**. This will take you to the page that allows you to download several files, which you will use to install ArcGIS Pro and supporting applications. You can also check system requirements, see additional products, and select language packs:

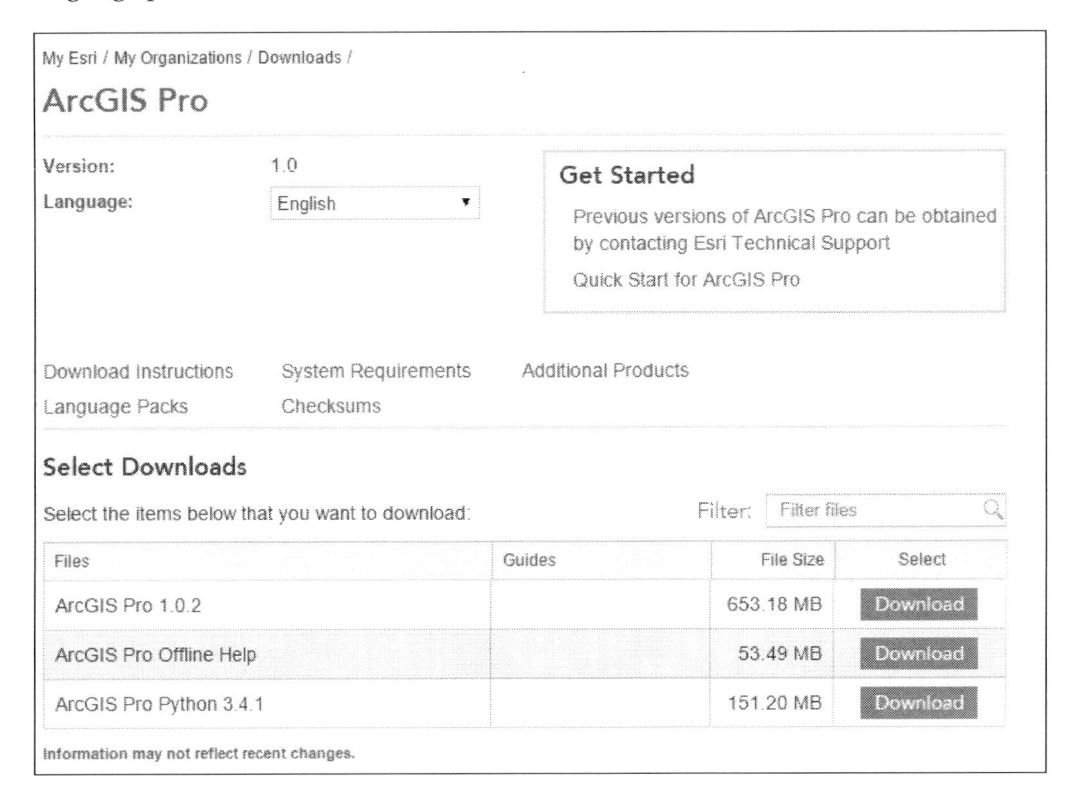

The ArcGIS Pro download is the primary install file for the main ArcGIS Pro application. You must download this file if you wish to install ArcGIS Pro. The other two files are optional but recommended.

The **ArcGIS Pro Offline Help** option will install help files on your local machine, so you can access them without having an internet connection. By default, ArcGIS Pro accesses help information located on the ArcGIS Pro website, `http://pro.arcgis.com`. This means that you will always see the most current help documentation for the version you are using. You will need to download and install this file if you will be using ArcGIS Pro in a disconnected environment and think you might need to access the **Help** documentation.

By default, ArcGIS Pro allows you to run Python scripts inside the application without the need to download additional resources. However, if you wish to create scripts that will run outside ArcGIS Pro, that interact with other applications, or can be scheduled to run automatically, you will need to download the Python install files from the **My Esri** website.

If you cannot access your organization's Esri customer account through the **My Esri** website, you can request a 60-day trial from the Esri website at `http://www.esri.com/software/arcgis/arcgis-for-desktop/free-trial`. This free trial includes ArcGIS for Desktop, ArcGIS Extensions, ArcGIS Online, and ArcGIS Pro. Once you activate your trial, you will be sent instructions to download the install files for ArcGIS Pro in addition to other applications included with the trial. These install files are the same ones that are available through the **My Esri** website linked to your organization.

Installing ArcGIS Pro

Now that you have downloaded the install files, you are ready to begin installing ArcGIS Pro. First, you will need to make sure that you have sufficient rights to install new software on your computer. If you do not, you will need to contact your IT department to see if they can provide assistance installing ArcGIS Pro. Then, perform the following steps:

1. To install ArcGIS Pro, click on the **ArcGIS Pro** file you downloaded. This will unzip the install files and start the installation process.
2. Once the install begins, you will first be asked to review and accept the license agreement from Esri. If you wish to install the software, you must accept the license.
3. Next, you will have to choose who will have access to the application, all users or just the current user. If you want to install such that all users can use ArcGIS Pro, the user installing the software must have full administrative rights.
4. Then, you will choose the install location. By default, ArcGIS Pro will be installed into `C:\Program Files\ArcGIS\Pro\`. It is recommended that you use the default location to avoid issues that could cause problems when running ArcGIS Pro after the installation.
5. Finally, you will be asked if you wish to take part in the **Esri User Experience Improvement** (**EUEI**) program. This will send Esri information about system crashes and other use information automatically if you choose to participate.

Once you have run the ArcGIS Pro installation and it is completed, you can install the **ArcGIS Pro Offline Help** and Python if needed. This installs in a similar manner as the main ArcGIS Pro.

Conducting a silent install for IT departments

ArcGIS Pro also supports a **silent install**, which IT staff can use to automate the install of ArcGIS Pro throughout an enterprise. This is done through the use of command line. To conduct a silent install of ArcGIS Pro for all users, use the following command line syntax:

```
msiexec.exe /i <setup staging location>\ArcGISPro.msi ALLUSERS=1
INSTALLDIR="C:\MyArcGISPro\" /qb
```

The setup staging location is the location of the install files, which you downloaded and unpackaged. This can be a mapped drive or a **Universal Naming Convention** (**UNC**) path. The following are parameter settings, which can be used in the command line.

- `INSTALLDIR`: This parameter is not required. If it is not included in the command line for the silent install, ArcGIS Pro will install to the default location of `C:\Program Files\ArcGIS\Pro\` if all users are specified.

  If a current user is specified it will install to `C:\Users\%UserProfile%\AppData\Local\ArcGIS\Pro`.

- `AllUSERS`: This parameter determines which users will be able to start and access the ArcGIS Pro application on the installed computer.
  - `AllUSERS: 1` equals all users will have access to ArcGIS Pro
  - `AllUSERS: 2` equals current user will have access to ArcGIS Pro
- `ENABLEEUEI`: This parameter controls the user's participation in the EUEI program. Participation in this program will send application usage and crash information to Esri automatically.
  - `ENABLEEUEI: 0` disables participation in the program
  - `ENABLEEEUEI: 1` enables participation in the program
- `qb` or `qr` or `qn`: This parameter controls the display of the UI when the install is being run:
  - `qb` equals show basic user interface during install
  - `qr` equals show reduced user interface during install
  - `qn` equals show no user interface during install

All these parameters are case sensitive.

Managing and assigning ArcGIS Pro licenses

Unlike ArcGIS for Desktop, ArcGIS Pro does not use traditional single use licenses or rely on a third-party license manager to manage concurrent use licenses. ArcGIS Pro licenses are managed through ArcGIS Online or Portal for ArcGIS.

The number and level of ArcGIS Pro licenses

Your ArcGIS Pro licenses are determined based on the number and level of the ArcGIS for Desktop licenses you or your organization own and have under current maintenance. So, if you have two licenses for ArcGIS for Desktop advanced, four licenses of ArcGIS for Desktop standard and 10 Licenses of ArcGIS for Desktop basic, you will have the same number and level of ArcGIS Pro licenses, which you can assign to users.

ArcGIS Pro has three license levels: basic, standard, and advanced. The functionality of the different license levels for ArcGIS Pro is very similar to that for the equivalent ArcGIS for Desktop license level. The following are a few of the functionality differences between the three license levels:

	Basic	**Standard**	**Advanced**
Visualize spatial and tabular data	Yes	Yes	Yes
Edit Shapefiles	Yes	Yes	Yes
Edit personal or file geodatabase	Yes	Yes	Yes
Edit workgroup or enterprise geodatabase	No	Yes	Yes
Perform spatial and attribute queries	Yes	Yes	Yes
Create and use geodatabase topology	No	Yes	Yes
Create and use geometric networks	No	Yes	Yes
Create, manage and update relationship classes	No	Yes	Yes
Perform overlay analysis	Limited to union and intersect	Limited to union and intersect	Yes
Perform proximity analysis	Limited to buffer and multi-ring buffer	Limited to buffer and multi-ring buffer	Yes

To download a complete functionality matrix, go to Esri's product webpage at `http://www.esri.com/software/arcgis/arcgis-for-desktop/pricing`.

ArcGIS Pro also has extensions like ArcGIS for Desktop. Licenses for these extensions are also matched with the extensions for ArcGIS for Desktop. So, if you have one license of Spatial Analyst extension, you will also have one Spatial Analyst for ArcGIS Pro extension license.

You cannot purchase additional ArcGIS Pro licenses independently. If you wish to have additional licenses, you must purchase new ArcGIS for Desktop licenses at the level you need to provide the functionality required.

ArcGIS Pro uses a subscription licensing model that is tied to your annual ArcGIS for Desktop software maintenance. If you do not stay under a current maintenance agreement with Esri, you will lose access to ArcGIS Pro.

Managing ArcGIS Pro licenses

As we mentioned earlier, ArcGIS Pro licenses are managed through ArcGIS Online or Portal for ArcGIS. In order to assign or manage licenses, you must be designated as an administrator. If you are, you can assign ArcGIS Pro and extensions licenses to named users.

To manage or assign licenses you must perform the following steps:

1. Login to ArcGIS Online (`www.argis.com`) or Portal for ArcGIS.
2. Once logged in, you will need to click on **Manage Licenses,** as shown in the following screenshot:

3. From the **Manage Licenses** page, you can then assign ArcGIS Pro and extension licenses to named users within your organization's ArcGIS Online account or Portal for ArcGIS by clicking on the link located under **Licensed For**:

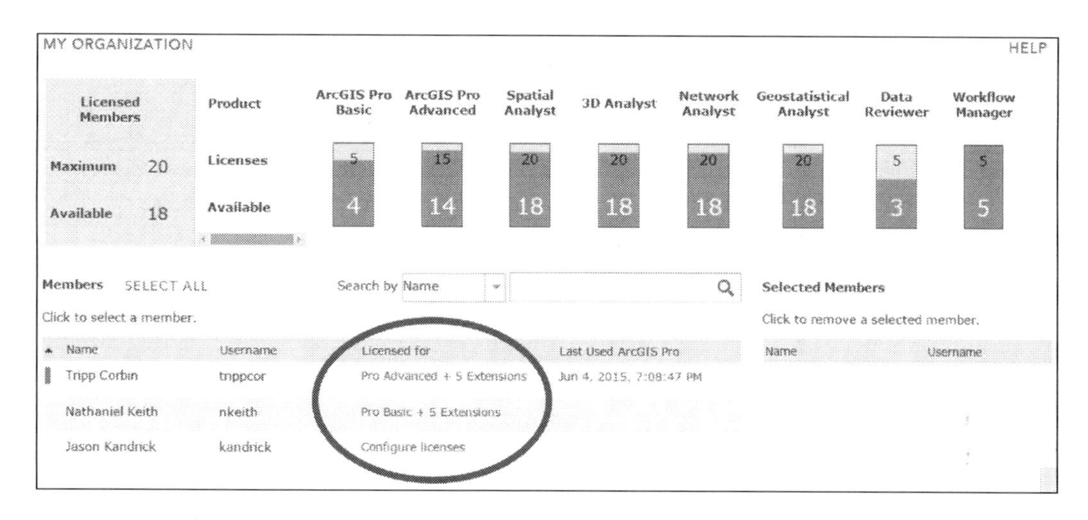

4. After clicking on the link, you are taken to a page that allows you to assign the ArcGIS Pro level and extensions to the user. To assign an ArcGIS Pro license to the user, simply click on the radial button next to the level you wish to assign. At a minimum, you will need to have basic in order for ArcGIS Pro to run on your computer:

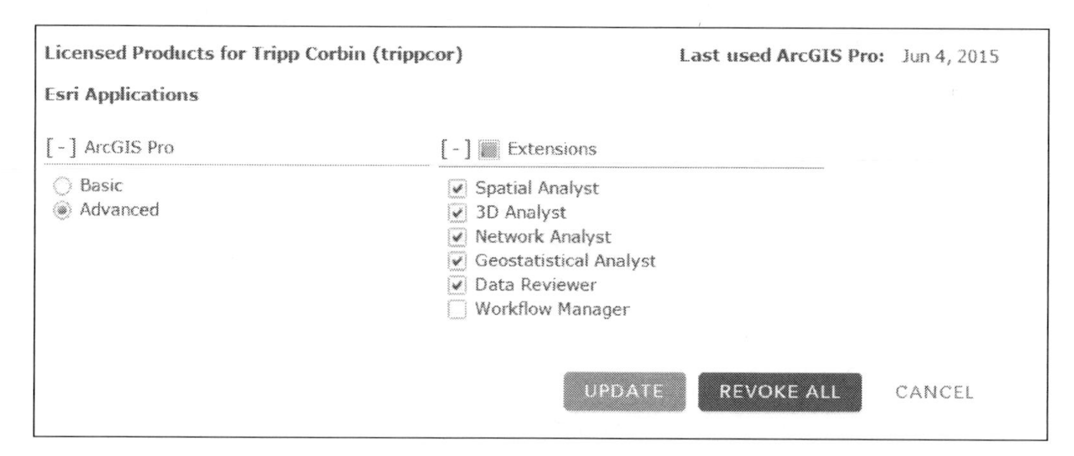

As you can see in the preceding figure, **Tripp Corbin** has been assigned an **Advanced** ArcGIS Pro license along with the **Spatial Analyst**, **3D Analyst**, **Network Analyst**, **Geostatistical Analyst**, and **Data Reviewer** extensions.

If ArcGIS for Desktop is installed on the user's computer, the ArcGIS Pro license level does not have to match the ArcGIS for Desktop license level, the user may be running. These are independent from one another.

As an administrator, you can change the ArcGIS Pro licenses and extensions assigned to users as needed. You can assign licenses as new users are added or revoke licenses as older users are deactivated or removed.

Launching ArcGIS Pro

Now that you have installed ArcGIS Pro and have a license, it is time to launch ArcGIS Pro for the first time. You need to ensure that you are connected to the Internet because when you launch ArcGIS Pro, it will need to connect to ArcGIS Online or Portal for ArcGIS to make sure that you have a valid license.

Let's open ArcGIS Pro:

1. Open ArcGIS Pro. How you do this will depend on your OS and whether you have added shortcuts to your desktop or taskbar.
2. Once you start ArcGIS Pro, you will need to sign in to your ArcGIS Online or Portal account. This allows ArcGIS Pro to verify your license. You do have the option to allow ArcGIS Pro to remember your login credentials for ArcGIS Online, so you will not have to log in every time you launch ArcGIS Pro.

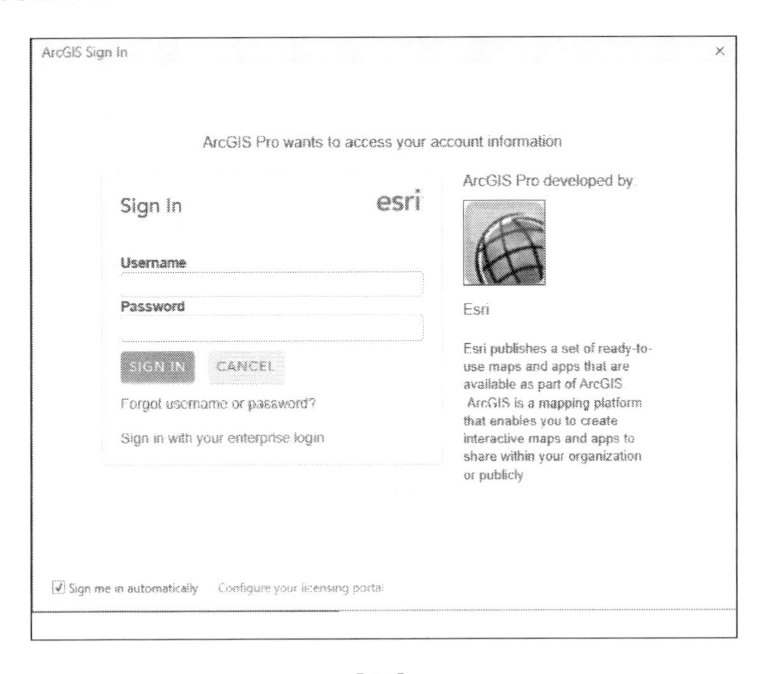

After ArcGIS Pro verifies your login and license, it will take you to the window that allows you to start a new project or open an existing project. You will now open an existing project.

Downloading and installing exercise data

Before you can proceed further in the book and complete the exercises, you will need to download and install the exercise data. To do this, follow these steps:

1. Download the exercise data from *Packt Publishing*. In most cases, this will be downloaded to your `Downloads` folder unless you specify a different location.
2. Open **Windows Explorer** or **File Explorer** depending on which OS you are using. If you are using Windows 7 or 8.1, you will see an icon located on your task bar, which looks like a small file folder in a holder. This will open **File Explorer**.
3. Navigate to the location you downloaded the exercise data to. If you downloaded the data to the standard windows default location, you should be able to click on the **Downloads** option under **Favorites** in the tree located in the left-hand side of the **File Explorer** interface.
4. Double-click on the `LearningArcGISPro.zip` file.

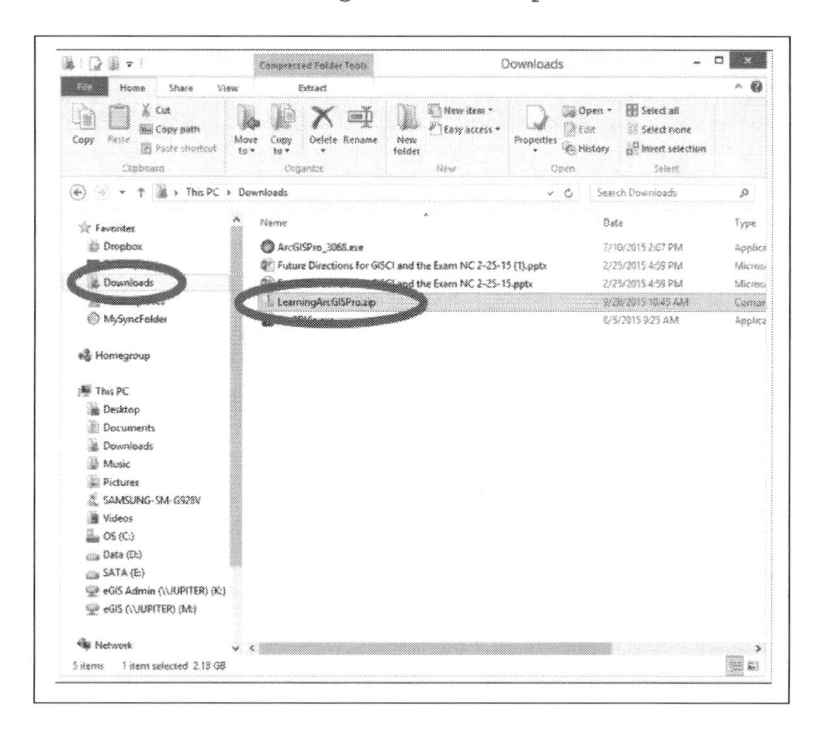

5. Right-click on the **IntroArcGISPro** folder. Then, select **Copy**.
6. In **File Explorer**, now navigate to your `C:\ drive` often named **Local Disk** or **OS**. The `C:\ drive` may be found under this PC in the tree on the left-hand side of the **File Explorer** interface.
7. Right-click on the `C:\ drive` and select **New | Folder**.
8. Name the new folder `Student`.
9. Right-click on the `Student` folder you just created and select **Paste**. This will copy the **IntroArcGISPro** folder to the `Student` folder you just created.
10. Close **File Explorer** once the copy is complete.

Opening an existing ArcGIS Pro project

ArcGIS Pro makes use of projects that can contain 2D and 3D data. You will now open an existing project and begin your first journey into ArcGIS Pro to help verify the successful install of ArcGIS Pro. You will get an opportunity to explore 2D and 3D maps, a layout, and more:

1. In the ArcGIS Pro Open Project window, click on **Open another project** option as shown in the following screenshot:

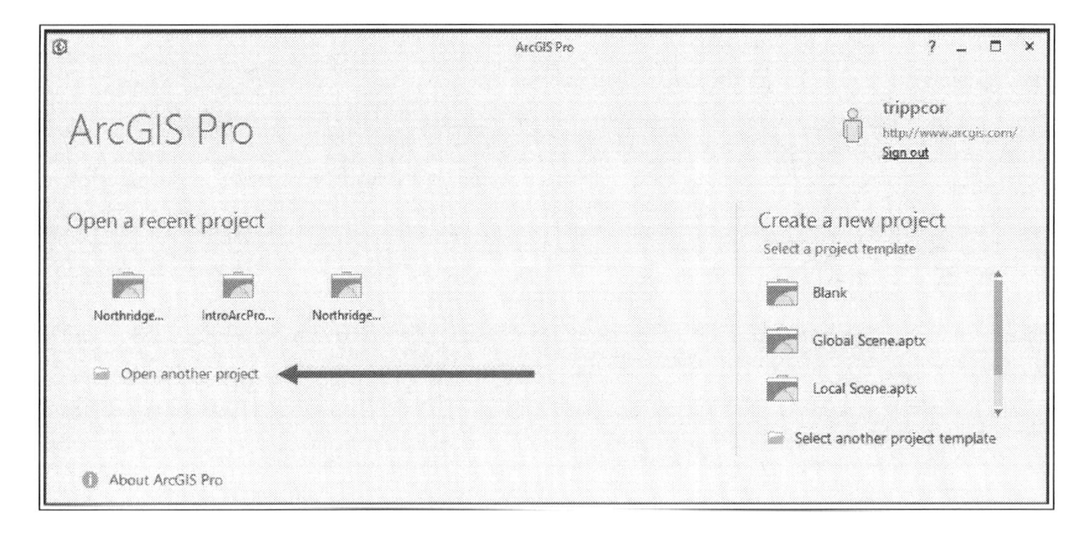

2. Click on **Computer** under the **Open** pane. Then, select **Browse** under the **Computer** pane.
3. Navigate to `C:\Student\IntroArcPro\Chapter1\IntroArcPro_Chapter1` or the location you installed the exercise data and select `IntroArcPro_Chapter1.aprx`.

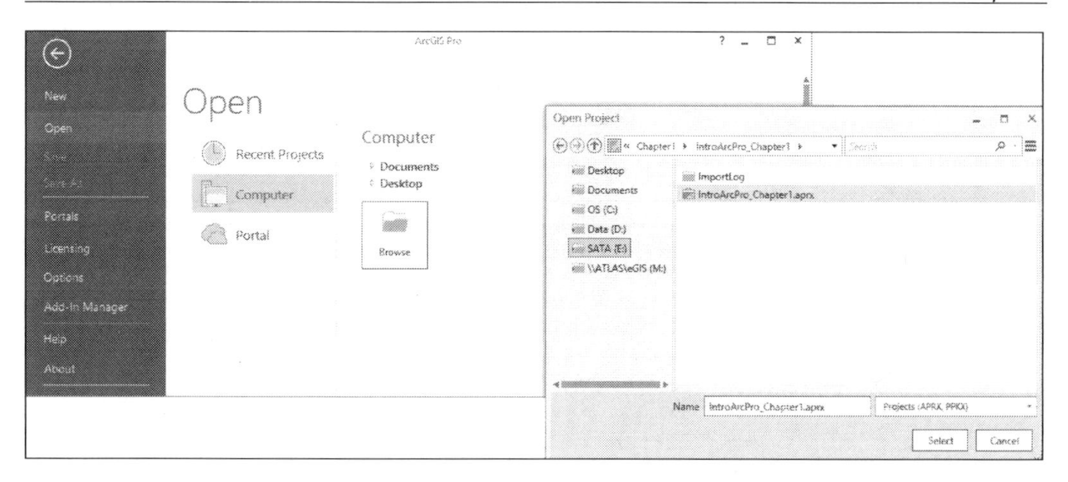

If the project opens successfully, you will see two map views located side by side showing a granite quarry. One is a traditional 2D map, and the other is a 3D map. These two views have been linked together, so when you zoom or pan in one, the other updates to match.

4. Click on the **Explore** tool.
5. Move your mouse pointer to one of the two map views and click your left mouse button and drag it to pan within the view. As you pan in one view, watch what happens in the other view.

> **Question**: What happens when you use the **Explore** tool within the views?
>
> ____
>
> ____

As you can see, the two views are linked together. This keeps them in sync with one another as you move between them. This is one of the functions you will find in ArcGIS Pro. You do not have to link views, but it can be useful. You will learn how to configure view linking in a later chapter.

6. Using the **Explore** tool once again, click on a parcel in either view.

> **Question**: What happens when you click on a parcel in one of the map views?
>
> ____
>
> ____

As you have now experienced, the **Explore** tool not only pans and zooms within the map views but will also retrieve information about the features in the map. Feel free to continue to investigate the ArcGIS Pro project and some of its functionality. When you are finished, close ArcGIS Pro without saving the project.

Summary

You successfully installed ArcGIS Pro. This new 64-bit hyperthreaded application from Esri provides an effective tool to visualize a range of data, including 2D and 3D. This increased capability does require a greater amount of system resources to run effectively compared with the earlier ArcGIS applications.

ArcGIS Pro makes uses of a new licensing model. It uses a subscription model as opposed to traditional single user or concurrent licenses. The ArcGIS Pro licenses are tied to the annual maintenance for your traditional ArcGIS for Desktop licenses and are managed through your organization's ArcGIS Online account or Portal for ArcGIS.

2

Using ArcGIS Pro – Navigating through the Interface

Now that you have successfully installed ArcGIS Pro, it is time to begin examining how it works. ArcGIS Pro is the first Esri desktop product to make use of a ribbon interface, which has become common in many current desktop applications, such as Microsoft Word, Excel, and AutoCAD. The new ribbons make finding and accessing tools and functionality quicker and easier than the traditional drop-down menu and toolbar interface used by ArcGIS for Desktop.

So, you may ask yourself how the new ribbon style interface makes accessing tools and functionality quicker. This new interface is smart. It will present you with tools and information relevant to your current operation. When working in a layout, tasks associated with creating and printing a layout are presented. If you switch your focus to performing analysis, then a new series of menu options will be presented automatically that allows you to access analysis tools.

This automatic behavior is referred to as **contextual menus** or **tabs**. These smart ribbons present you with the tools you need when you need them. When you move on to other tasks, they automatically update accordingly. This not only puts the tools you need at your fingertips, but it also removes the need to have a large number of toolbars and windows open at one time, thus reducing screen clutter and opening up real-estate for other uses.

The interface can be further customized to meet your specific needs. You can group tools you use most commonly, so they are readily available at all times. You can make changes to the **Quick Access Toolbar** located at the top of ArcGIS Pro. You can also share these customizations with others in your organization. So, ArcGIS Pro can be even easier to use.

In this chapter, you will:

- Learn terminology associated with ArcGIS Pro ribbon menus
- Work with the ribbon menus and understand how the contextual tabs work
- Locate several commonly used tools in the ArcGIS Pro interface
- Learn how to move between maps, scenes, and layouts
- Learn how to customize the ArcGIS Pro interface

Understanding the new interface terminology

Anyone who has been working with GIS for any length of time knows that GIS has its own language. As your experience with GIS grows, this new language becomes second nature. You find yourself using words such as topology, projection, and buffer in conversations without realizing that you are doing it. While you may completely understand what you are saying non-GIS users may not. The new ArcGIS Pro ribbon interface will require you to expand your GIS dictionary.

The new interface used by ArcGIS Pro has a whole new set of words and phrases associated with it. In order to successfully use, navigate, and understand this interface, you need to make sure that you know what some of these are and what they mean.

The ArcGIS Pro interface terminology

The following table gives the definition of various terms used in ArcGIS Pro:

ArcGIS Pro terms	Definition
Ribbon	This is the rectangular menu area across the top of ArcGIS Pro. The ribbon is divided into tabs, which group related tools and functions. Some tabs are consistent, whereas others are contextual, which means that they will change depending on tasks being performed.

ArcGIS Pro terms	Definition
Tab	This is a collection of related tools and functions accessed via the ribbon, and it is similar to toolbars in ArcGIS for Desktop. Tabs can be core or contextual. Core tabs are those that stay constant on the ribbon such as project and share. Contextual tabs will change and appear depending on the tasks users are currently performing. Examples of contextual menus include **MAP**, **LAYOUT**, **APPEARANCE**, and **LABELING**.
Pane	This is a dockable window that allows users to access information or tools. It is also similar to the dockable windows in ArcGIS for Desktop, such as **CATALOG**, **SEARCH**, **ArcToolbox**, and **TABLE** windows. Like the windows in ArcGIS for Desktop, panes can be opened and closed, pinned in place or set to auto hide.
Group on a tab	This is a subset of related tools located within a tab. An example of a group on a tab is **LAYER**, which contains multiple tools in order to add layers to a map and is located on the **MAP** tab.
Quick Access Toolbar	This is a collection of shortcuts to commonly used tools and commands. The Quick Access Toolbar is typically located in the top left-hand corner of the ArcGIS Pro interface though the position can be changed. Similar to the taskbar in Windows.
View	This is a window that allows you to view data and perform tasks. A view might be a 2D map, a 3D scene, a layout, or some other visualization of your data. This is similar to the **Data** and **Layout** views in ArcGIS for Desktop. However, in ArcGIS Pro, you can have multiple views and see them displayed at the same time.

These are some of the key terms you will need to understand as you begin to use ArcGIS Pro and navigate through its interface. As you start using the interface, I believe that you will find it much more intuitive than the one used by ArcGIS for Desktop, which hides much of its functionality through a myriad of toolbars and menus.

Using the interface

Now that you understand the key terminology associated with the new interface, it is time to start using it. You will start with using some of the tools in the core tabs on the ribbon. Then, you work with different views and the contextual tabs associated with them. Finally, you will investigate some of the panes.

Navigating the ribbon

You will now begin to investigate the ribbon. It is the rectangular area located at the top of ArcGIS Pro. It includes a series of tabs, group tabs, and tools as shown in the following screenshot:

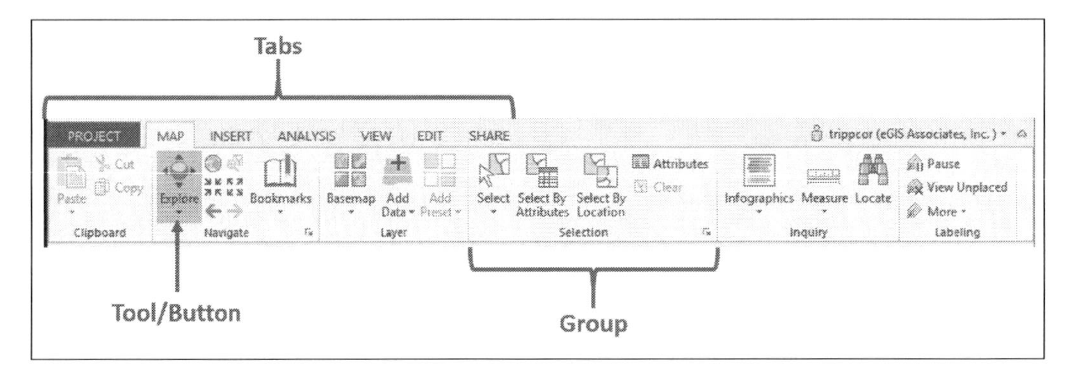

When you first open a new project, you will typically see the **PROJECT**, **MAP**, **INSERT**, **ANALYSIS**, **VIEW**, **EDIT**, and **SHARE** tabs on the ribbon. Each tab contains tools associated with the name of the tab. For example, the **MAP** tab includes tools for navigating within the map, adding layers, selecting features, and accessing information about the features in the map.

Let's now begin taking a closer look at a couple of the most used tabs and some of the key tools located within that tab. You will look at others as you move through this book. You will start with the **PROJECT** tab.

The PROJECT tab

The **PROJECT** tab provides tools needed to manage the project and configure options. This tab allows you to create new projects, open existing projects, save your current project, and make a copy of your current project. You can also connect to different ArcGIS Online accounts or Portals for ArcGIS from this tab.

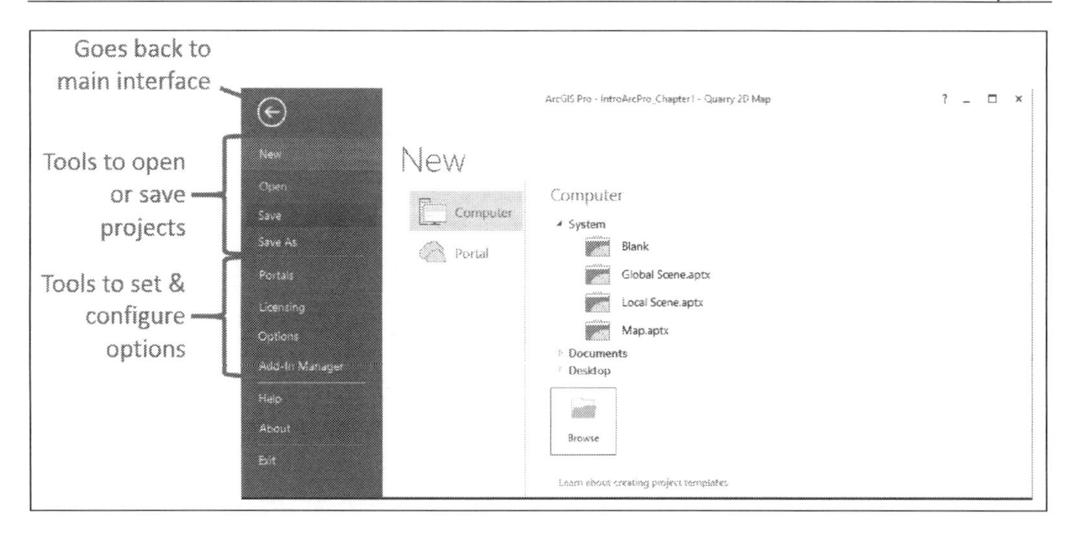

New

The **New** tool will open a brand new project using the template you select. The template controls what is included within the project just as a map document template did within **ArcMap**. The default project templates included with ArcGIS Pro are **Blank**, **Global Scene.aptx**, **Local Scene.aptx**, and **Map.aptx**.

The following table gives their description:

Template	Description
Blank	This starts a new empty project. The new project will not contain any maps, scenes, or layouts. It will be a blank canvas, so you can build the project completely from scratch.
Global Scene.aptx	This starts a new project with a single 3D scene built on the entire world. This is similar to a new blank globe created with ArcGlobe, which is part of the 3D Analyst extension for ArcGIS for Desktop. A new empty project geodatabase will also be created within the folder containing the new project file.
Local Scene.aptx	This starts a new project with a single 3D scene, which is contained in a localized area. This is similar to scenes created with ArcScene including the 3D Analyst extension for ArcGIS for Desktop. It also creates a new blank project geodatabase within the folder containing the project file.
Map.aptx	This starts a new project containing a single 2D map and creates a blank project geodatabase in the same folder containing the project file.

Once you create and open a project, you can add whatever maps, scenes, or layouts you wish regardless of which template you start with. These templates just provide you with a helping hand. You can create your own custom templates as well. You will learn how to do that in a later chapter.

Save

The **Save** tool saves your project including any changes you have made to maps, layer properties, and layouts. This does not save edits you have made to features in your map. Edits you make to create new features or update existing features must be saved using a **Save Edits** tool located on the **EDIT** tab.

Save As

The **Save As** tool creates a copy of your project file with a new name. It does not create a copy of the data referenced by the original project. The new project will continue to point back to the same data sources used by the original project it was created from.

Portal

The **Portal** tool allows you to manage which portals you are connected to. This can include connection to ArcGIS Online or Portal for ArcGIS. The **primary portal** is used by ArcGIS Pro to determine if you have a valid license and what level and extensions you have the ability to use. Other portals can be used to access data, services, and basemaps, which can be used to create maps and perform analysis within ArcGIS Pro supplementing your own standard datasets.

Licensing

The **Licensing** tool allows you to see what ArcGIS Pro license level and extensions you have been granted. Once you know what license level and extensions are available, you can then determine what functions you will be able to perform in ArcGIS Pro. Remember that the ArcGIS Pro license level and extensions assigned to you do not have to be the same as the ArcGIS for Desktop licenses you use.

After your initial login to use ArcGIS Pro, you do have the option to use your license offline. This checks out the license to the computer you are currently using and no longer requires you to be logged in to ArcGIS Online or Portal for ArcGIS to use ArcGIS Pro. Do be careful when using this option. If the computer is lost or fails while the license is checked out for offline use, there is no easy way to get it back. You will need to contact Esri Support and have them resolve the issue.

Options

The **Options** tool allows you to set various ArcGIS Pro user options giving you the ability to customize it. You can set different options for the specific project you are working in, such as the units or for the application in general such as a default basemap for all new maps or scenes added to a project. As you explore each option, you can click on the **Learn more about** link located at the bottom of the window for more information about the settings associated with that option.

The MAP tab

The **MAP** tab on the ribbon provides access to tools used to work with both 2D and 3D maps. From this tab, you can add new layers to your map, select features, change your basemap, and so on. This tab combines functionality that was found on the **Standard and Tools** toolbars in ArcGIS for Desktop.

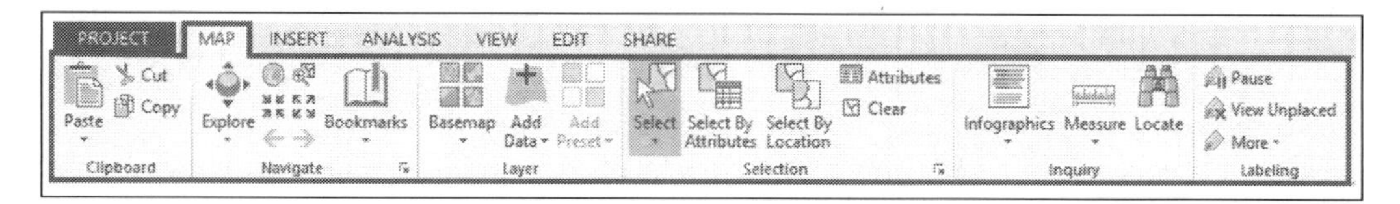

The **MAP** tab contains six groups: **Clipboard**, **Navigate**, **Layer**, **Selection**, **Inquiry**, and **Labeling**. Each group then contains tools associated with the name of the group. For example, the **Selection** group tab includes several tools that allow you to select features in the map using various methods. You can select features directly from the map or based on specific attribute values or based on the spatial relationships between features in one or more layers. You will now take a closer look at a few of the most commonly used tools.

The Navigate group

The **Navigate** group contains tools that allow you to navigate to locations within the map. This includes tools that allow you to zoom in to areas so that you can see more detail or to zoom out to see a greater area. It also has tools that allow you to return to specific areas within the map.

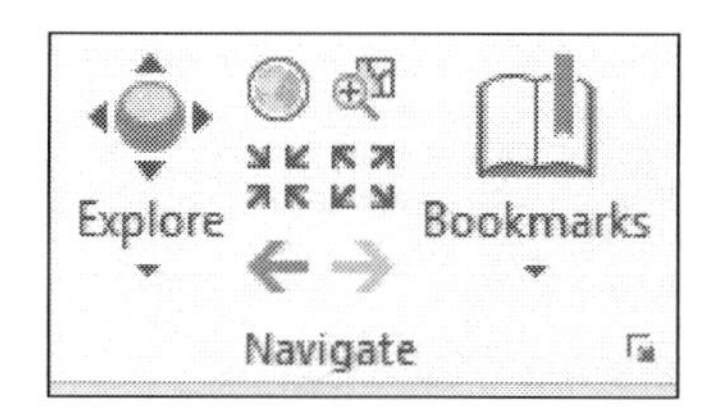

The Explore tool

The **Explore** tool is a multifunction tool that allows you to zoom and pan within the map. It also allows you to click on features within the map and retrieve attributes linked to the feature. This tool combines the functions of the **zoom in**, **zoom out**, **pan**, and **identify** tools found in ArcMap into a single tool.

You use this tool to navigate within your map using your mouse as follows:

- **Left click on a feature**: This opens the **Identify** pop-up window showing attributes linked to the feature.
- **Left or right drag**: Clicking and holding your left mouse button down and then dragging your pointer to the left and right pans your map view in that direction.
- **Moving scroll wheel**: This zooms map in and out. Scrolling the wheel away from you causes the map to zoom in. Scrolling the wheel toward you causes the map to zoom out.
- **Right click and drag**: This continuously zooms the map in or out depending on the movement of your mouse.
- **Single right click**: This opens a menu. The menu that appears will vary depending on where you right-click.
- **Hold down scroll wheel and drag (3D map/scene only)**: This rotates and tilts the map view along 3D axis.

As you can see, this tool works best with a mouse that has a scroll wheel in between the right and left button. Other types of pointing devices can be used, but the functionality may be different depending on your hardware. Some of the buttons associated with this tool can be altered under the **Options** tool located on the **PROJECT** tab. For example, you can change the zoom direction of the scroll wheel.

The Bookmarks tool

The **Bookmarks** tool allows you to zoom to save spatial locations within your active map. This allows you to quickly return to important locations, such as project areas, special event locations, key parcels, and so on. Each map or scene within your project will have its own set of unique bookmarks.

The drop-down arrow below the primary tool will allow you to access the bookmarks that have been saved with the active map or scene. This also provides access to tools in order to create new bookmarks and managing existing ones. There is no limit to the number of bookmarks each map or scene can have.

The Layer group

The **Layer** group contains tools in order to add new layers or data to your map. Some tools work with both 2D and 3D data. Others work best with one or the other. There are also tools in order to add nonspatial data to your map. These tools use either coordinate values or addresses to show the location of points.

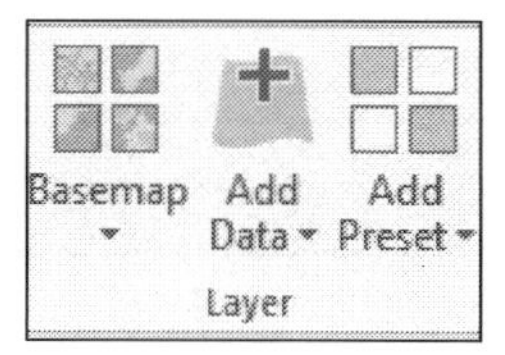

The Add Data tool

This tool allows you to add new 2D or 3D layers to your maps. Added layers can reference various types of data and can come from different locations. It also allows you to add tabular data.

This tool is basically the same as the **Add Data** button in ArcMap. However, Esri has incorporated some additional options that the old button in ArcMap did not have. The options included with this tool in ArcGIS Pro are:

Options	**Descriptions**
Data	This option adds spatial or tabular data to a map. It can be used to add layers, which reference Shapefiles, geodatabase feature classes, and ArcGIS Server web services. Standalone tables can be added to a map from various sources as well, including dbase files, CSV files, database tables, and Excel spreadsheets.
X, Y Event Data	This allows you to add points to a map based on coordinates stored in a Standalone table. Each point will also be linked to the additional attributes or information associated with that record or row.
Route Events	This shows location of route between points of interest along a linear network.
Query Layer	This creates a query layer from an enterprise database based on an SQL select statement. This does not work with file or personal geodatabases.
Address Layer	This geocodes a table of addresses using a specified address locator creating a new geodatabase feature class or a Shapefile.

The Add Preset tool

This tool is used to add new 3D layers to a scene with a predefined set of symbology settings.

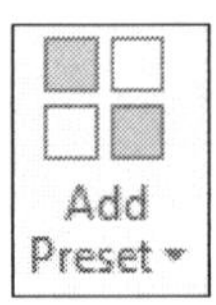

Predefined properties are determined from a gallery of defined symbology. ArcGIS Pro comes with galleries for realistic trees, thematic trees, realistic buildings, ground (terrain), icon points, and thematic shapes. Using this tool to add layers makes visualizing data in a 3D environment easier, especially for those that may be new to working with 3D data.

The Selection group

The **Selection** group contains various tools used to select data including **Select**, **Select By Attributes**, **Select By Location**, and **Clear**. These tools all work very much like their cousins found in ArcMap though with some subtle differences.

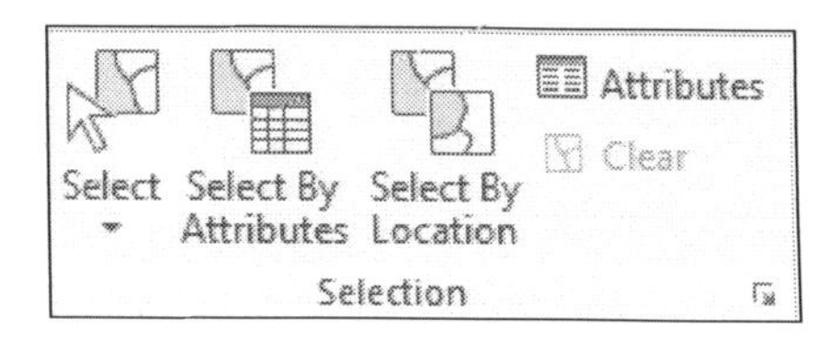

The Select tool

The **Select** tool allows you to select features from within a map or scene by clicking on a feature or drawing an area.

The drop-down arrow located at the bottom of the tool lets you select what type of area you want to draw. You have options to draw a rectangle, polygon, freehand lasso, circle, and line. This is similar to the **Select Features** tool in ArcMap.

The Select By Attributes tool

The **Select By Attributes** tool allows you to select features based on their attribute values. For example, you might want to select all parcels owned by Tripp Corbin or all sewer pipes made of ductile iron. This tool will allow you to do that.

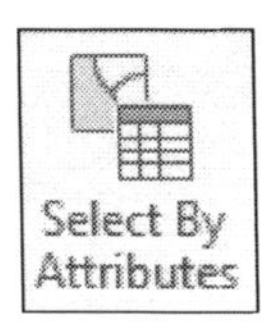

Like the tool by the same name in Esri's older ArcMap application, it creates a SQL `where` clause. You create the clause in a tool pane normally located on the right-hand side of the interface.

The Select By Location tool

This tool allows you to select features in one or more layers based on spatial relationships.

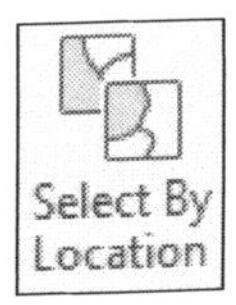

For example, you might select all parcels located within 100 feet of a road which you know is going to have work done on it, so you can get a list of all those that live along the road. You might select all the roads that intersect or cross a floodplain so you know not to include those in an emergency evacuation plan.

The Inquiry group

The **Inquiry** group contains tools that allow you to retrieve some basic information about your map. You can measure lengths and areas in different units or retrieve some basic demographic data or even find a point of interest using the tools in this group.

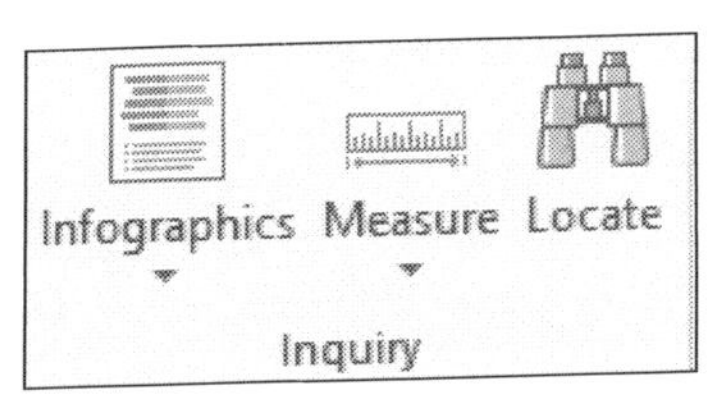

The Infographics tool

This is a completely new tool that allows you to retrieve basic demographic data for the area you select in the map. It will retrieve information and graphs showing average household income or population ethnic makeup.

This tool accesses Esri's **GeoEnrichment** service, which is connected through your ArcGIS Online login. So, this tool will require you to have an internet connection, be connected to ArcGIS Online, and have available ArcGIS Online credits to work. **Using this tool will also cost you ArcGIS Online credits**. You should always keep that in mind when using this tool. This tool can also be customized to show specific geographic locations, data collections, and how the data is aggregated.

The Measure tool

This tool allows users to measure length and area by simply clicking on locations within a map. If you are working within a 3D scene, you also have the option to measure vertical distances.

You can change the units used by this tool, so they are different from the units assigned to your map. Even if your map is in meters, you can have the **Measure** tool provide distances in feet or miles or kilometers. Area units can also be set. Unlike ArcMap, you can still measure areas when your map is in a geographic coordinate system.

The Locate tool

This tool allows you to locate a place using an address or common name. By default, this tool uses the Esri **World Geocoder** service from ArcGIS Online. This service requires you to have an active connection to ArcGIS Online to work. However, unlike the **Infographics** tool, it does not use credits. You can add and use your own address locators to your project, and they will then be available for use with this tool.

While the **Locate** tool uses an icon that is very similar to the **Find** tool in ArcMap, it is limited to just locating an address or common name. It does not have all the other functionality of the old **Find** tool. The **Locate** tool is tied directly to the address locators connected to the project. It is those locators that control what the **Locate** tool can locate or not.

Exercise 2A – working with the MAP tab

You will now get to take some of the tools we discussed for a test drive. You will see how they work within the new ribbon interface and some of the options associated with those tools first hand. You will start by opening an existing project and using the navigation tools.

Step 1 – opening a project

In this step, you will open a project that has already been created. This project includes a single 2D map with several layers for the imaginary City of Trippville:

1. Start ArcGIS Pro. Remember that you will need to make sure that you are connected to the Internet, so your license can be validated. ArcGIS Pro should remember your user login from *Chapter 1, Introducing ArcGIS Pro*. If not, log in to ArcGIS Online when asked to.
2. Once ArcGIS Pro starts, click on **Open another project**.

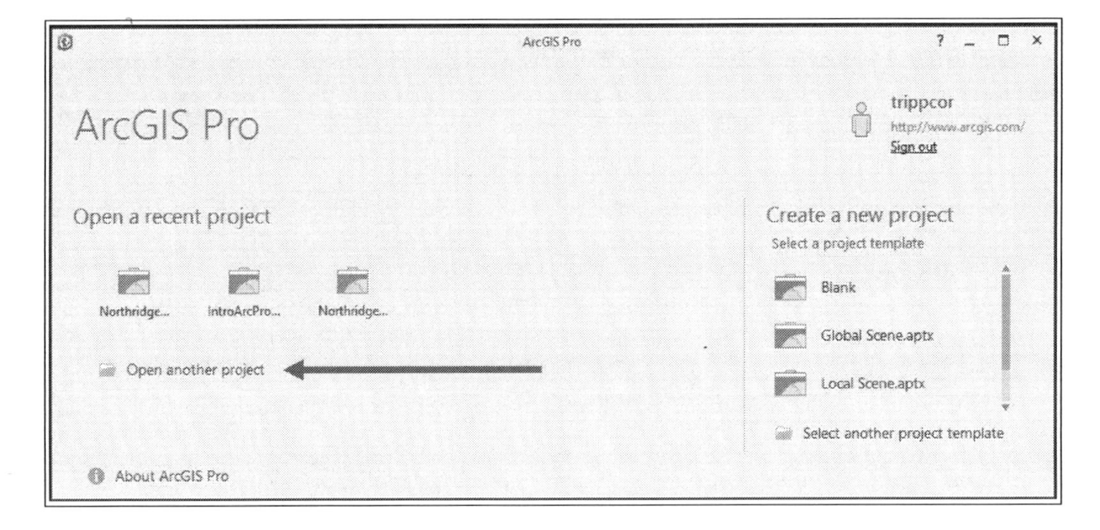

3. Click on **Computer** under the **Open** pane. Then, select **Browse**.
4. Navigate to the `C:\Student\IntroArcPro\Chapter2` folder or the location you installed your exercise data.
5. Select the `Ex 2A.aprx` file and click on **Select**.

If you have successfully opened the project, ArcGIS Pro should look as follows:

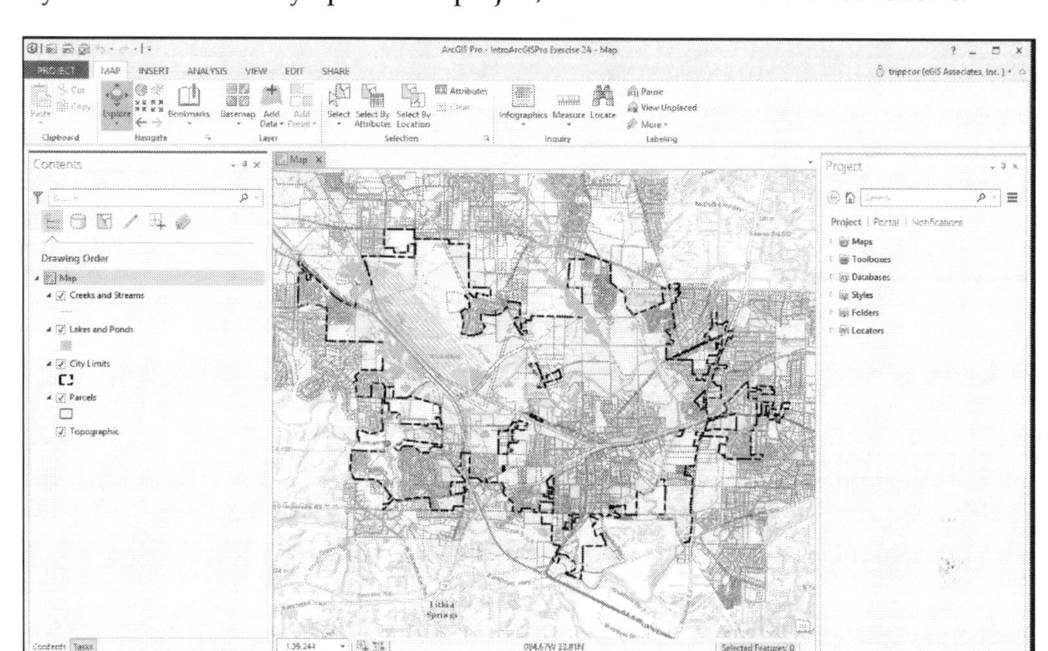

As you can see in the preceding image, this project contains a single 2D map with layers showing city limits, creeks, lakes, and parcels. Above the map is the ribbon that contains the tabs, group tabs, and tools. To the right of the map view is the **Project** pane. This pane allows you to access various items associated with the project including maps, databases, folders, and address locators. You will work with the **Project** pane throughout this book.

Step 2 – navigating in the map

Now that you have the project open, you will begin to explore the navigation tools on the **MAP** tab:

1. Click on the **Bookmarks** tool and select the Washington Park bookmark. This will zoom you to the location of Washington Park automatically.
2. Click on the **Explore** tool to make it the active tool.
3. Click on the parcel for Washington Park with your mouse's left button. Continue to hold the mouse button down and drag your mouse toward the lower left-hand corner of the map view until Washington Park is located in the lower left-hand corner. Then, release the mouse button.

You have just used the **Explore** tool to pan the map view. As you learned earlier in this chapter, the **Explore** tool has many uses. Now you will explore some of the other functions of this tool:

4. With the **Explore** tool still active, click on any parcel within the map view.

Question: What happens when you click on a parcel using the **Explore** tool?

____

____

5. Close the pop-up window that appears.
6. Click on the small drop-down arrow located under the **Explore** tool and select **Visible Layers**.

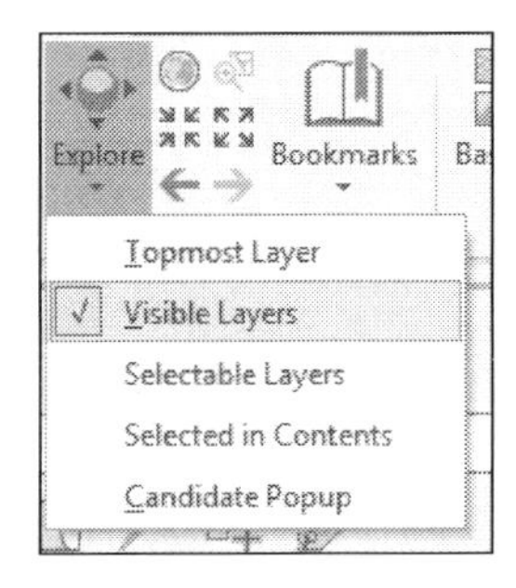

7. Click on the same parcel you did in task 4 earlier.
8. Look at the bottom of the information pop-up window. You should see 1 of 2 located in the lower left-hand corner.
9. Click on the small drop-down arrow located to the right of 1 of 2. You should see a list of all visible features that overlap the location you clicked on.

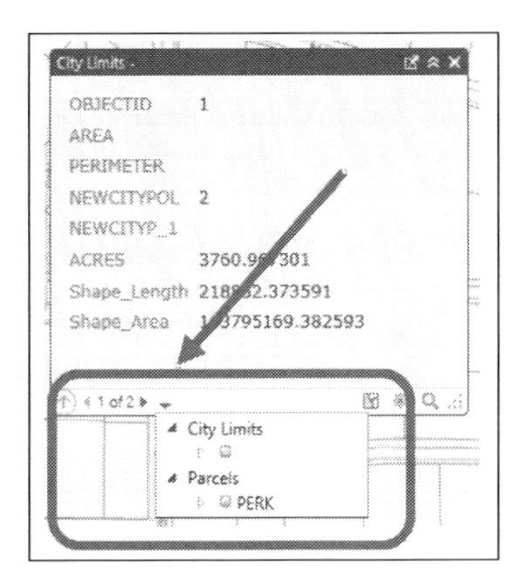

[The list of features you see and the values shown will depend on where you clicked in the map. So, do not be surprised or worried if what you see in your ArcGIS Pro is different than the screenshot.]

10. Click on one of the other values in the list and see what happens in the information pop-up window. You will need to click on one of the unique values listed and not just the layer name. For example, using the screenshot as a guide, you would need to click where it says **PERK**, not on **Parcels**.

[**Question**: What happens when you click on one of the other values in the drop-down list?

____

____]

The information window can also be resized and moved if needed. You can even move to another monitor or display. One of the interesting features of the information pop-up window is the ability to pin to the screen. This locks in the current values for that window and opens a new information pop-up when you click on other features with the **Explore** tool. Let's see how that works:

11. Close the information pop-up window.
12. Turn off the **City Limits** layer in the **Contents** pane by clicking on the checkmark next to the layer name. This ensures features from this layer will not show up in the information pop-up when you use the **Explore** tool.
13. With the **Explore** tool active, click on a parcel to open the information pop-up window.
14. Click on the *pin to screen* button located in the upper right-hand corner of the information pop-up window, as indicated in the following screenshot:

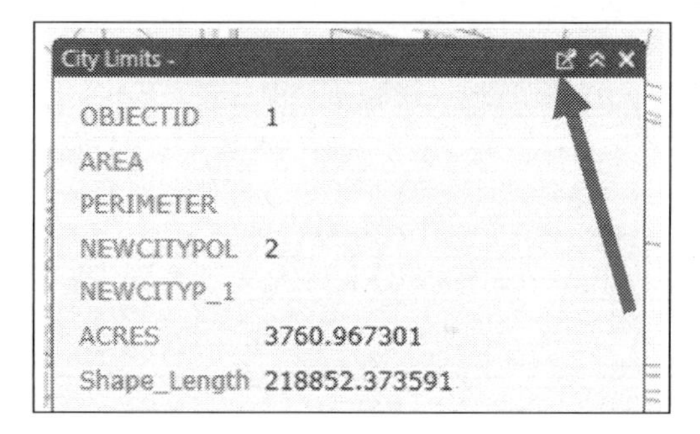

15. With the **Explore** tool still active, click on another nearby parcel.
16. Click on the blue title bar of the information pop-up and drag it to another location in the ArcGIS Pro interface.

You should now see two information windows, as illustrated in the following screenshot. Again, your actual values will most likely differ based on the parcels you selected in the map.

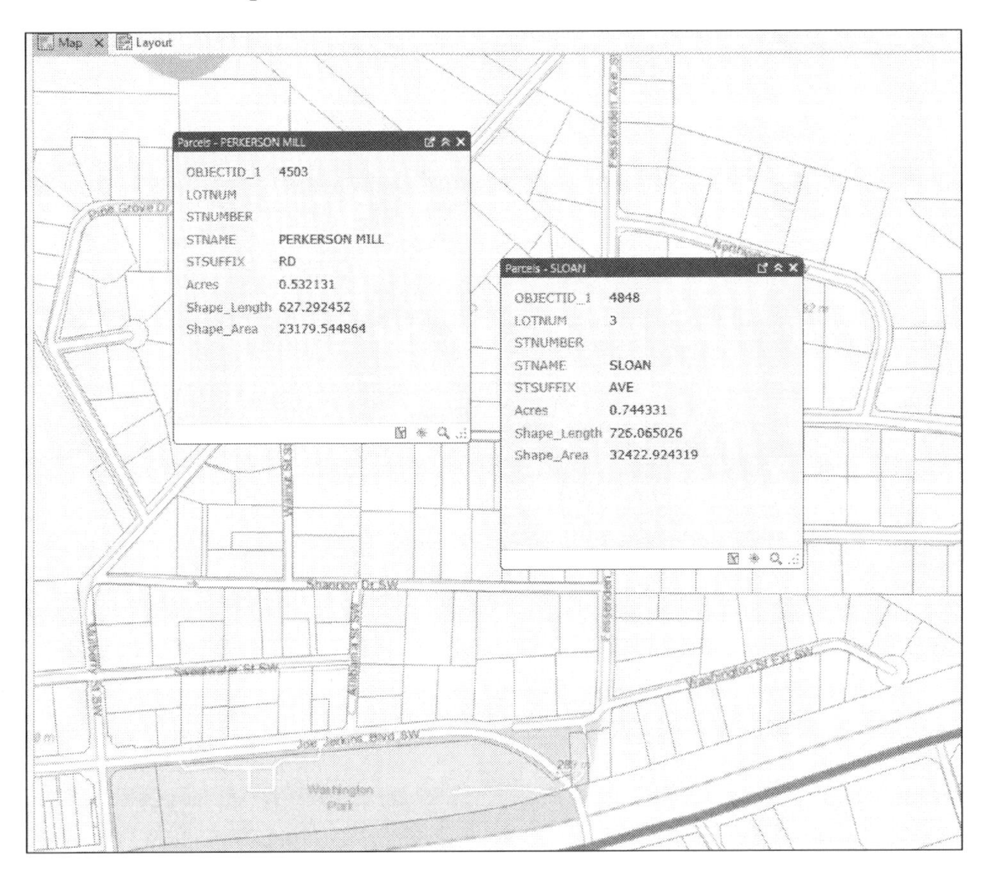

Pinning an information window to the screen allows you to open multiple information windows at one time, so you can easily see the attributes of multiple features at the same time. Feel free to continue to experiment with this functionality. Also, examine the buttons located in the lower right-hand corner of the information pop-up windows. Close all information popups once you are done experimenting.

You will continue to explore some of the other functions associated with the **Explore** tool:

17. Click on the **Full Extent** tool located in the **Navigate** group on the **MAP** tab. It has an icon that resembles a small *globe*. This will zoom to the full extents of the map.
18. With the **Explore** tool active, place your mouse pointer in the middle of the map view and roll your scroll wheel away from you.

Question: What happens when you roll the scroll wheel away from you?

____

____

19. Now roll the wheel back toward you to zoom back out to see a larger area.
20. Click on the **PROJECT** tab located within the ribbon and select **Options**.
21. In the **Project** pane in the **Options** window, select **Navigation** located under **Application**.
22. Under the **Navigation** option, you will see different settings for navigation tools. Set the mouse wheel to roll forward to zoom out and click **OK**.
23. Click on the return arrow located in the upper left-hand corner of the **PROJECT** tab.

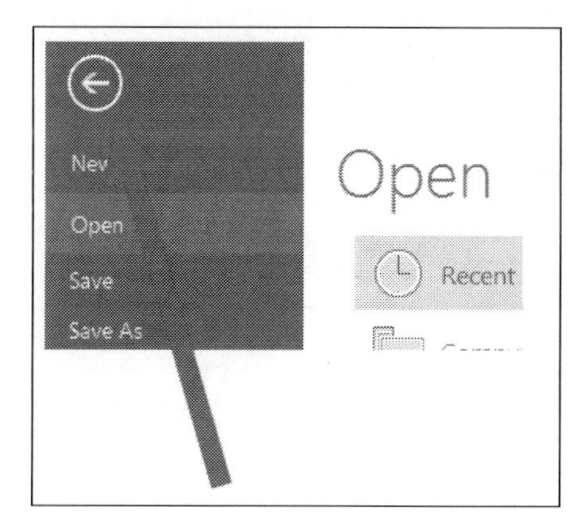

Move your mouse pointer to the center of the map view and roll your scroll wheel away from you and then back toward you again. The zoom has now been reversed because you changed the option. Feel free to change the zoom back to the original settings by repeating the process you used to change it. If you preferred this setting then you can keep it.

Step 3 – using the Infographics tool (optional)

Now, let's take a quick look at the **Infographics** tool. As mentioned earlier, this tool allows you to retrieve demographic information using Esri's GeoEnrichment Service from ArcGIS Online. **This does use ArcGIS Online credits**. However, the amount used for this step will be minimal. You should use less than one credit by the time you complete this step:

1. Zoom back to the full extent of the map using the **Full Extent** tool.
2. Click on the **Infographics** tool in the **Inquiry** group tab.
3. Click near the middle of the map.

The **Infographics** window will open displaying demographic information for the area.

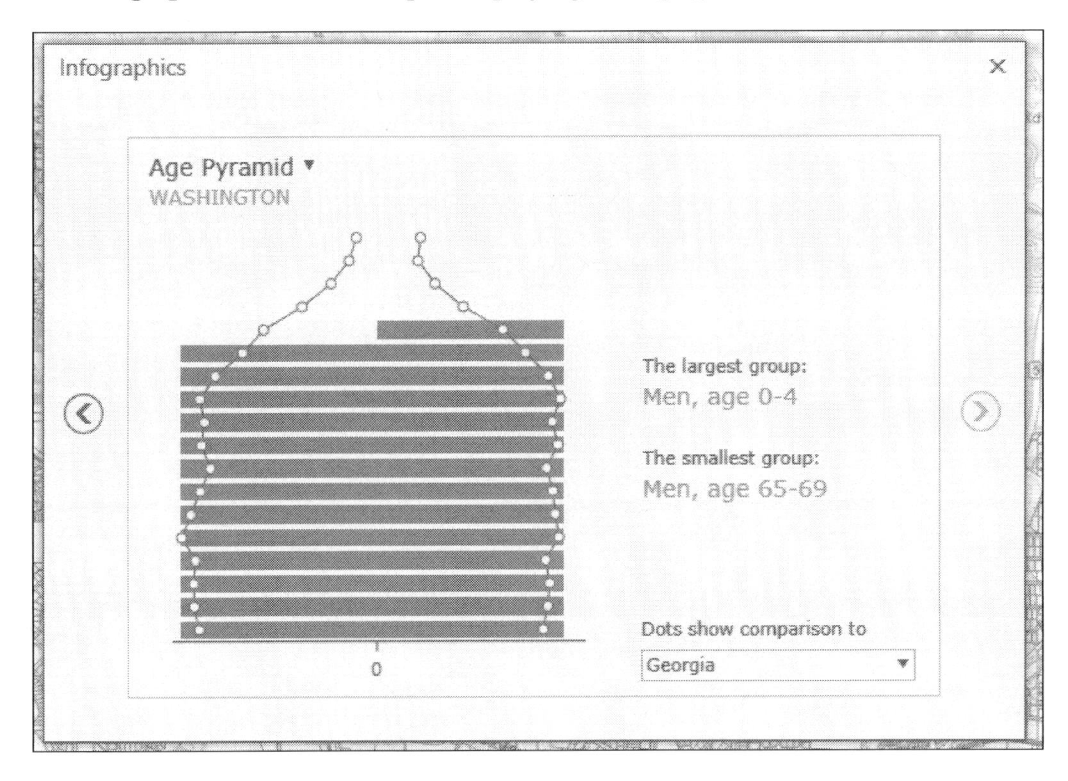

The first thing you will see is the **Age Pyramid**, which shows different age groups compared with the information for the geographic area indicated in the lower right-hand drop-down cell:

4. Move your mouse pointer over some of the blue or pink bars to see what happens. Then, move your mouse pointer over one of the dots located along the curved line located toward the end of the colored bars.
5. Now click on one of the arrows located on the right- and left-hand sides of the window. This will page you through different demographic data associated with the area shown in the map.
6. Once you are done examining the **Infographics** tool, click on the **PROJECT** tab and select **Save As**.
7. In the **Save As** window, navigate to `C:\Student\IntroArcPro\My Projects`. Name your project `your name_Ex2A` and click on **Save**.
8. Exit ArcGIS Pro.

So, you have now successfully opened an ArcGIS Pro project, retrieved data, and navigated within the map using the ribbon interface. Now you will move on to contextual tabs.

Contextual tabs

Now that you have had a chance to gain some experience using ArcGIS Pro and the ribbon, you hopefully have begun to understand the concepts of tabs. You will now begin to explore contextual tabs.

Contextual tabs are smart tabs that appear when you select an item within ArcGIS Pro. This could be a layer, a map, a layout, a table, and so on. When you select one of these items, a tab will appear on the ribbon, which contains tools specific to working with that item. For example, if you select a layer from the **Contents** pane, the **FEATURE LAYER** contextual tab will appear. You will examine the **FEATURE LAYER** contextual tab next. You will explore other contextual tabs throughout other chapters in the book.

The FEATURE LAYER contextual tab

The **FEATURE LAYER** contextual tab appears when you click on a layer within the **Contents** pane. It includes three tabs: **APPEARANCE**, **LABELING**, and **DATA**. Each of these allows you to access various properties associated with the layer you have selected.

The APPEARANCE tab

The **APPEARANCE** tab contains tools that control the display of the layer. You can change the symbology, set visibility scales, apply transparency, and so on.

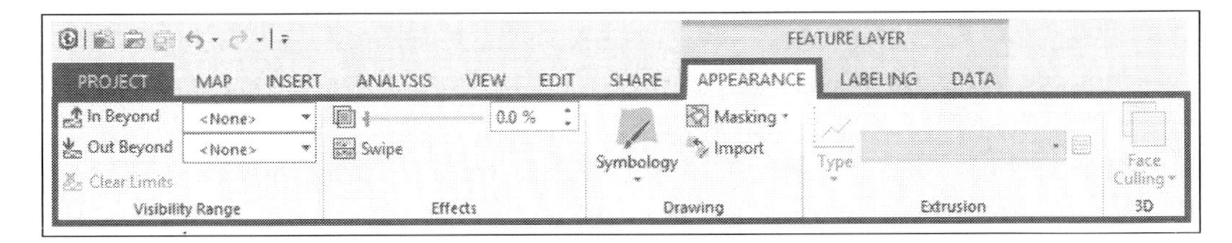

You can see that there are five groups within the **APPEARANCE** tab. They include **Visibility Range**, **Effects**, **Drawing**, **Extrusion**, and **3D**. **Visibility Range**, **Effects**, and **Drawing** will be available for all layers.

Extrusion and **3D** groups will only be available when working with a 3D scene. **Extrusion** allows you to extend 2D features above a 3D surface, such as building footprints or light poles. The extruded layer must be in the **3D Layers** category in the **Contents** pane. **Face Culling** allows you to see through parts of a 3D feature.

The LABELING tab

This tab allows you to access tools to add text labels to the features within the selected layer. The tools presented in this table will vary depending on what type of layer you have selected.

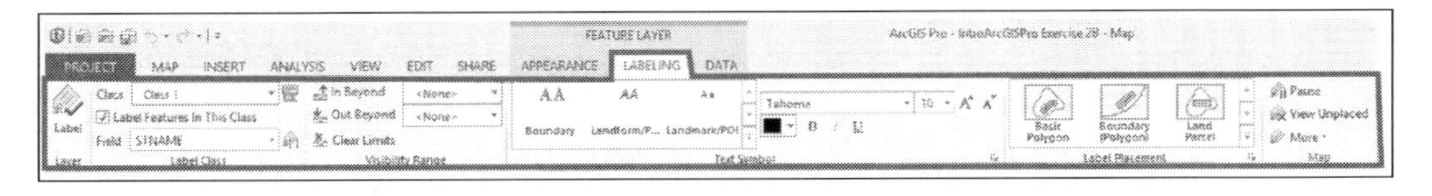

This tab contains six groups. These include **Layer**, **Label Class**, **Visibility Range**, **Text Symbol**, **Label Placement**, and **Map**. The **Label** tool in the **Layer** group will turn labeling on or off for the selected layer.

The **Label Class** group allows you to create different label classes for a group of features based on a SQL query statement. Each label class can have its own unique label settings. For example, you could create a label class based on the type for each road so that highways could be labeled using a different setting than a local city road. This can greatly expand the flexibility of labels.

The **Text Symbol** group contains tools and settings, which control how a label is visualized. Here, you can change the font, size, and color. You can also choose to make the label bold, italicized, or underlined. ArcGIS Pro also includes several predefined label styles, which you can also choose to use from this group tab.

The **Label Placement** group controls the general placement of the label in relation to the feature being labeled. The options here will vary depending on whether you are working with points, lines, or polygons. For example, for a line layer, you will have the options to place the label above, below, or on the line. The options for a point or a polygon layer will be different.

The last group is **Map**. The tools located here control the drawing of the labels within the map. You can pause labeling to help speed up navigation within the map. You can view unplaced labels. One of the most important tools located in this group sets the labeling engine used by ArcGIS Pro. Like ArcMap, ArcGIS Pro uses one of the two labeling engines: **Esri Standard Labeling Engine** or **Maplex**. By default, ArcGIS Pro uses Maplex, which provides much greater control over labeling than the Standard Esri Label Engine.

Exercise 2B – working with 2D and 3D layers

It is now time to see how the **FEATURE LAYER** contextual tab works for you. In this exercise, you will use tools on the tab to make changes to layers within an existing project. You will change a layers symbology, labels, and more.

Step 1 – opening the project and adding layers

In this step, you will open a project that has already been created. You will then add some additional layers to a map:

1. Using the skills you have already learned, start ArcGIS Pro and open the `Ex 2B` project. It is located at `C:\Student\IntroArcPro\Chapter2` or the location you installed the course data.
2. Click on the **Add Data** tool located in the **Layer** group of the **MAP** tab.
3. Expand the **Project** folder in the pane located on the right-hand side of the **Add Data** window. Then, select the `Databases` folder so you see two geodatabases: `IntroArcGISPro Exercise 2A` and `Trippville_GIS`.
4. Double-click on the `IntroArcGISPro Exercise 2A` geodatabase.

5. Select the `bldg._footprints`, `RR_Tracks`, and `Street_Centerlines` feature classes and click on the **Select** button. You can hold down your *Ctrl* key while clicking on each feature class to select and add multiple feature classes to a map at one time.

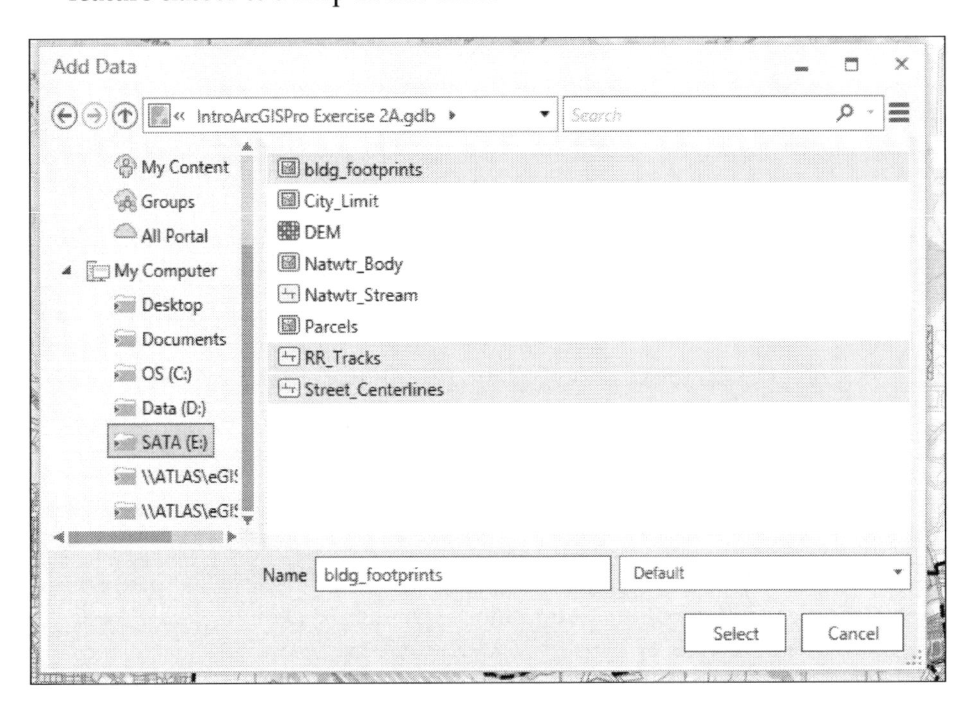

Step 2 – changing symbology

You have just added three new layers to the map. Now you need to adjust the symbology for the layers you added. You will change the railroad layer you added to use a common railroad symbol. You will similarly change the building footprints. Finally, you will adjust the road centerlines, so they are symbolized based on their condition:

1. Right-click on the `bldg_footprints` layer and select **Properties**.
2. Click on **General** located in the pane on the left-hand side of the **Properties** window.
3. Change the name to `Buildings` and click on **OK**. Note what happens to the layer in the **Contents** pane.

When you right-clicked on the `bldg_footprints` layer, did you notice what happened to the ribbon? The **FEATURE LAYER** contextual tab automatically appeared providing access to the tools it contains. You will now use those tools to make adjustments to the `Building` layer you just renamed:

4. In the **FEATURE LAYER** contextual tab, select the **APPEARANCE** tab.
5. Click on the **Symbology** tool and, if necessary, select **Single Symbol**. This opens the **Symbology** pane on the right-hand side of the ArcGIS Pro interface. The **Symbology** pane allows you to make changes to the symbology settings for a layer.

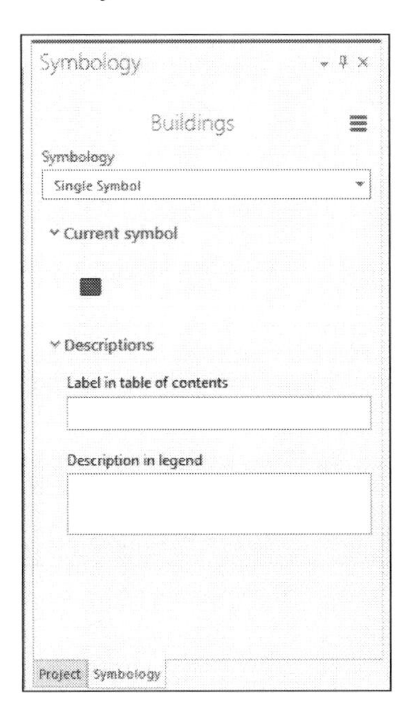

6. Click on the small rectangular symbol located under **Current symbol** to open the **Symbol Gallery**.
7. Select the `Gray Building Footprint` symbol that is located next to the `Airport Runway` symbol. You can use another symbol if you prefer.

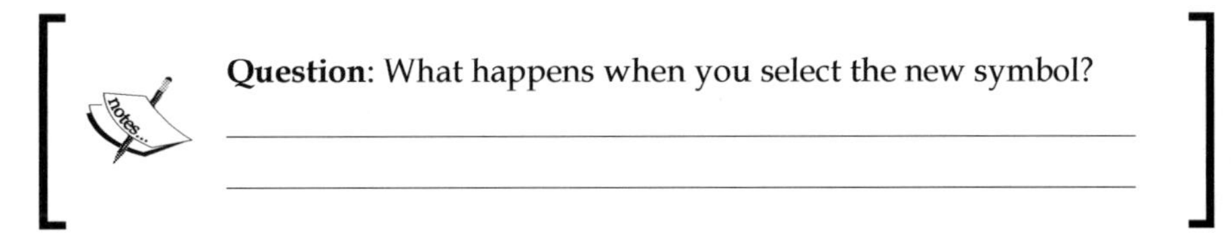

Question: What happens when you select the new symbol?

____

____

8. Now select the `RR_Tracks` layer in the **Contents** pane and watch what happens in the **Symbology** pane.
9. Using the same process mentioned earlier, change the symbology to the `Railroad` symbol.
10. Rename the `RR_Tracks` layer to `Railroad` using the same process you used to rename the `Buildings` layer.
11. Click on the **PROJECT** tab and select **Save As**.
12. Navigate to the `C:\Student\IntroArcPro\My Projects` folder and name your project `<your name> Ex2B`.

Now you will change the symbology for the road centerlines to reflect their condition. This will be a bit more challenging than changing a single symbol:

13. Using the skills you have already learned rename the `Street_Centerline` layer to `Streets`.
14. Select the **APPEARANCE** tab and click on the small arrow located below the **Symbology** tool. Choose **Unique Values** from the drop-down options.

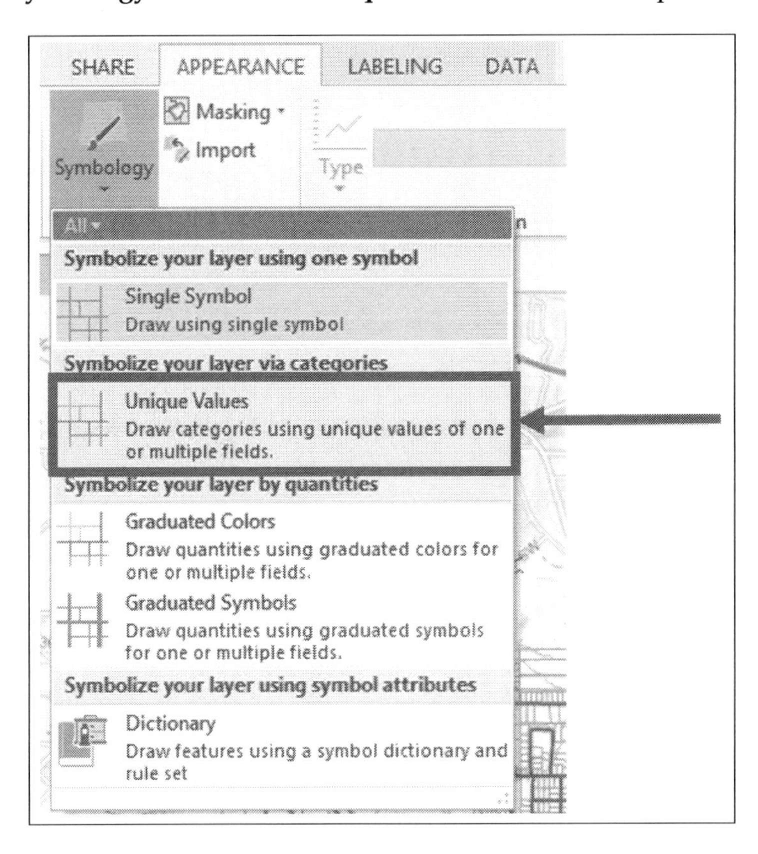

The **Symbology** pane will once again be opened on the right-hand side of ArcGIS Pro. It will look different than what it did when you were working with single symbols. It contains many more options and settings. These allow you to configure your symbology based on the attributes found in one or more fields:

15. In the drop-down box located next to **Field 1**, select **Condition**. Note that ArcGIS Pro automatically adds all values found in this field and assigns them a symbol.
16. Click on the symbol located next to **Fair** to open the **Symbol Gallery**.
17. Click on the **Properties** tab located at the top of the pane.
18. From the **Color** drop-down selection, set the color to an *orange* tone or another you prefer and set the width to `1 pt`. Then, click on **Apply**.
19. Click on the return arrow located at the top of the **Symbology** pane to return to the symbology settings for the entire `Streets` layer.
20. Using the same method mentioned earlier make the following adjustments to the symbols for **Good** and **Poor** values.
    - **Good**: *Bright green* color and width of `1.5 pt`
    - **Poor**: *Bright red* color and width of `3 pt`

    When you are done, your map should look similar to this:

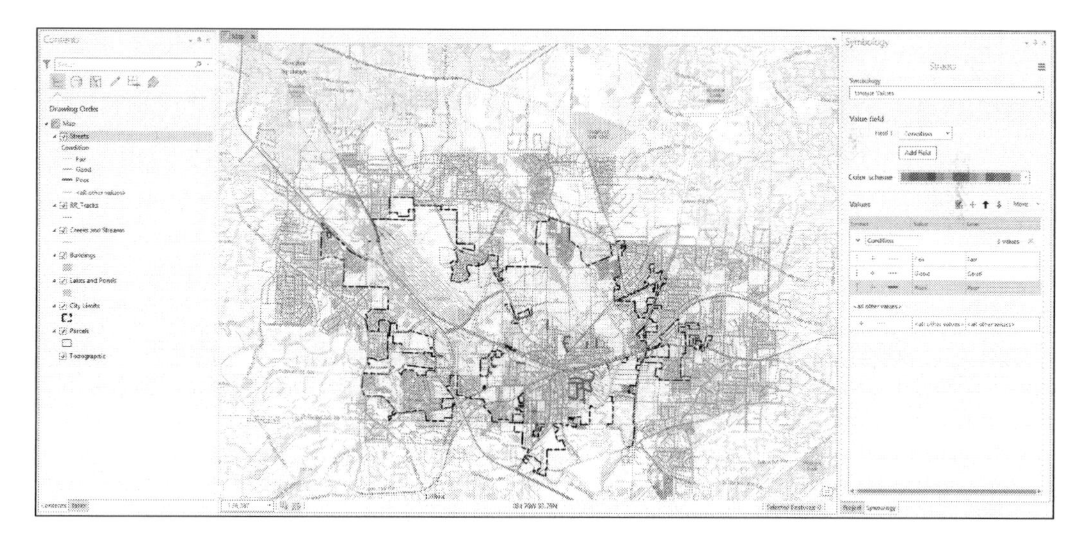

21. Close the **Symbology** pane when you are done changing the symbology. Feel free to make changes to other layers in the map.

22. Save your project by clicking on the **Save** button located on the **Quick Access Toolbar** located at the very top left-hand side of the ArcGIS Pro interface.

After adjusting symbology using the **FEATURE LAYER** contextual tab, let's add some labels to the maps.

Step 3 – labeling

In this step, you will use the **LABELING** tab in the **FEATURE LAYER** contextual tab to add street names to the map. You will configure the labeling for the `Streets` layer to accomplish this task:

1. Select the `Streets` layer in the **Contents** pane.
2. Click on the **LABELING** tab in the **FEATURE LAYER** contextual tab.
3. Click on the **Label** tool to turn on labels for the `Streets` layer. You should see labels appear in the map once you click on the tool.
4. Within the **Label Placement** group tab select the **North American Streets** option. ArcGIS Pro will then use the placement properties defined by this placement style such as the road name appearing above the centerline and be curved to follow the road.

   When you enabled labeling for the `Streets` layer, the **Label Class** pane also opened in the right-hand side of ArcGIS Pro in the same place the **Symbology** pane did. This pane allows you to refine various label settings.

5. In the **Label Class** pane, click on the **Fitting Strategy** button. It is the one that looks like a knight piece from chess. If you do not see the **Label Class** pane, you can right click on the layer and select **Label Properties** from the context menu.

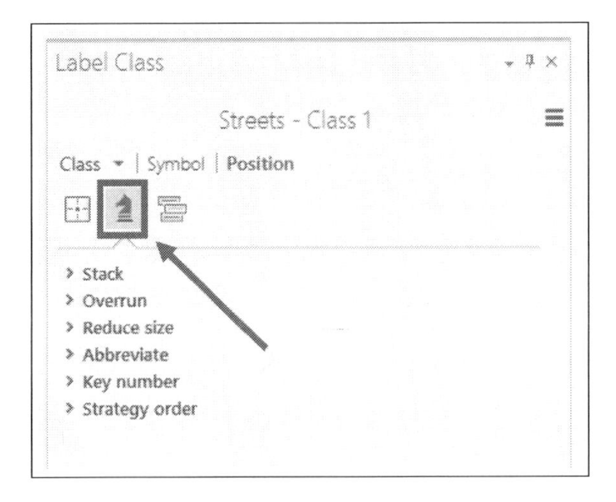

6. Click on the small arrow next to **Reduce size** to expand the options for that strategy.
7. Click on the check box next to the **Reduce font** size to enable this fitting strategy.
8. Set the **Lower Limit** to `5.0 pts` and the **Step interval** to `1.0 pts`.

This will allow ArcGIS Pro to automatically adjust the font size in order to get a road name to fit on the map. This will allow more road names to be displayed.

9. Zoom in and out to see how well your labels appear. Turn off the **Reduce font** size option and then zoom in and out again to see the impact that option has.
10. Save your project and close the **Label Class** pane.

Step 4 – working with a 3D layer

ArcGIS Pro also allows you to work with 3D data even if you don't have the 3D Analyst or Spatial Analyst extensions. In this step, you will add a layer containing building footprints to a 3D scene and then extrude that layer, so you can see the heights of the building:

1. In the **Project** pane located in the same place as the **Symbology** and **Label Class** panes, click on the little arrow next to **Maps** located at the top of the pane to expand its contents. If you don't see the **Project** pane, you may have accidentally closed it when you closed the other panes. You can reopen it by clicking on the **VIEW** tab and then the **Project** button.
2. Once you expand the **Maps** folder, you should see two maps: **Map** and **Scene**. Right-click on **Scene** and select **Open Local View**. This opens the **3D Scene**.

3. Press the scroll wheel on your mouse down and push your mouse away from you slightly to rotate the map view on the 3D plane until it looks similar to the following image:

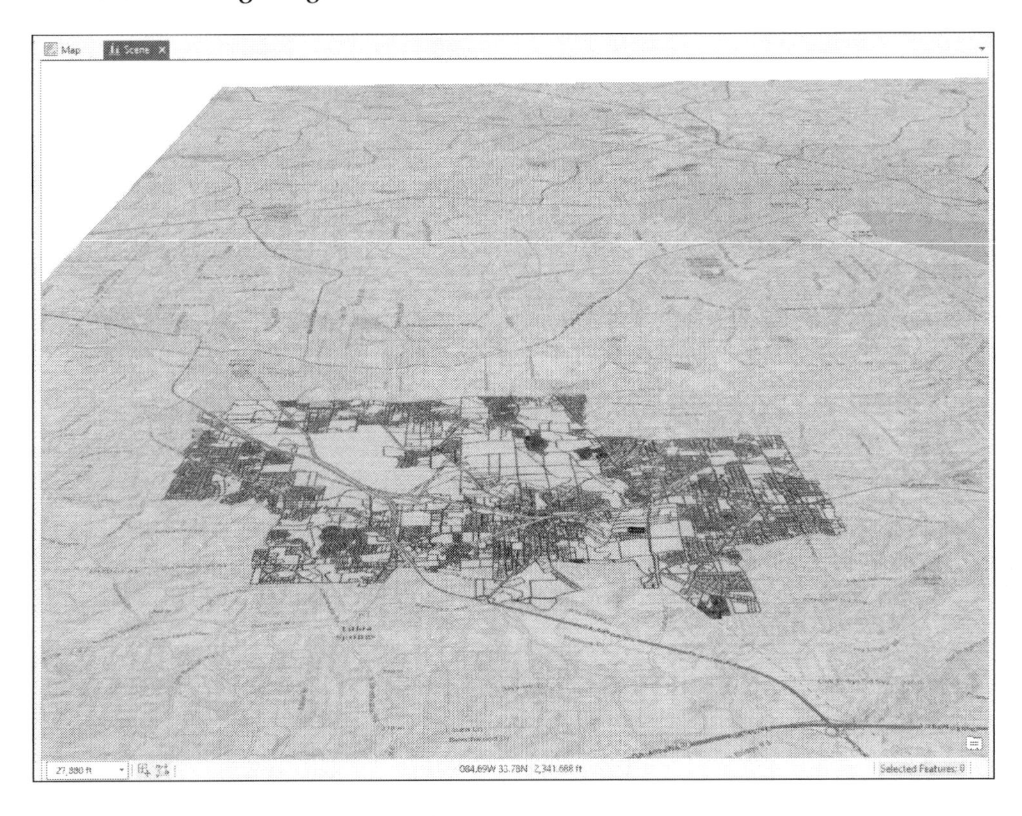

4. Place your mouse pointer near the center of the parcels layer and roll your wheel away from you to zoom in. Continue to work with the scroll wheel until you get comfortable with how it allows you to navigate within a **3D Scene**.
5. Click on the **Add Data** tool and navigate to the `IntroArcGISPro Exercise 2A.gdb` file. Click on `bldg_footprints` feature class and click on **Select**.
6. Rename the `bldg_footprints` layer to `Buildings` using the same process you learned in previous steps.
7. Select the `Buildings` layer and drag it up to 3D layers. This will allow you to apply 3D symbology to this layer.
8. Click on the **Bookmarks** tool and select the `Buildings` bookmark. This will zoom you out, so you can see the buildings as they overlay the ground elevation.

9. Select the `Buildings` layer once again and click on the **APPEARANCE** tab. Note that it looks quite a bit different from what you saw when working with a 2D map.
10. On the **Extrusion** group tab, click on **Type** and select **Base Height**. This will allow you to extrude the buildings based on their estimated height, so you can see the differing heights of each building in a 3D view.
11. In the drop-down window next to **Type**, set that to **Est_HGT**. This is the estimated height of each building.

Your map should now look similar to the following image. The colors of your buildings may be different because ArcGIS Pro assigns random colors to newly added layers:

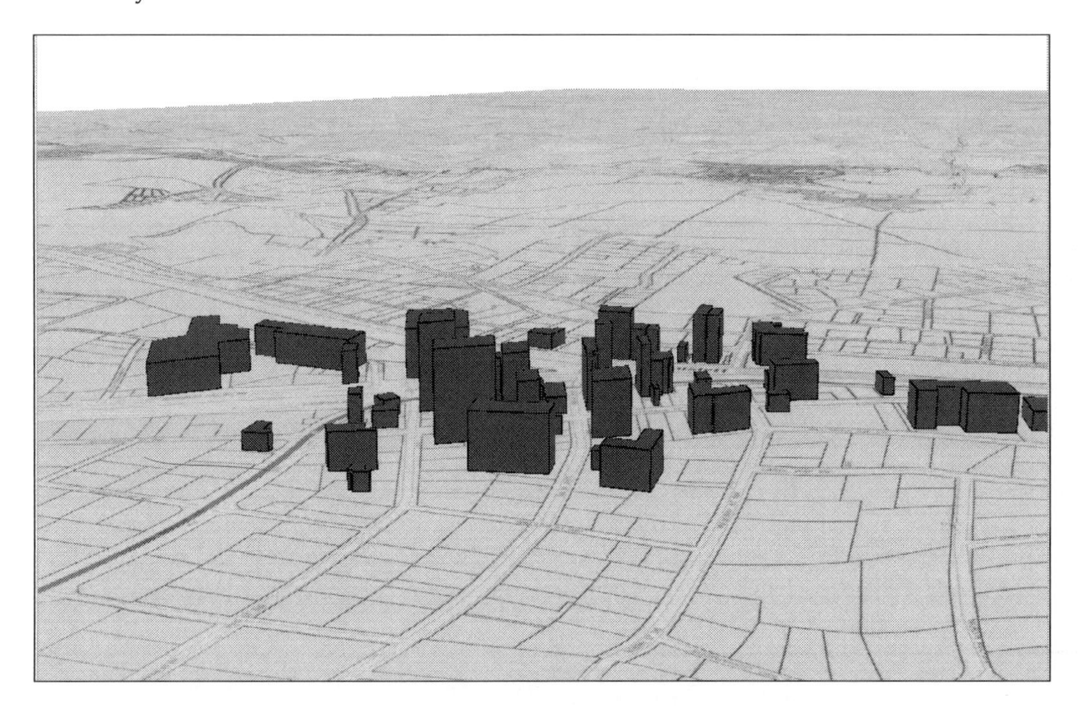

You have just added and symbolized a 3D layer in ArcGIS Pro. You can now see the 3D relationships between the different buildings within the city.

> **Challenge**: Try to change the symbols for the buildings to one of the `Gray Building Footprint` symbols found in the **Symbol Gallery**.

Summary

In this chapter, you learned how to use the ArcGIS Pro ribbon interface and the terminology associated with it. As you saw that the ribbon provides easy and quick access to tools. The contextual tabs make specific tools associated with tasks and items available to you with minimum clicking or searching.

Many of the tools in the ribbon work in conjunction with panes that appear on the right-hand side of the interface. These panes provide access to settings and inputs used by the selected tool to refine its function.

Putting these tools to use, you were able to work with both 2D and 3D data. You were able to label features and change their symbology. You learned how the **Explore** tool has many functions, from navigating in the map to retrieving data.

3
Creating and Working with ArcGIS Pro Projects

ArcGIS Pro reintroduces the concept of projects to Esri users. Esri first introduced the concept of managing your GIS using project files with its ArcView GIS application. ArcView GIS used projects to manage map layouts, views, tables, data connections, and more within a single **APR** file. This allowed you to access everything associated with a project from a single location.

When Esri released ArcGIS in 1999, this was dropped. In its place, Esri decided to use separate individual files. You have map documents (**MXD** files) for ArcMap, which contain a single map with one or more data frames. So, for each map you needed, you would have to create a separate map document. You have scenes (**SXD** files) for **ArcScene**, which contained one local 3D view. All of this means you often have tens to hundreds of files to manage and keep track of which makes management much harder.

ArcGIS Pro projects greatly simplify GIS management. Projects allow you to consolidate all of the resources needed to complete a project in a single place. No longer will you need to search for the correct map document. You will just need to open the ArcGIS Pro project file for the project you are interested in, and all the maps, data connections, layouts, tools, and more will be there.

In this chapter, you will:

- Be introduced to ArcGIS Pro projects
- Understand how to create a project and what can be included within one
- Learn how to use and create project templates

Working with an ArcGIS Pro project

ArcGIS Pro utilizes project files in place of map documents and scenes to access GIS data, make maps, perform analysis, and share work. Project files have an `.aprx` file extension and include all the GIS resources that are needed for a given project. Projects can be created and shared on your local network so that others in the project team can quickly access the resources they need to complete their assigned tasks. You can even create a project package that can be shared via ArcGIS Online or Portal for ArcGIS or other cloud storage solutions.

Each project will have components and resources, which are associated with each specific project. Some of these will be common to multiple projects, whereas others will be unique to individual projects. So, even if a component or resource is linked to one project, it can be used in another project.

Projects also have their own terminology associated with them. It will be important to understand the terminology when working with projects. You will now begin exploring the terminology associated with projects and some of the component or resources that can be included in a project.

Understanding project terminology

Like the ArcGIS Pro application, projects in ArcGIS Pro have a vocabulary associated specifically with them. We will quickly define some of these terms, so you can begin to understand what they mean as you see them throughout this book, hear them at conferences, see them in Esri publications, and so on.

A project item

A **project item** is any item, component, or resource stored within an ArcGIS Pro project. Project items include maps, scenes, folder connections, database connections, ArcGIS server connections, **Web Mapping Service** (**WMS**) connections, toolboxes, locators, and more. These items are accessed within ArcGIS Pro from the **Project** pane or **Project** view.

A portal item

A **portal item** is any item that is accessed via a designated portal. A portal is a connection to ArcGIS Online or Portal for ArcGIS. Portal items might include web maps, feature layer, tile layer, layer package, map package, and project package.

A project package

A **project package** is similar to a layer or map package. It bundles or zips all items associated with a project into one consolidated file. A project package can be used to share projects with others that may not be connected to your network or to archive a project. Project packages will have a **PPKX** file extension and may be stored in a local folder or uploaded to your portal.

The home folder

A home folder refers to the folder where the project file is stored. The home folder is where ArcGIS Pro creates the project-specific content unless the users specify a different location. The home folder can be changed only by saving a project to a different folder. The default home folder for new projects is `C:\Users\%user name%\Documents\ArcGIS\Projects`.

The default geodatabase

The default geodatabase is used by the geoprocessing tools as the automatic default workspace for the resulting output items of the tools. This means new feature classes resulting from the use of analysis, conversion, and other tools will automatically be saved to the default geodatabase unless you specify a different location. Any geodatabase can be used as the default geodatabase. When a new project is created, ArcGIS Pro will automatically create a new file geodatabase for the project in the home folder.

Using the Project pane

When you open a project or create a new project in ArcGIS Pro, the **Project** pane will be one of the first things you see. By default, it will be located on the right-hand side of the ArcGIS Pro interface opposite to the **Contents** pane.

The **Project** pane provides access to project items. From there, you can access maps, scenes, folder connections, layouts, database connections, and more and are all associated with the project. You can view the **Project** pane in two different modes: **Outline** or **Gallery**. The **Outline** mode is default and similar to looking at **Details in Windows** file explorer. The **Gallery** mode provides a thumbnail view of items along with the title.

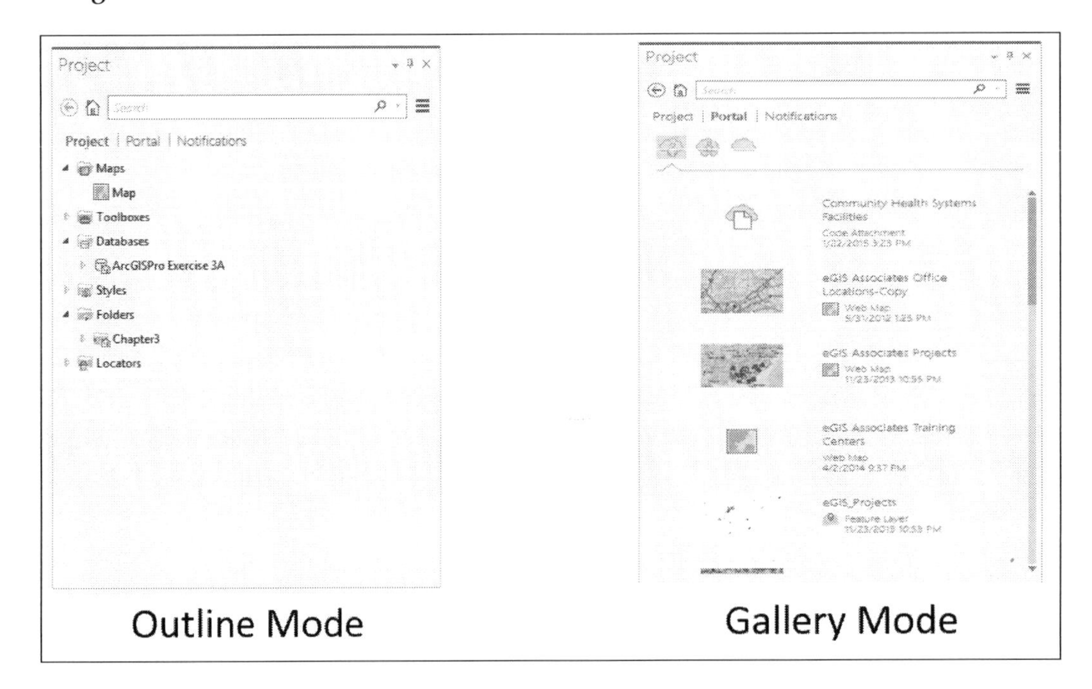

The **Project** pane contains three tabs at the top: **Project**, **Portal**, and **Notifications**. The **Project** tab allows you to access project items that are stored either within the project, such as maps or layouts, or that you have established a connection to for the project, such as a folder, a database, or a web connection. It should be mentioned that ArcGIS Pro does not support personal geodatabases. However, file, workgroup, and enterprise geodatabases are supported.

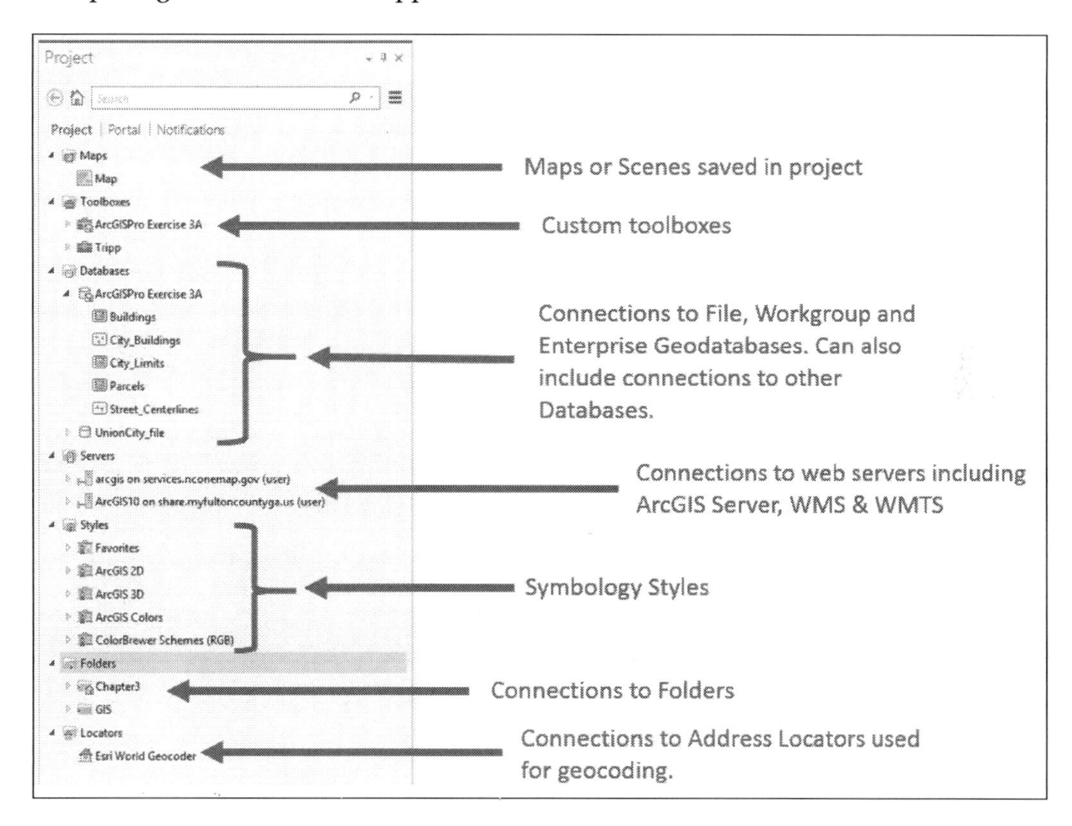

The **Portal** tab provides access to connections you have established to either ArcGIS Online or Portal for ArcGIS. Using this tab, you will then be able to access content that has been published and that you have permissions to use. This can include web maps, feature layers, and web apps. The **Portal** tab has three buttons located on it. The first button is **My Content**. This allows you to access content that was created by you in ArcGIS Online or Portal for ArcGIS. The second button is **Groups**. This button allows you to see the various groups you are a member of and content shared to those groups. The last button is **All Portals**. This allows you to see content shared with you in all portals you have connected other than your primary portal.

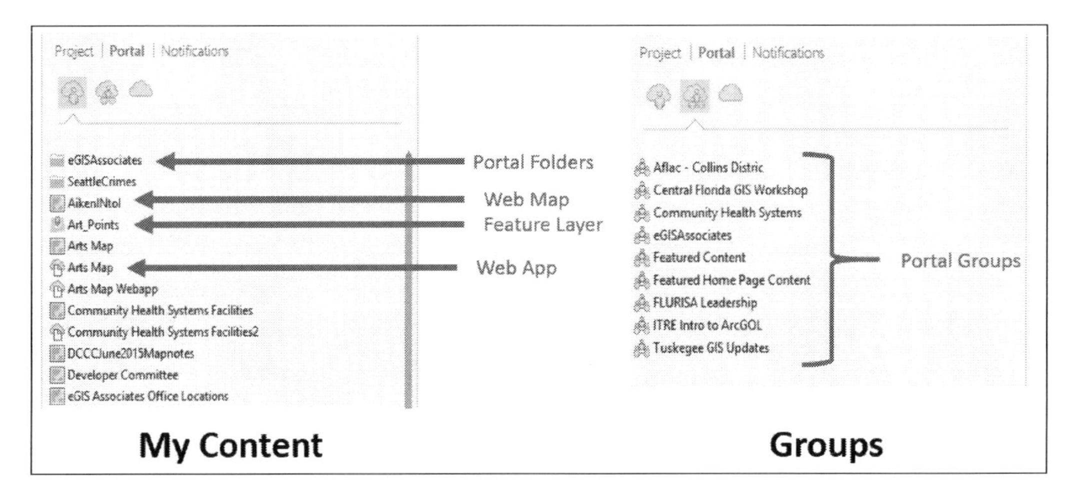

The **Notifications** tab contains messages from processes and tools used while you are working in your project. These messages are not saved and will be cleared if you open a new project or close ArcGIS Pro.

Exercise 3A – using the Project pane

Now that you have learned some of the terms associated with a project and been introduced to the **Project** pane, it is time for you to see it in action. In this exercise, you will work with **Create a Map** which shows the floodplains and drainage basins for the city of Union City. You will need to use the **Project** pane to establish several connections including a geodatabase and server to access additional information, which is needed to create the map. You will start with an existing project that already contains a basic map.

Step 1 – open and save a project

In this step, you will open an existing project and save it. This will be the project you will continue to work with in this exercise:

1. Start ArcGIS Pro. Again remember that you will need an Internet connection and an access to ArcGIS Online.
2. Using the skills you have learned in past exercises, open the `Ex 3A` project located at `C:\Student\IntroArcPro\Chapter3`.
3. Using the skills you have learned, save the project as `%your name% Ex3A` (that is, `Tripp Ex3A.aprx`). Save the project to `C:\Student\IntroArcGISPro\My Projects`.

If you have successfully opened the correct project and saved it, you should see the following:

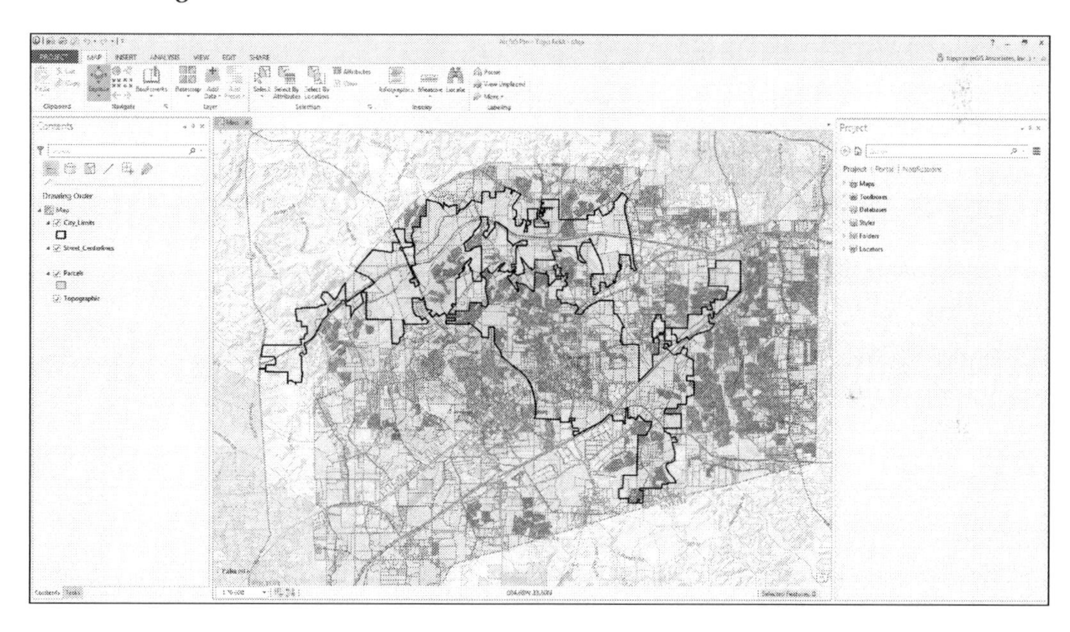

Step 2 – navigating in the Project pane

Now that you have opened the project, you should see the **Project** pane located on the right-hand side of the ArcGIS Pro interface. In this step, you will see what content is already available in the **Project** pane:

1. In the **Project** pane, click on the small arrow head located to the left of **maps** to expand that section.

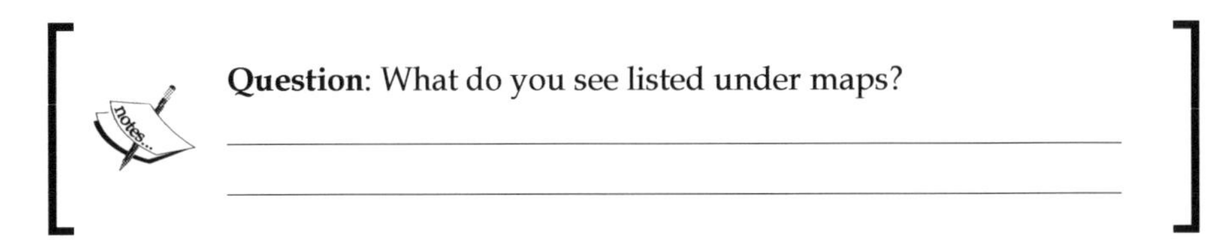

Question: What do you see listed under maps?

2. Expand the **Database** section using the same process.

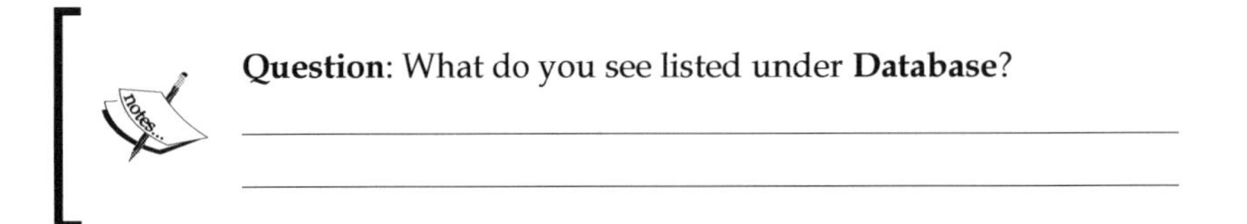

Question: What do you see listed under **Database**?

3. Expand the ArcGIS Pro `Exercise 3A` geodatabase to see what feature classes it contains.
4. Compare the layers in the map to the feature classes in the geodatabase for the project.

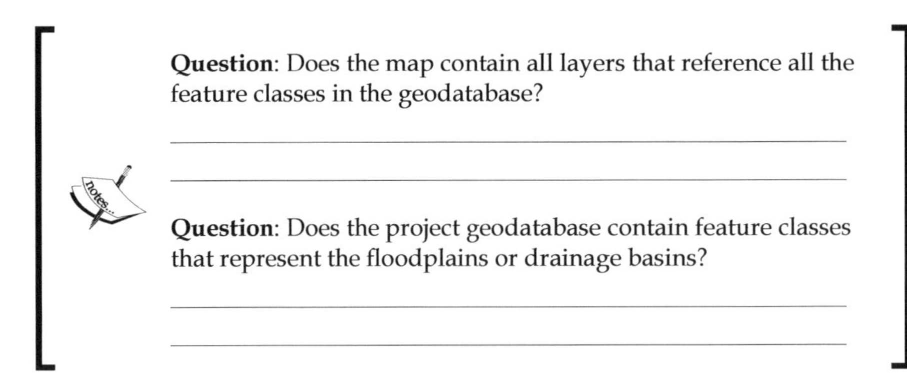

Question: Does the map contain all layers that reference all the feature classes in the geodatabase?

Question: Does the project geodatabase contain feature classes that represent the floodplains or drainage basins?

5. Expand the remainder of the sections within the **Project** pane to see the rest of the project items you currently have access to.

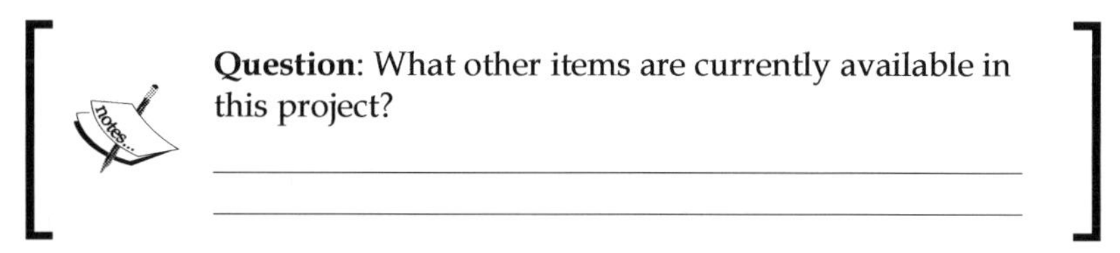

6. Click on the **Portal** tab in the **Project** pane. By default, you start on the **My Content** button.
7. Explore the portal items that are available to you in **My Content** by moving your pointer over some of them.

It is possible that you may see nothing if you have never published any content to ArcGIS Online or Portal for ArcGIS. Later in this book, you will publish a map and project to ArcGIS Online. These will then show in your portal and **My Content**.

8. Click on the **Groups** button located on the **Portal** tab to see what groups you are a member of or have access to.

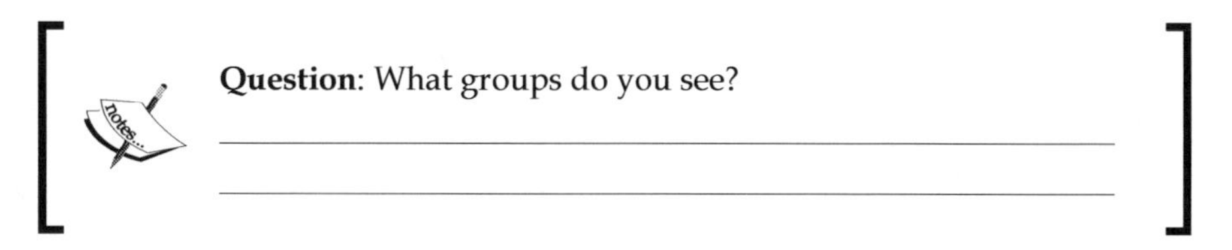

Again you may see nothing here if no one has created groups for your organization in ArcGIS Online or Portal for ArcGIS. As your organization makes more use of those technologies, expect the list of groups to grow.

Step 3 – adding a database connection

In this step, you will add a connection to another geodatabase that contains additional data you need for this project:

1. Click back on the **PROJECT** tab in the **Project** pane.

2. Right-click on **Databases** and click on **Add Database**.

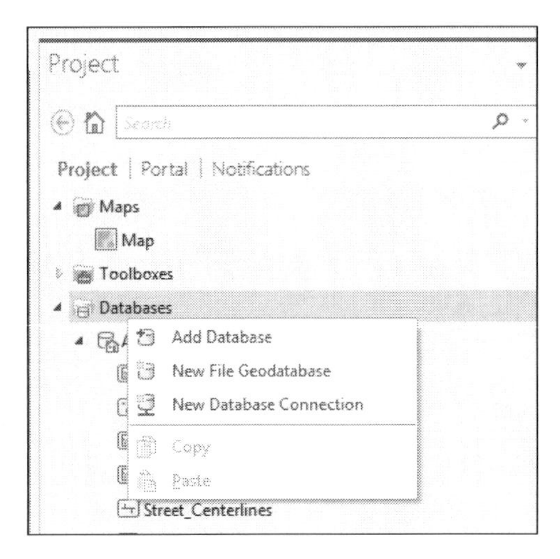

3. In the **Select Existing Geodatabase** window, navigate to `C:\Student\IntroArcPro\Databases`. Select the `UnionCity.gdb` file and click on **Open**.

   You should now see the `UnionCity` geodatabase in the **Project** pane below the ArcGIS Pro `Exercise 3A` project geodatabase. The `UnionCity` geodatabase contains additional layers, which you will need to add to your map.

4. Expand the `UnionCity` geodatabase, so you can see what feature classes it contains.
5. Select the `floodplains` feature class. Holding down your *Ctrl* key, also select the `Watersheds` feature class.
6. Right-click on one of the two selected feature classes and choose **Add to Current Map**.

Your map should look similar to this now. The colors maybe different depending on what colors ArcGIS Pro assigned to the new layers you add. Remember that ArcGIS Pro assigns a random color to new layers added to the map.

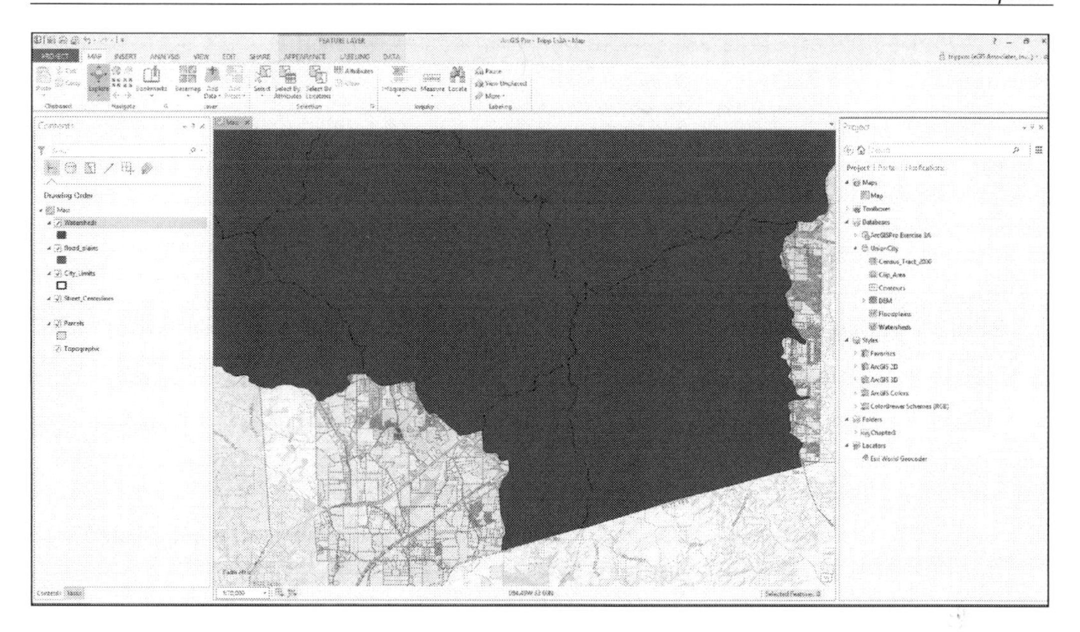

You have just added two layers from a different geodatabase to the map after you made the connection to it. Although the layers are now visible in the map, you will need to make some adjustments to the layers to improve the readability of the map.

Step 4 – adjusting the layers

You will now make changes to the symbology and layer orders so that the map is easier to read. First, you will make changes to the layer draw order:

1. Select the `Watersheds` layer in the **Contents** pane located on the left-hand side of the ArcGIS Pro interface.
2. While holding your left mouse button down, drag the `Watersheds` layer, so it is below the `Parcels` layer in your layer list.
3. Now drag the `Floodplains` layer, so it is below the `Parcels` layer, but above the `Watersheds` layer.
4. Right-click on the `Floodplains` layer and select **Zoom to Layer**.

Now that you have the layers drawing in the order you want them, it is time to work on the symbology for each. You will need to add a new style to our project first which contains the symbol for the floodplains:

5. In the **Project** pane, right-click on **Styles** and select **Add Style**.
6. In the **Add Style** file window, navigate to `C:\Student\IntroArcPro\Chapter3`. Then, choose `ESRI.stylx` and click on the **Select** button.

You have just added this new style to your project, which will allow you to use the symbols it contains for the layers in your map. Styles store symbols, colors, north arrows, scale bars, and other map elements, which are used to create maps and layouts:

7. Click on the symbology patch for the `Parcels` layer to open the **Symbology** pane. A symbology patch is the small sample symbol located just below the layer name in the **Contents** pane, as illustrated here:

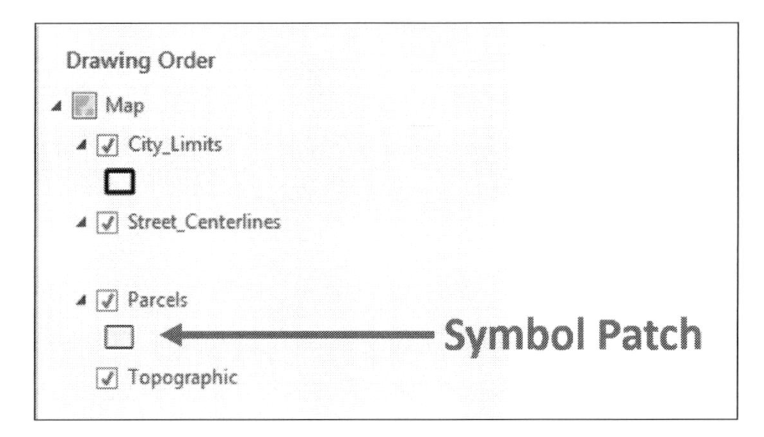

8. Click on the **Properties** tab and set the **Color** to *no color* using the drop-down arrow.
9. Make sure that the **Outline** color is set to black and click on **Apply**.
10. Select the `Floodplains` layer in the **Contents** pane.
11. Select the **APPEARANCE** group tab from the ribbon.
12. Click on the **Symbology** drop-down arrow and select **Unique Values**. This will open the **Symbology** pane on the right-hand side of the ArcGIS Pro interface.
13. For **Field 1**, select **SFHA**. This field is used to specify if an area is in or out of the **Special Flood Hazard Area** (**SFHA**), which is also commonly referred to as the 100-year flood zone. You want to symbolize the areas that are in the SFHA.

14. Click on the symbol patch located next to **IN** in the symbol grid.
15. At the top of the **Symbology** pane in the `search` function, type `Flood` and press the *Enter* key. You will need to be in **Gallery** and not **Properties** to do this.
16. Select the symbol named `100 Year Flood Overlay`. Note that this is in the Esri style you added to the project. It is only available for you to use because you added it.
17. Click on the back arrow located in the top-left corner of the **Symbology** pane to return to the settings for the `Floodplain` layer.
18. Click on the symbol patch next to **OUT** in the symbol grid and set it to **Hollow**. Use the `search` function again if needed.
19. In the symbol grid, select where it says **SFHA** and type `100 Year Floodplain` to replace it.
20. Click on **More** located in the right-hand side just above the symbol grid and deselect **Show all other values**. This removes that from appearing in the **Contents** pane and the legend.

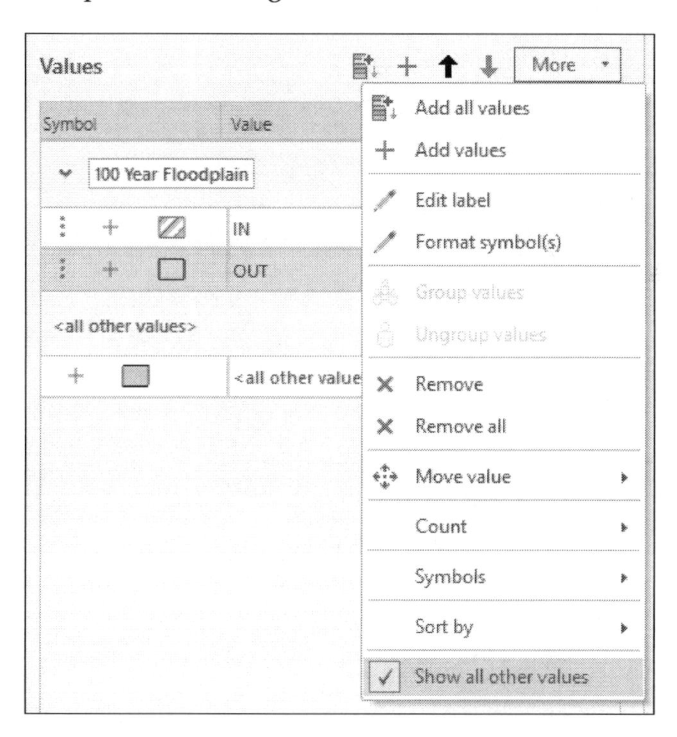

So, now you have added a new style to the project and used a symbol contained in that style to update the symbology for the `Floodplains` layer. You have one last layer to adjust, the `Watershed` layer. You will now apply symbology to this layer using an existing layer file. A layer file stores the properties used to display a layer. You will learn more about layer files later in this book:

21. Select the `Watersheds` layer in the **Contents** pane.
22. Select the **APPEARANCE** tab in the ribbon.
23. Click on the **Import** tool. This will open the **Geoprocessing** pane in the right-hand side of the interface.
24. Click on the browse button located next to the **Symbology** layer.
25. Click on **Folders** located under **Project** in the left pane.
26. Select the `Watersheds.lyrx` file and click on the **Select** button.
27. Make sure that the following are set under the **Symbology** fields in the **Geoprocessing** pane:
    - **Type**: **Value_Field**
    - **Source Field**: **WATERSHED**
    - **Target Field**: **WATERSHED**
28. Once you have verified the **Symbology** fields, click on the **Run** button located at the bottom of the **Geoprocessing** pane.
29. Save your project.

Step 5 – connecting to ArcGIS server

You are almost done. At the last minute, it has been decided that you need to add an aerial photograph to the map. You have been instructed to use aerials served up by Fulton County using ArcGIS for Server. So, you will need to first add a connection to the County's ArcGIS Server and then add the new aerial layer to the map:

1. Right-click in the **Project** pane somewhere that is blank.
2. Click on **New ArcGIS Server** connection.
3. For the server URL: type `http://share.myfultoncountyga.us/ArcGIS10/rest/services`.
4. Leave **User Name** and **Password** blank and click on **OK**.
5. You should now see that servers have been added to your **Project** pane. Expand the **Servers** section to reveal the new connection you just added.

6. Expand the **ArcGIS10 on share.myfultoncountyga.us (user) (1)** connection to reveal the map services that can be used via this connection.

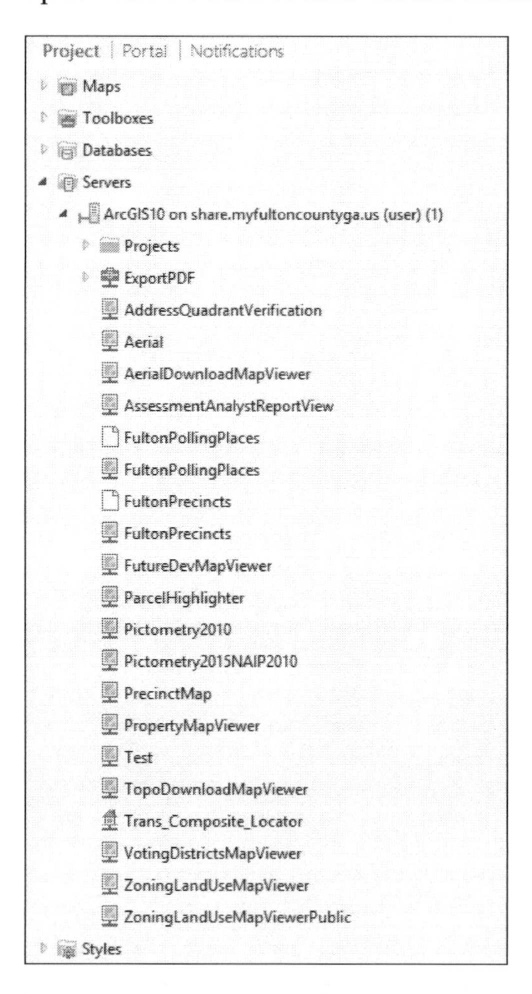

As you can see, the County provides access to a wide range of map services that you can access and use in your maps. By publishing these map services to the Internet, the County is easily able to share its data with others that may need it including the cities within the County. By establishing a connection to this data within your project, you can now begin to use this published information as well. You will now do just that by adding the County published aerials to your map:

7. Right-click on the **Aerial** map service under the new server connection you just added.
8. Select **Add to Current Map**.

9. Now, let's adjust the `Watersheds` layer, so you can see the aerials underneath by applying a transparency. Select the `Watersheds` layer in the **Contents** pane.
10. Click on the **APPEARANCE** tab in the ribbon.
11. In the **Effects** group tab, set the transparency to **65.0%**.

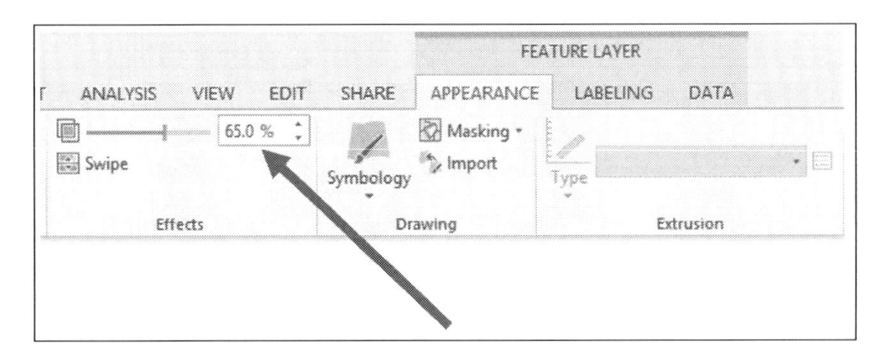

12. Save your project and close ArcGIS Pro.

So, you have experienced how the **Project** pane works. You added several connections to your project and used them to add more information to your map.

Creating a new project and project templates

As you have seen, projects are the core of ArcGIS Pro. They contain all things a user needs to complete GIS tasks required for the project. ArcGIS Pro will not even allow you to open the application without designating a project to open. So far, you have been working with existing projects in ArcGIS Pro.

These projects have already been set up and are ready for you to use. They have included maps, 3D scenes, and various connections. So, how do you create a new project from scratch? In this section, you will learn how to create a new project using one of the Esri-provided templates. Then, you will learn how to create your own template, which you can use to create new projects.

Creating a new project with Esri templates

When you open ArcGIS Pro, you have the option to open an existing project or create a new one. If you choose to create a new one, you first select a template. The template automatically generates content and connections within the new project being created. The template can create maps, scenes, layouts, folder connections, database connections, and more.

In addition to what is created by the template, a new project-specific file geodatabase and toolbox are created in the home folder. Remember that the home folder is the folder you create the new project in. Your new project will set these as the default geodatabase and toolbox used by the new project automatically.

Esri has already created some templates you can use to begin creating new projects as soon as you install ArcGIS Pro. This includes four templates each with a specific use in mind:

Esri template name	General description
Map	This is for use with the 2D data primarily and automatically includes a 2D map that has the Esri topography basemap layer.
Global scene	This is for use with the 3D data that covers a large area primarily and automatically includes a 3D Global scene.
Local scene	This is for use with the 3D data that covers a smaller area primarily and automatically includes a 3D Local scene.
Blank	This creates a new completely empty project with no maps, scenes, or layouts.

To use one of these templates, you simply select it from the **Create a new project** pane in the ArcGIS Pro startup screen. You also have the option to select another project template, which you will learn about later. Let's now see how one of these work.

Exercise 3B – using an Esri template to create a new project

In this exercise, you will create a new project that will contain a local scene using some of the same data you used in the last exercise. You have been asked to create a 3D scene for the city of Union City showing the elevation change of the city and the locations of the basins.

Step 1 – creating a new project using the Local scene template

In this step, you will create and explore a new blank project that you created using the Local scene template:

1. Open ArcGIS Pro.
2. On the right-hand side in the **Create a new project** pane, select the `Local Scene.aptx`.
3. In the **Create a New Project** window, name your project `%Your Name% Ex3B` and set your location to `C:\Student\IntroArcPro\My Projects`. Your window should look similar to the following image:

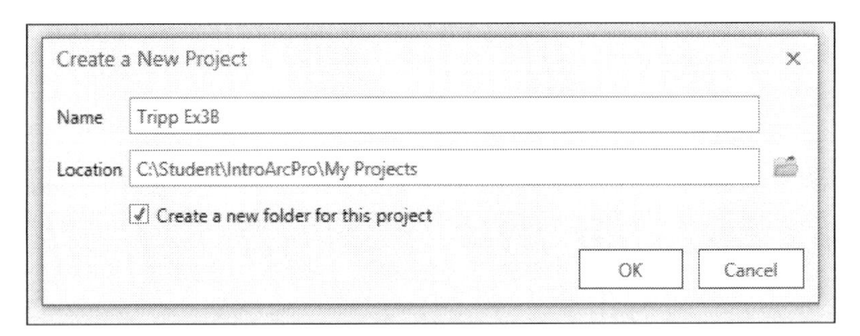

4. Click on **OK** to open your new project.

Your new project should have opened with a new scene, which only contains a 2D `Topographic` basemap layer. You will now take a moment to explore what else was done automatically by using this Esri template:

5. Explore the **Project** pane and answer the following questions:
    - **Question**: What is available or listed under **Maps**?

      ____

      ____

    - **Question**: What is available or listed under **Toolboxes**?

      ____

      ____

    - **Question**: What is available or listed under **Databases**?

      ____

      ____

6. Continue to explore the remaining items in the **Project** pane until you have become completely familiar with the new project you created and what you currently have access to.

Now you will take a look at the backend of what happened when you created your new project, so you can see what ArcGIS Pro does when a new project is created. You will look at your project using your operating systems file explorer.

7. Open your file explorer application. If you are using Windows 7, 8, or 8.1, it is called **file explorer** and uses an icon that resembles a file folder in a file organizer. You can normally find it on the task bar or desktop. If you are using Windows XP, it is called **My Computer**.
8. Once you have opened your file explorer, navigate to `C:\Student\IntroArcPro\My Projects`. You should see a folder inside the `My Projects` folder, which is named `%Your Name% Ex3B`. This folder was automatically created by ArcGIS Pro when you created the new project. This is your projects home folder.
9. Open the home folder, so you can see its contents. The home folder should contain several folders and files all of which were automatically generated by ArcGIS Pro when you created the new project.

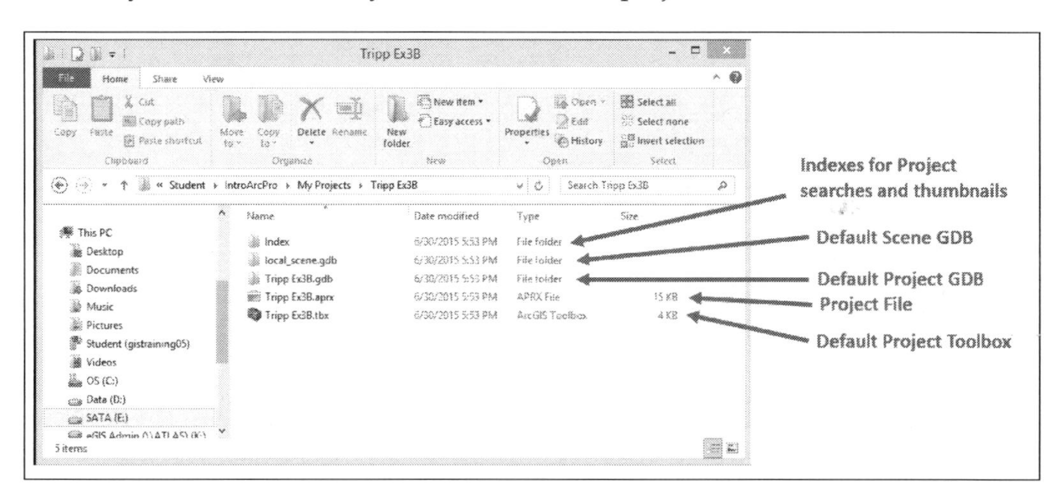

Different templates will create different files and folders. Remember templates are created with specific purposes in mind, so each will be different to some extent.

10. Feel free to explore some of the folders more fully if desired. Once you are done, close your file explorer and return to ArcGIS Pro.

Step 2 – connecting to data sources and adding your surface

In this step, you will connect to the `UnionCity` geodatabase, which you used in the last exercise. You will then set up your scene to use the **Digital Elevation Model (DEM)** in that geodatabase as the elevation surface. The elevation surface serves as the ground level or height all other layers are overlaid on to within the 3D views provided for in a scene:

1. Right-click on your scene in the **Contents** pane and select **Properties**.
2. Click on the **General** option in the left-hand side window. Rename the scene `Union City`.
3. Click on the **Coordinate** system option in the left window. Then, click on the **Import Coordinate System** button located next to the **Search** cell.
4. In the **Import Coordinate System** window, navigate to `C:\Student\IntroArcPro\Databases\UnionCity.gdb`. Select the DEM raster and click on the **Open** button. This sets your scene, so it will use the same coordinate system as the DEM and other layers you will add to the scene during this exercise.
5. Click on **OK** in the **Map Properties** window.
6. Using the skills you learned in the last exercise, add a database connection to the `UnionCity.gdb` in the **Project** pane. As you learned before, this allows you to use the data in this geodatabase within your project.

Before you add any layers to your scene, you need to define the elevation surface. By default, ArcGIS Pro uses Esri's world elevation model from ArcGIS Online. However, you can use your own if you have one. In this case, you do have your own digital elevation model, which is more accurate than Esri's model. So, you will now configure your scene to make use of your local more accurate DEM.

7. Right-click on the `Union City` scene and open its properties.
8. Go to the **Elevation Surface** option.
9. Expand the **Ground** surface, and you will see the default surface from ArcGIS Online.
10. Click on the add elevation surface button. This button is located to the far right of the **Ground**. It looks like a yellow folder with a **+** on top of it.

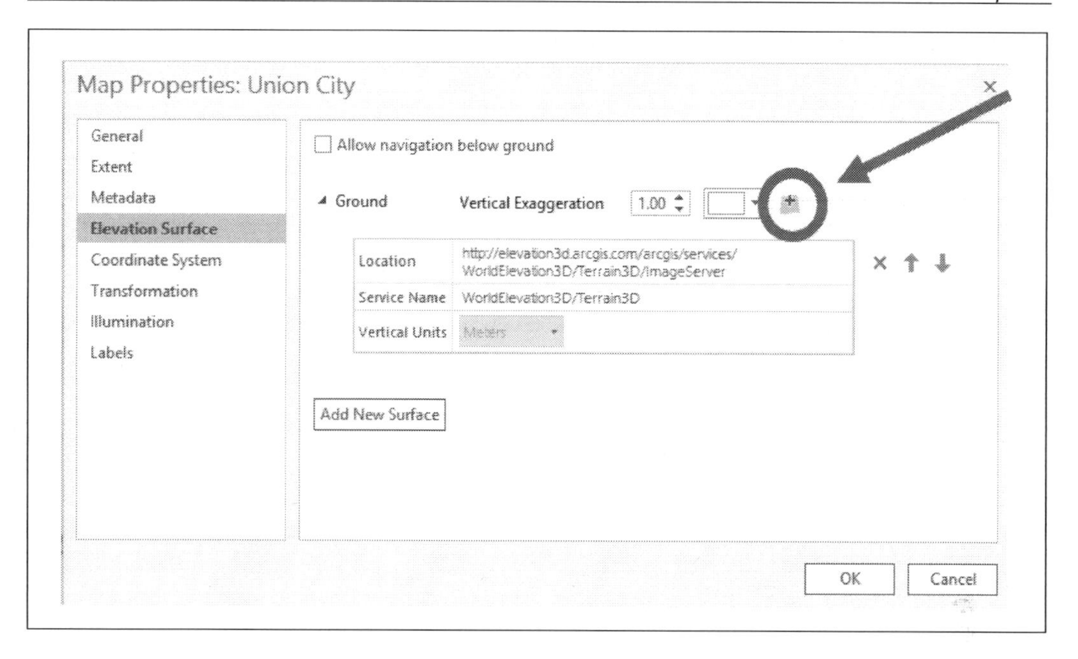

11. In the add elevation source window, click on **Databases** in the left-hand pane.
12. Double-click on the `UnionCity.gdb`. Select **DEM** and click on the **Select** button.

You should now see the `Union City` DEM above the default ArcGIS Pro ground elevation. Since we are not concerned with areas outside the immediate Union City area in this scene, you will remove the default ArcGIS Pro ground surface:

13. Click on the small red **X** located next to the default ArcGIS Pro surface to delete it.
14. Click **OK** to close the **Map Properties** window.
15. Using the skills you learned in previous exercises, add the `Floodplains` and `Watersheds` feature classes to the scene.
16. Using the same process you followed in the previous exercise, adjust the symbology, so they are symbolized the same way they were in the *Exercise 3A – using the Project pane* section.

Hint: You may need to connect to the `Chapter3` folder and add the Esri style.

17. Save your project.

18. Click on the **MAP** tab in the ribbon.
19. Click on the **Add Data** button and navigate to the `Chapter3` folder

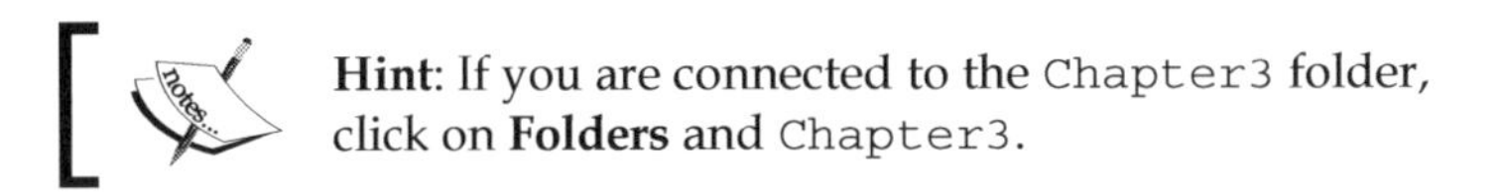

Hint: If you are connected to the `Chapter3` folder, click on **Folders** and `Chapter3`.

20. In the `Chapter3` folder, choose the `3D Buildings.lyrx` file and click on **Select**.

You have just added a new layer to your scene using a layer file. As you can see, the layer was automatically assigned a specific symbology. This was because you used a layer file to add the data instead of going directly to the geodatabase and adding the feature class. A layer file contains predefined settings for a layer, including symbology, labeling, transparency, source, and more. They allow you to standardize layers for use in multiple maps:

21. Right-click on the **Buildings** layer you just added and select **Zoom to Layer**.
22. Using the **Explore** tool along with the scroll wheel on you mouse zoom into the area where the Whitewater Creek, Morning Creek, and Deep Creek `Watershed` basins intersect. Zoom in until your map looks similar to the following image:

You can now see the buildings in 3D with a realistic facing applied. You can see what basin they are in, and if you pan around, you can also see which ones are in the 100-year floodplain.

23. Save your project and close ArcGIS Pro.

So, you have just created a new project using Esri's Local scene template and then built out the scene with layers and a local elevation surface. How do you create your own project templates for use by ArcGIS Pro users in your organization? You will investigate that next.

Custom project templates

In the previous exercise, you saw how useful a template can be to start a new project. Templates can standardize all projects you or your organization creates in ArcGIS Pro, so they access the correct databases, contain the appropriate styles, and have the proper layout elements. Project templates are really specialized versions of a project package that can be used to create new projects. They have an **APTX** file extension.

Project templates can be saved to several locations. You can save them to your computer, a network share, ArcGIS Online, or Portal for ArcGIS. The location where you save them will help determine who can use them.

Project templates saved to your computer generally can only be used by you. This can limit their value to an organization. However, since they are stored locally on your computer, they can be better suited for your personal use. This is because they can contain connections to local resources, which exist on your computer and others may not have access to. Also, if you include connections to secure data sources or files, you do not have to worry if other users will have permissions to access those locations.

You save project templates to a network share such as a folder on your file server or to ArcGIS Online, which can be used by others in your organization when creating new projects assuming that they have sufficient privileges to access the location. This allows you to standardize projects created by ArcGIS Pro users throughout your organization, which has several advantages:

- Templates can ensure common map layouts, which include standard title blocks, north arrows, logos, and legal disclaimers.
- Templates can ensure that everyone accesses the correct data sources, folders, styles, and toolboxes.

- Templates can ensure that data and files are saved to the proper locations by standardizing a common project structure.
- Templates can ensure that everyone uses the correct basemap when creating maps or scenes.

Those are just a few of the advantages of using templates within an organization. If you are going to save your templates to ArcGIS Online, there is a security concern you need to make sure to pay careful attention to, who you choose to share with.

If your organization has enabled sharing outside your organization, it is possible that others who are not in your organization may get access to your templates. When you create and save a template to ArcGIS Online, you are asked who you wish to share the template with. If you select everyone, then you will be sharing your template with all users that have access to ArcGIS Online. This includes those that are not part of your organization. This means ArcGIS Pro users not affiliated with your company or group will be able to use your template. This also means that they will be able to see everything you included in the template, such as database connections, logos, folder connections, layouts, and more. Access to your templates could represent a security breach waiting to happen if you are not careful. So, when you save a template to ArcGIS Online, make sure to pay attention to who you choose to share it with.

Exercise 3C – creating a custom project template

Now that you understand *why* you would want to create custom project templates, let's look at the *how*. In this exercise, you will create a custom project template, which you will save to your computer. This template will include a map, database connections, and more. Once you create the template, you will then create a new project that uses the template.

Step 1 – create a project

All templates start as a project. So to create a new template, you must first have a project that contains all the standard settings to include in the template. So, you will start in this step by creating a new project using the blank template. This provides you with a blank slate upon which to build your template:

1. Start ArcGIS Pro.
2. On the right-hand side in the **Create a new project** pane, select `Blank.aptx`.
3. In the **Create a New Project** window, name your project `%Your Name% Ex3C` and set your location to `C:\Student\IntroArcPro\My Projects`.
4. In the **Project** pane, right-click on **Database** and select **Add Database**.
5. In the **Select Existing Geodatabase** window, navigate to `C:\Student\IntroArcPro\Databases` and select `Trippville_GIS.gdb` to add it as a connected database.
6. Using the skills you have learned, add the `ESRI.stylx` style to the project.

Step 2 – setting up a default map in the template

Now that you have opened a new project and established some default connections to databases and styles, you will now create a default map, which will be included in the template. You will configure a new basemap and then set a default view extent:

1. Right-click on the **Maps** folder in the **Project** pane and select **New Map**.
2. On the **MAP** tab in the ribbon, click on the small drop-down arrow under **Basemap** and select **Imagery**.
3. In the **Project** pane, expand the `Trippville_GIS` geodatabase you connected to.
4. Expand the **Base** feature dataset and right-click on the `City_Limit` feature class. Select **Add to Current Map**.

Your map should zoom to the new `City_Limit` layer you just added, so it should look similar to the following image. Remember that the color for the `City_Limit` layer maybe different in your map because ArcGIS Pro assigns a random color to new layers unless you add the new layer using a layer file. Adding the `City_Limit` layer also assigned the coordinate system you wish to use for the map and in your template. Just as with ArcMap, the first layer you add to a map in ArcGIS Pro assigns the coordinate system to the map.

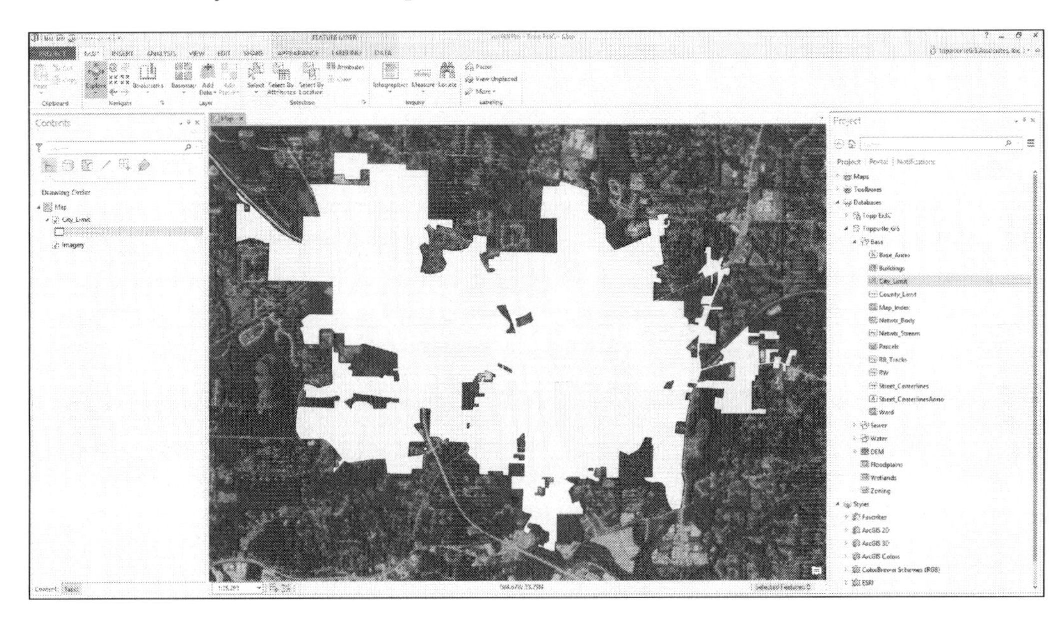

Now you will set this view extent to be the default for this map in your project:

5. Right-click on **MAP** in the **Contents** pane and select **Properties**.
6. In the **Map Properties** window, select **Extent** from the left-hand pane.
7. Click on the radial button next to **Custom extent**.

8. Click on the drop-down menu located to the right of **Calculate from** and select **Current visible extent** so that your **Map Properties: Map** window looks like this:

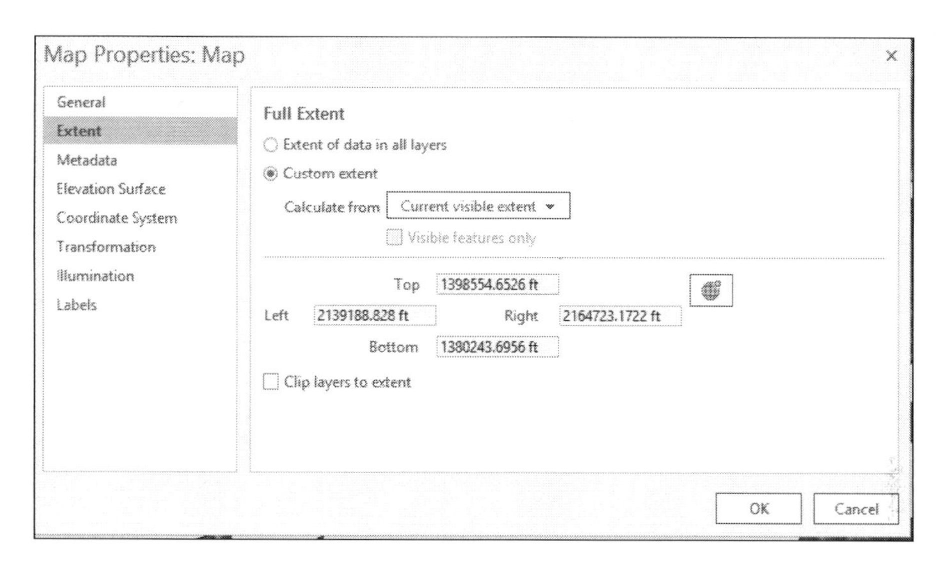

9. Once you have verified that the extent has been properly set, click on **OK**.
10. Using the skills you have learned, make the following changes to the `City_Limit` layer:
    - Symbology to a hollow fill
    - Outline to `Mars Red`
    - Outline width to `2 pt`
    - Rename the layer `City Limits`
11. Save your project.

Step 3 – adding a layout from an existing map document

You have one last thing you wish to add to your project before you save it as a template. You want to add a layout that you have been using in your map documents you created with ArcMap, another Esri application. This will allow your maps to look the same whether they are created in ArcGIS Pro or ArcMap:

1. Click on the **INSERT** tab in the ribbon.

2. Click on the small drop-down arrow located next to **New Layout**, and select **Import a layout file...**, which is located near the bottom of the window.

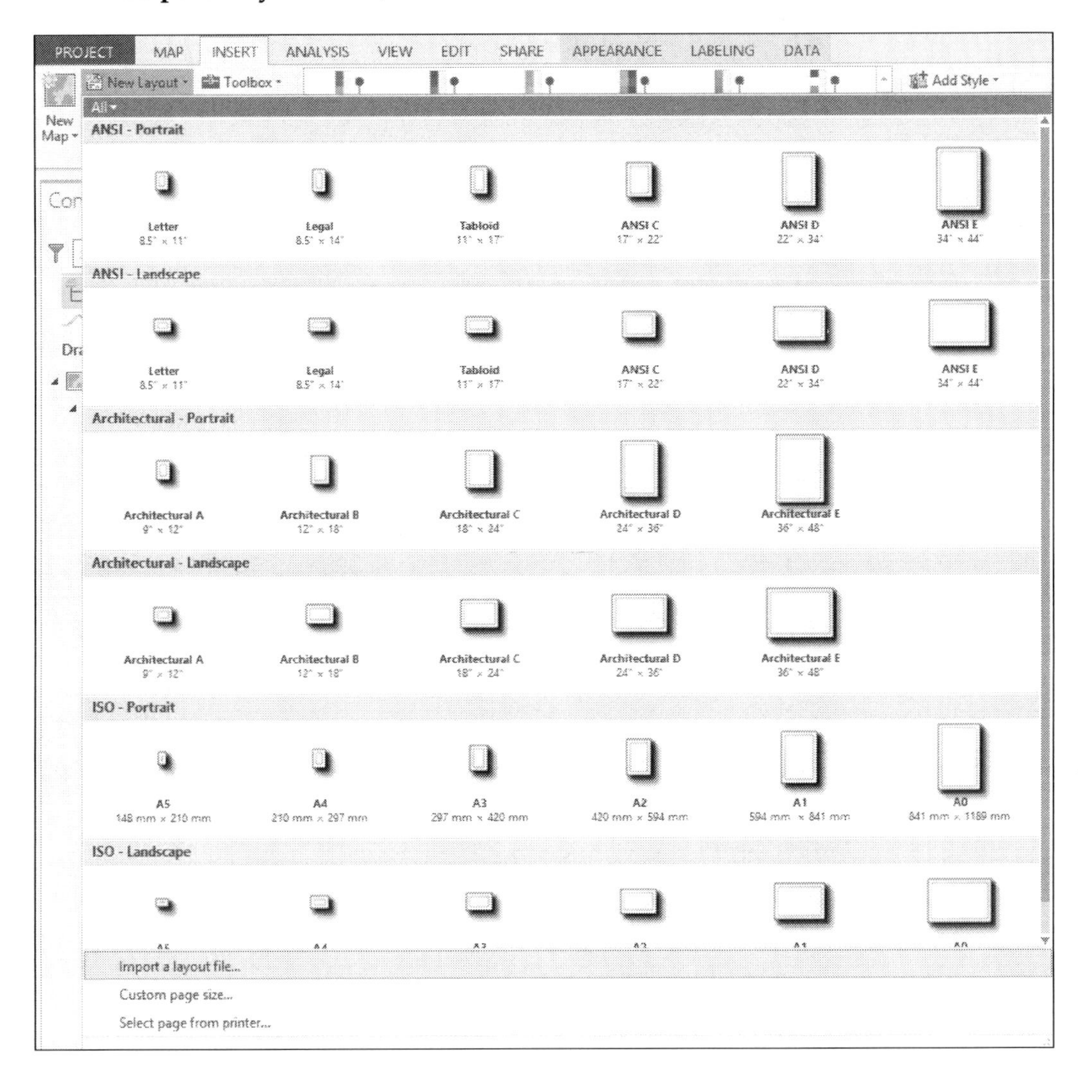

3. In the **Import** window, navigate to the `C:\Student\IntroArcPro\Chapter3` folder. Choose the `Sample Layout.mxd` file and click on **Select**.

You have just imported the map layout, which was in the map document. This layout was created with ArcMap and is used by your organization as a standard layout for all maps you print. You will need to make some adjustments to this layout for it to work properly in ArcGIS Pro:

4. Right-click on **Map Frame** in the **Contents** pane and select **Properties** to open the **Element** pane on the right-hand side of the interface.
5. Make sure that the **Options** button is selected in the pane. It is the first one under the word **Properties**.
6. You should see two options, **General** and **Map Frame**. In the **General** option, rename the frame to `Map Frame`.
7. In the **Map Frame** option, click on the drop-down menu and select **Map**. When you do this, you should see that the map you created in Step 2 should appear in the layout. It should look like this now:

8. Now, we need to make adjustments to the legend. Select the legend in the **Contents** pane.
9. The **Legend Tools** contextual tab will appear on the ribbon. Select the **Design** tab.
10. Click on the small drop-down arrow next to **Layers** in the **Map Frame** group tab and click on the box next to `City Limits` to turn this layer on in the legend.
11. Set the **Resize Behavior** to **Adjust columns** using the drop-down menu.
12. In the **Elements** pane located on the right-hand side of the interface, select the **Placement** button located below the word **Properties**.

13. Make the following adjustments to the legend placement properties:
    - **Width**: `1.75 in` (**Note**: the height should automatically adjust to `.55 in`).
    - **X position**: `8.65`.
    - **Y position**: `2.00`.

    Your layout should look very similar to this when you are done:

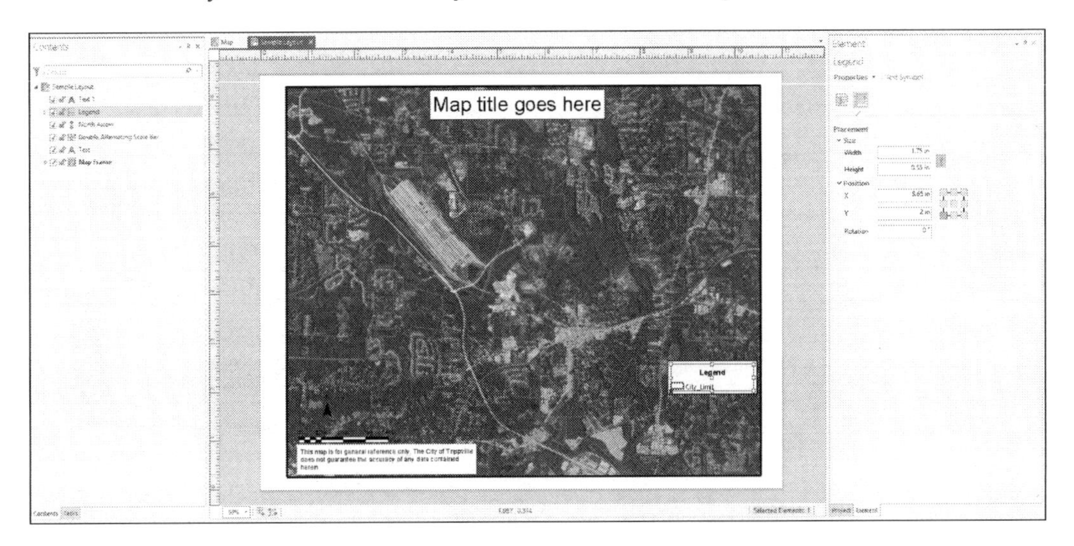

14. Close the **Elements** pane.
15. You will save a bookmark for this layout, so you can easily return to this location and scale. Select the **Layout** tab on the ribbon.
16. Choose **Create Bookmark**. In the **Create Bookmark** window, name your bookmark `Print Layout` and click on **OK**.
17. Close the **Sample Layout** and save your project.

Step 4 – saving a project as a template

Now that you have configured the project to have your standard database, style, and folder connection plus included a sample map and layout, which will be the common basis for new projects, you are ready to save your project as a template:

1. Click on the **SHARE** tab on the ribbon.
2. Select the **Project Template** button in the **Save As** group tab.

3. The **Create Project Template** pane is displayed in the right-hand side of the interface. Under the **Start Creating** section, select **Save template to file**.
4. Click on the **Browse** button located under **Name and Location** to specify the location you will save the template at and the name you will give it.

The default location is located in your user profile and the `ProjectTemplates` folder. You will accept the default location so that the template appears in your list when you start ArcGIS Pro:

5. Name your template `%your name% 2D Project Template` (that is, `Tripp 2D Project Template`) and click on **Save**.
6. You will now need to fill out the item description information for the template. The item description is a shortened form of metadata. Complete the following as indicated here:
    - **Summary**: This serves as a template for 2D ArcGIS Pro projects and includes a single 2D map and a layout. It will also automatically connect to the `Trippville_GIS` geodatabase and add the Esri symbol style.
    - **Tags**: 2D, Trippville, layout, map, template.
7. Click on the **Analyze** button to make sure that your template does not contain any issues that would prevent it from working properly. The **Analyze** button will automatically take you to the **Messages** tab in the pane. You should not see any errors or warnings. If you find any error, correct them as recommended by ArcGIS Pro.
8. Once you have no errors in the **Messages** tab, click on the **Create** button in the pane.

    When ArcGIS Pro is done creating the template, you will see a message letting you know that it is complete.
9. Close ArcGIS Pro. You can choose to save the project if you desire, but it is not required.

Step 5 – creating a new project using a custom template

Now that you have created a custom project template, you need to know how to use it when you create a new project. In this step, you will use the project template you just created to create a new ArcGIS Pro project.

1. Start ArcGIS Pro.
2. In the **Create a new project** pane, you should see the new template you just created at the bottom of the list of available templates. Select the template you create in the last step.

3. Name your new project `%your name% template test`.
4. Set the save location to `C:\Student\IntroArcPro\My Projects` and click on **OK**.

   Your new project should open with a map that looks very familiar. You should see the `City Limits` layer with the **Imagery** basemap.

5. Take some time to explore the **Project** pane. Look at what connections, maps, layouts, and styles are present.
6. When you are done exploring your new project, save it and close ArcGIS Pro.

Summary

In this chapter, you learned how important the projects are when using ArcGIS Pro. They are the very core of the application. You have seen how they contain all the items associated with a project, including database connections, maps, styles, toolboxes, layouts, server connections, and more. By having all these in a project, it is much easier to access required resources.

You have created a new project and seen how you can expand it by adding new resources. A project is not limited to a single database, map, folder, or style. You can choose to make as many connections as you wish. The **Project** pane in ArcGIS Pro provides the tools and methods to access, add, view, or remove those resources.

Finally, you learned how a project template can be used to make creating new projects more efficient by preconfiguring various project items. You have the option to use one of the predefined ArcGIS Pro templates, or you can create your own custom templates.

4
Creating 2D Maps

One of the key functions of a GIS is to visualize data through the use of maps. Maps allow us to see and analyze the spatial relationships between features in one or more layers. Maps are created for many different reasons. Often, they are used to show the location of specific assets or to highlight attributes associated with features such as a parcel's zoning classification or a pipe size. Sometimes, they are used to show the results of analysis we have performed.

So, each map you create in ArcGIS Pro has a reason to exist. This purpose will help to dictate what layers you choose to include and how those layers are visualized. There is no limit on the number of layers a map can contain. However, it is possible to have so many layers in a map that it distracts from the purpose. As the author of the map, you must strike a balance between the amount of information you include in a map and its overall purpose.

Each layer in a map has a set of properties. This includes things such as the source, symbology, visibility ranges, coordinate system, and so on. When you create a map and add layers, you must configure many of these settings manually to ensure that your map is legible and supports the purpose of the map.

As you have already experienced, ArcGIS Pro allows you to create both 2D and 3D maps within a project. In this chapter, you will focus on creating a 2D map. These have been the mainstay of GIS since it was first developed over 50 years ago. In this chapter, you will learn how to:

- Create a new 2D map within a project
- Use various methods to add new layers to your map
- Configure layer symbology using various methods
- Configure label settings
- Configure other layer properties

Creating and configuring 2D maps

ArcGIS Pro provides several methods to create new 2D maps within a project. You have already seen how creating a new project using a template can automatically create a new map. However, that is not the only way to create new maps within a project. There are several other methods to create or add maps to your projects. Remember that an ArcGIS Pro project can contain multiple maps.

A map only provides a canvas. You must add the content. This normally means adding layers, which represent the features you wish to display within your map. Just as there are multiple methods to create new maps, there is also more than one way to add new layers to a map.

You will now investigate different methods to create new maps and add layers.

Creating a new map

Maps are the primary way to visualize data in a **geographical information system** (**GIS**). With each project, there may be a need to have multiple maps. You may need a map that shows the general location of the project in relation to a larger area. For example, you may be working on a project to determine the location of a new water reservoir. You need to show the location of the proposed new reservoir within your county or city, so others can know where it will be generally located. Although this map is good at showing where the reservoir is located, it is at a scale that would make it hard to see the level of detail that might be needed by an engineer, planner, or surveyor. So, you might need another map that focuses on the site of the new reservoir, which shows its boundary, the parcel or parcels of land it is on, the elevation contours for the site, and so on.

To keep from making a single map too cluttered, you might actually need to create multiple maps that focus on one or two primary layers. Remember that a map should have a specific purpose. All layers within maps should support that purpose. It is possible to have too much information.

So, how do you create new maps within ArcGIS Pro? One way is using the **INSERT** tab on the ribbon. From there, you can choose the **New Map** button as shown in the following screenshot:

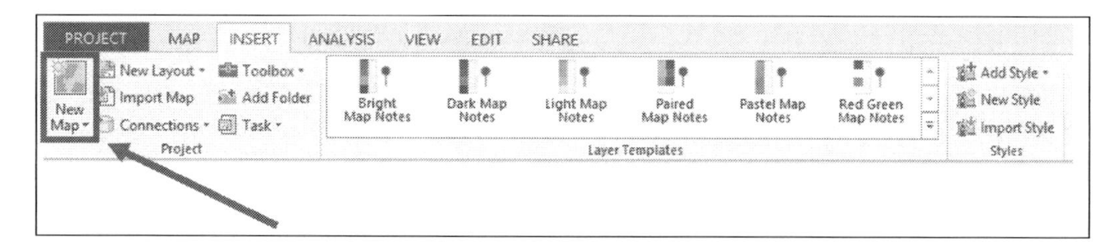

Another way is to right-click on the **Maps** folder within the **Project Pane** and select **New Map**, as shown in the following screenshot:

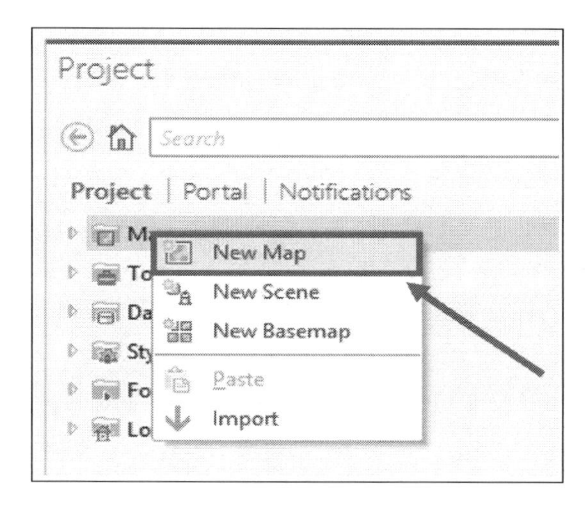

Another way is to import an existing map. ArcGIS Pro allows you to import maps from map documents, map packages, and map files. Each of these has unique file extensions, as shown in the following table:

Item	File extension	Software used to create
Map document	MXD	ArcMap
Map packages	MPK or MPKX	ArcMap, ArcCatalog, or ArcGIS Pro
Map files	MAPX	ArcGIS Pro

When you import an existing map from one of these files, the new map it creates in your project will contain the layers along with all of their settings, such as symbology and labels, that were in the map you imported. This provides a way to help populate a new project with items quickly, so you can get to work faster. This is especially true for those that are migrating to ArcGIS Pro from ArcGIS for Desktop.

Exercise 4A – adding and configuring layers in a map

You have been asked by the Community and Economic Development Director for the City of Trippville to prepare several maps for a business that wishes to open a location in the city. He needs a map that shows the location of all commercial property in the city. Then, he wants a map that shows all commercial parcels over 1 acre but less than 3 acres in size.

In this exercise, you will create a new project using the `Map` template. You will create the first map requested by the Director. This will require you to add and configure several layers using various methods. You will add the second map to the project in another exercise.

Step 1 – creating the project

In this step, you will create a project that will be used to develop the maps requested by the Community and Economic Development Director:

1. Start ArcGIS Pro.
2. In the **Create a new project** pane, which is located on the right-hand side of the ArcGIS Pro opening window, select the `Map.aptx` file.
3. Set the project location to `C:\Student\IntroArcPro\My Projects` by clicking on the **Browse** button located on the right-hand side of the **Location** cell and navigating to the **My Projects** folder as specified.
4. Name your project `%your name% Ex4A` (that is, `Tripp Ex4A`) and click on **OK**.

Now you need to connect to the folders and databases you need to use for this project. You will connect to the `Trippville_GIS` geodatabase and the folders that contain other files you need:

5. Expand **Databases** in the **Project** pane, so you can see the list of all connected databases.

> **Question**: What databases are currently connected to this project?
>
> ____
>
> ____
>
> **Question**: Which geodatabase is the default geodatabase?
>
> ____
>
> ____

You will not need the `Map` database for this project, so you will remove that connection. Removing unneeded connections will help ensure that your project performs efficiently and responds faster.

6. Right-click on the `Map` database connection in the **Project** pane. Select **Remove**.
7. Now you need to connect to the `Trippville_GIS` database. Right-click on **Databases** in the **Project** pane and select **Add Database**.
8. In the **Select Existing Geodatabase** window, navigate to `C:\Student\IntroArcPro\Databases` and select `Trippville_GIS.gdb`.
9. Finally, you need to connect to a folder that will contain files you will need to complete this project. Right-click on **Folders** in the **Project** pane. Select **Add Folder Connection**.
10. In the **Add Folder Connection** window, navigate to `C:\Student\IntroArcPro\`. Then, select the `Chapter4` folder and click on the **Select** button.

    You have now connected to all the resources you will need to complete this project. Your **Project** pane should look like this if you have connected to all the project items as directed:

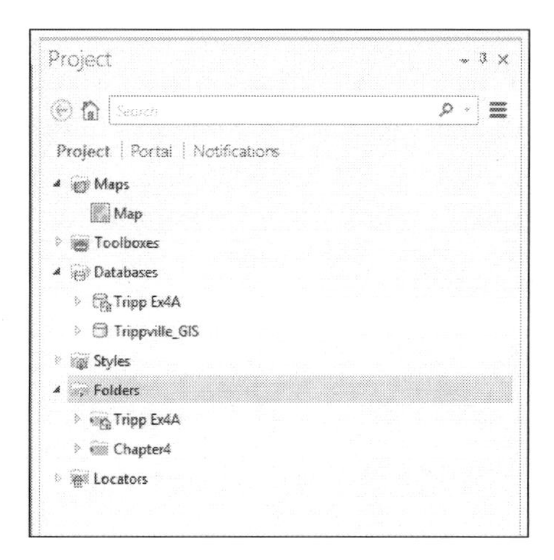

11. Save your project.

Step 2 – adding layers

Now that you have created your project, it is time to begin on the maps requested by the Community and Economic Development Director. You will start with the first map he requested, showing the location of all the commercially zoned property within the city.

You will use the map that was created when you generated your new project using the project template. You will add and configure a few layers using various methods:

1. The first layer you want to add is `City Limits`, so everyone will know which areas are in the City and which are out. In the **Project** pane, expand **Databases** and then the `Trippville_GIS` geodatabases.
2. Expand the **Base** feature dataset and select the `City_Limits` feature class. Holding your mouse button down, drag the `City_Limits` feature class into the map view. Once your mouse pointer is in the map view, release your mouse button.

`City Limits` have now been added to your map using a drag and drop method. This is one of the methods you can use to add a new layer from resources connected in the **Project** pane. Now you will need to change the symbology for the `City_Limits` layer:

3. Click on `City_Limits` in the **Contents** pane until a blue box highlights the layer name. Then, type `City Limits` to give the layer a new name, which will appear better in a legend.
4. Click on the symbol located below the `City Limits` layer name in the **Contents** pane to open the **Symbology** pane on the right-hand side of the interface.
5. Click on the **Properties** tab in the **Symbology** pane.
6. Make the following changes to the **Appearance** settings:
    - **Color**: No color
    - **Outline Color**: Black
    - **Outline Width**: 3 pt

The **Appearance** setting window looks as follows:

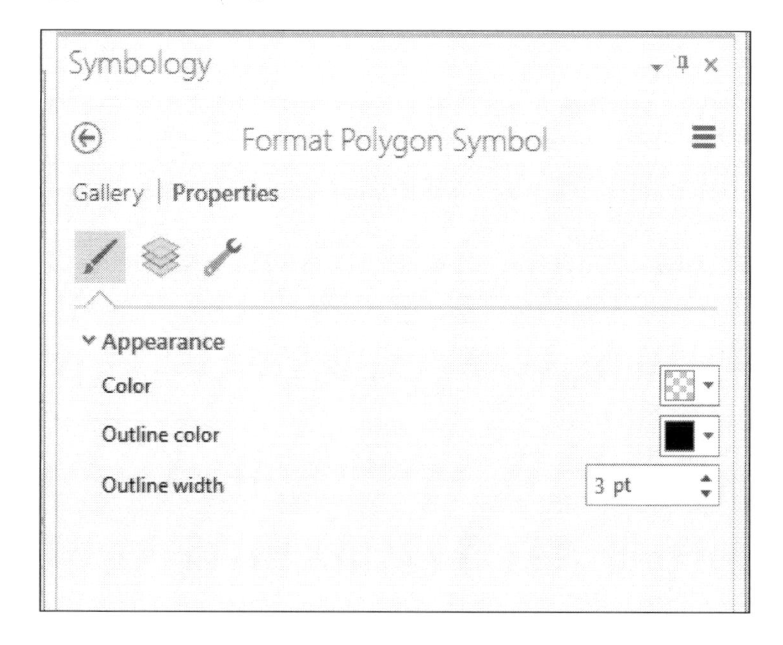

7. Once you verify that the settings are correct, click on **Apply**.
8. Click on the **MAP** tab on the ribbon.
9. Your map now has a single configured layer. Click on the **Add Data** button to add a few more layers.
10. Click on **Databases** in the **Add Data** window and then double-click on the `Trippville_GIS` geodatabase.
11. Double-click on the **Base** feature dataset.
12. Holding down your *Ctrl* key, select the `County_Limit`, `Natwtr_Body`, `Natwtr_Stream`, and `RR_Tracks` feature classes.
13. Click on the **Select** button to add these new layers to your map.
14. Make the following changes to the layers you just added using the skills you have learned so far:
    - `RR_Tracks`

      Name: Railroad

      Symbol: Railroad (Hint: Look in the gallery)

- `Natwtr_Stream`

  Name: Streams & Creeks

  Color: Cretan Blue

  Line Width: 1 pt

- `County_Limit`

  Name: County Limit

  Color: 60% Gray

  Line Width: 2 Pt

  Move County Limit layer below City Limits in the **Contents** pane

- `Natwtr_Body`

  Name: Lakes & Ponds

  Color: Yogo Blue

  Outline Color: Cretan Blue

  Outline Width: 1 pt

  Move the Lakes & Ponds layer above the Streams & Creeks layer in the **Contents** pane

  Your map should look like the following image:

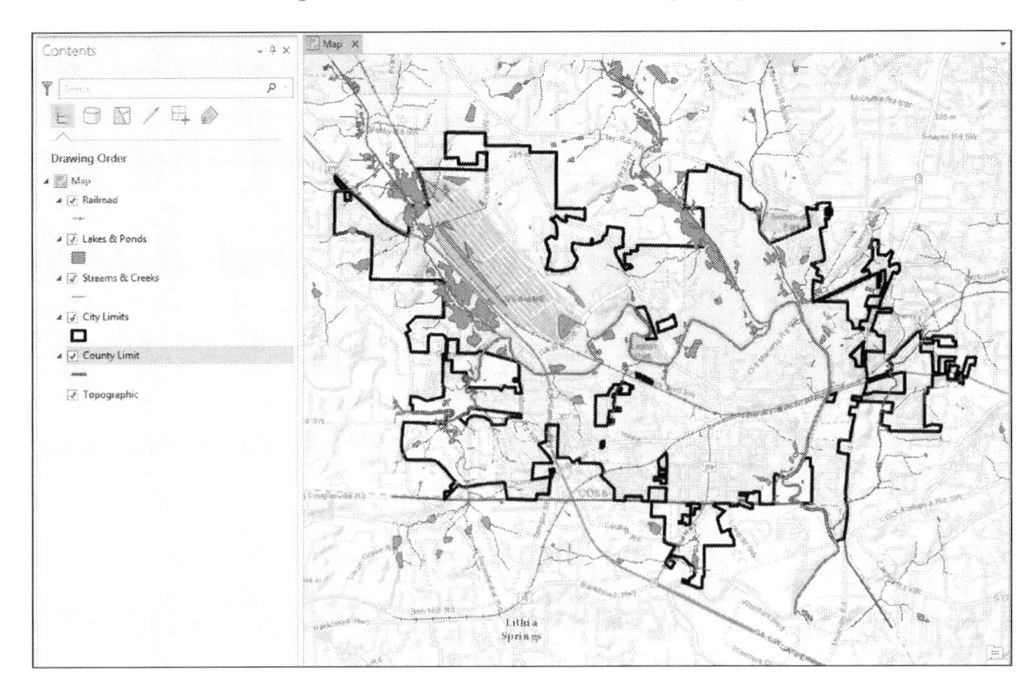

15. Save your project.

The last layer you need to add to this map to meet the request from the Director is a layer that shows all parcels zoned commercially. Luckily, another user has created a zoning layer file. As you have seen earlier, layer files can be used to set the symbology for a layer you already have in your map. It can also be used to add a new layer, which is already symbolized. You will use this layer file to add a zoning layer to the map and then limit it to only show those parcels with a commercial zoning classification. The zoning classes, which delineate commercial zoning, include GC, GC-S, and CBD. GC indicates a zoning of General Commercial. GC-S indicates General Commercial with special stipulations, and CBD is the central business district.

16. Expand folders in the **Project** pane.

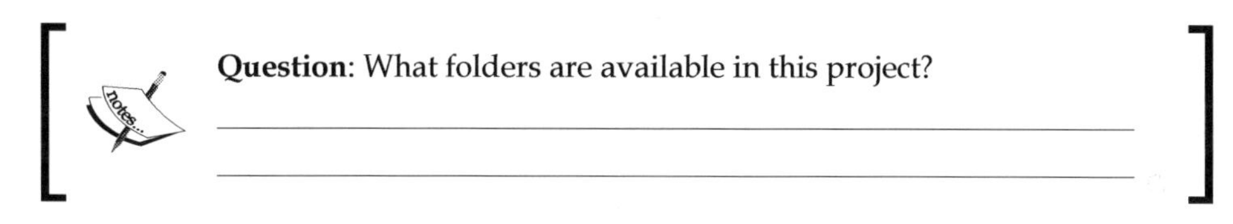

Question: What folders are available in this project?

17. Expand the `Chapter4` folder and right-click on the `Trippville Zoning.lyr` file. Select **Add to Current Map**.

You have added a layer that shows the zoning of all parcels in the City. Now it is time to limit the layer to only the commercial zoning classes.

18. Expand the symbology for the zoning layer in the **Contents** pane by clicking on the small arrow located to the left of the zoning layer. You can now see all the different zoning classifications and their corresponding symbology.
19. Select the **APPEARANCE** contextual tab in the ribbon.
20. Click on the **Symbology** drop-down menu and select **Unique values** to open the **Symbology** pane on the right-hand side of the interface.
21. Select the row for the PUD symbol settings in the grid.
22. Holding down your *Shift* key, click on the row containing Mobile Home Park to select all values from PUD to Mobile Home Park.
23. Right-click on one of the selected rows and select **Remove**.
24. You have just removed the selected values from the legend for your layer, so parcels with those values are no longer shown in the map. Repeat the same process to remove all values located below GC-S.

You have now isolated the commercially zoned parcels within the City as requested by your Director. Your map should now look like this:

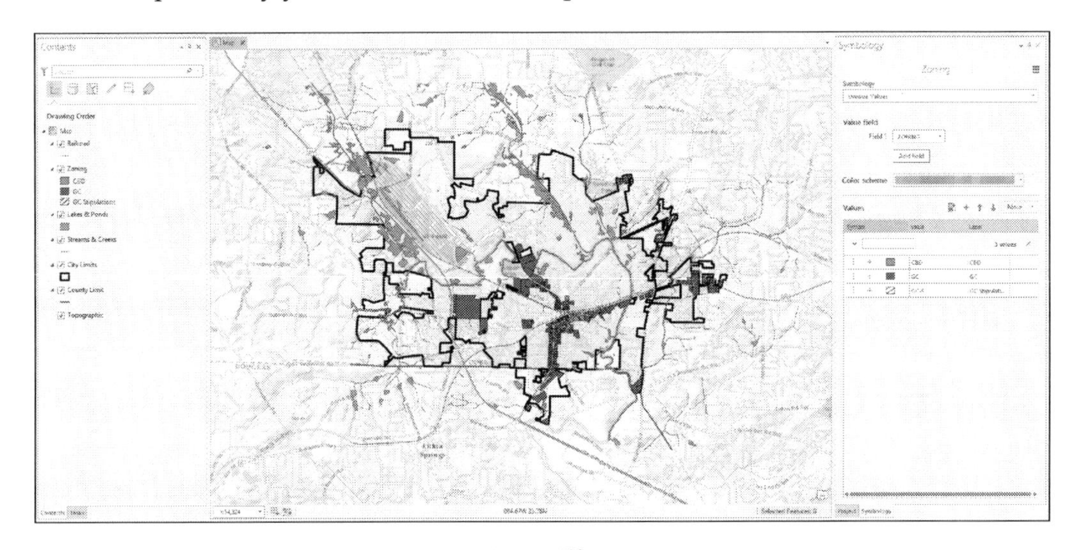

25. Once you have verified that your map is correct, expand `Maps` in the **Project** pane.
26. Right-click on `Map` and in the **Project** pane select rename.
27. Rename the map `Commercial Zoned Prop`.
28. Save your project.

Exercise 4B – adding a new map to a project

In the previous exercise, you created the first map asked for by the Community and Economic Development Director. Now you need to create the second map. This map needs to show those commercial properties which are over 1 acre and less than 3 acres in size. You will use the map you created in the last exercise as a starting point by copying it.

Step 1 – creating the new map

In this step, you will create a new map in your project by copying the map you created in the last exercise:

1. Save your project as `%Your Name% Ex4B` in `C:\Student\IntroArcPro\My Projects\Tripp Ex4A`.
2. In the **Project** pane, expand **Maps**. Then, right-click on the `Commercial Zoned Prop` map you created earlier and select **Copy**.
3. Right-click on **Maps** in the **Project** pane again and select **Paste**, as shown in the following screenshot:

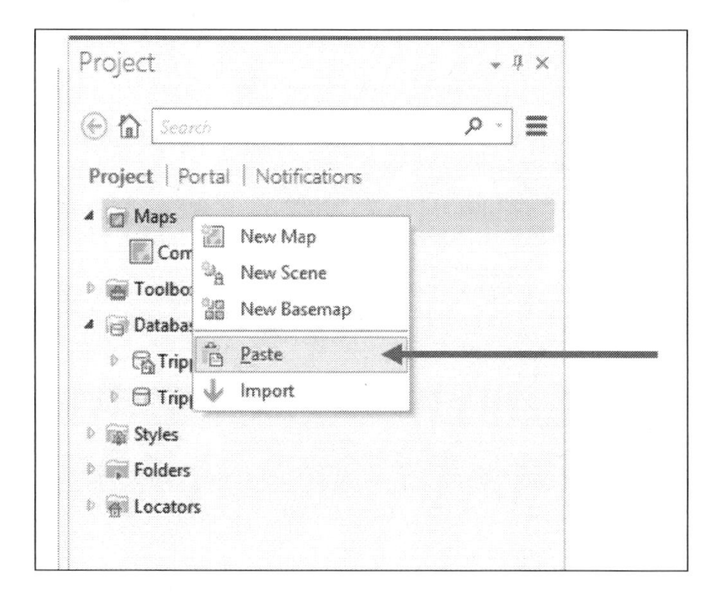

4. A copy of the `Commercial Zoned Prop` map should now appear under **Maps** in the **Project** pane. Right-click on the new map copy you just added and select **Rename**.
5. Rename the map `Commercial Between 1 to 3 acres`.
6. Right-click on the map you just renamed and select **Open**. This will open the new map you added in the view area, so you can see it
7. Save your project.

Step 2 – filtering a layer with a definition query

Now that you have created the new map in your project, you need to filter the **Zoning** layer, so it only shows those commercially zoned properties that are more than *1* acre and less than *3* acres:

1. Right-click on the **Zoning** layer in the **Contents** pane and select **Properties**.
2. Select **Definition Query** in the right pane in the **Layer Properties** window.
3. Click on the **Add Clause** button.
4. Verify that **Values** is selected in the upper-right-hand side of the window.
5. Set the **Query** field to **ACRE_DEEDE** by clicking on the small drop-down arrow and picking the field from the list.
6. Set the operator to **is Greater Than** again by clicking the small drop-down arrow and selecting the operator from the list.
7. Type `1` in the next cell after the operator.
8. Verify that your query looks like the graphic shown in the following screenshot and click on **Add**:

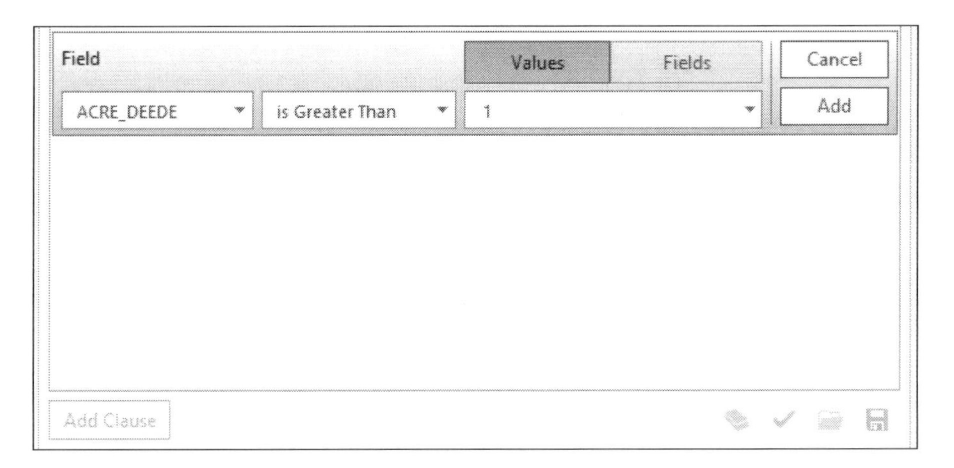

So, you have added the first query, which will limit the displayed properties, to only those greater than 1 acre. Now you need to add a second query, which limits it to those that are less than *3* acres.

9. Using that same process, add another query that states `ACRE_DEEDE is Less Than 3`. It should look like the following image:

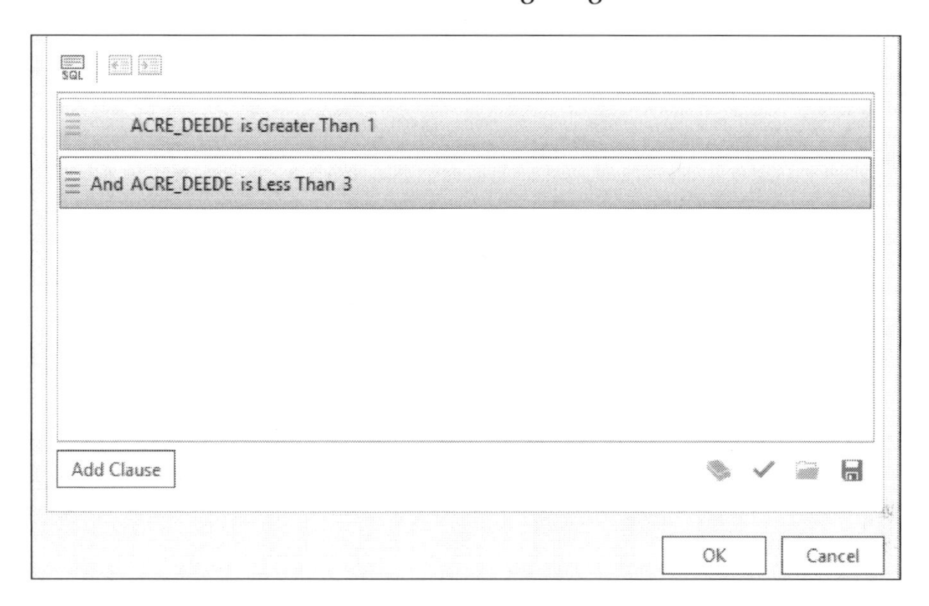

10. Once you have verified that your query has been built correctly, click on **OK** to apply the definition query

You have just added a filter to the **Zoning** layer, so only those commercially zoned parcels that are over *1* acre and less than *3* acres in size are shown in the map. However, you need to add some more layers to place the zoning information into context.

11. Save your project.
12. Expand **Databases** in the **Project** pane.
13. Expand the `Trippville_GIS` database and the **Base** feature dataset.
14. Right-click on the **Parcels** feature class and select **Add to Current map**.
15. In the **Contents** pane, drag the newly added parcels layer below the `County Limit` layer. You need to ensure that you are on the **List by Drawing** order in the **Contents** pane to do this. This is the first button at the top of the **Contents** pane.

You can now see the commercially zoned parcels that are between *1* and *3* acres and how they relate to the rest of the parcels in and around the City.

> Since ArcGIS Pro assigns random colors to new layers, you may need to adjust the symbology for the parcels layer so that the commercial zoning stands out more. Using the skills you have learned so far, adjust the symbology for the parcels so that it does not overpower the other layers, specifically the **Zoning** layer.

16. To also help the zoning layer stand out more, you need to change your basemap. Click on the **Basemap** button on the **MAP** tab. Select the **Light Gray Canvas**.
17. Click on the name for the **Zoning** layer, so the edit box appears around it. Rename the layer `Commercial Properties 1 to 3 AC`.
18. Right-click on the `City Limits` layer in the **Contents** pane and select **zoom to layer**.
19. In the scale box located on the bottom-left side of the map view, type `24000`.

Your map should look similar to the map shown here. The color of the parcels may be different depending on what you chose.

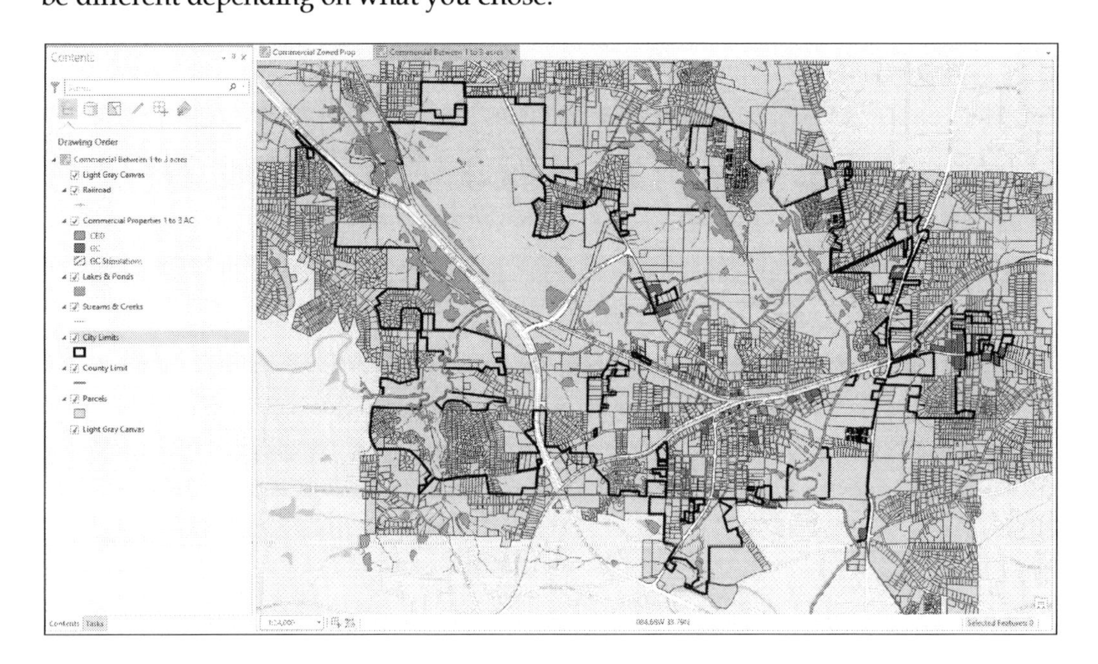

20. After you have verified that your map looks similar, save your project.

Step 3 – adding annotation to the map

You have now successfully created the map the Director asked for. However, after reviewing it with you, he has decided that he wants to see road names and points of interest added to the map as well.

Luckily, you already have an annotation feature class in the `Trippville_GIS` database which contains these. You will now need to add that annotation geodatabase to your map.

1. Expand **Databases** in the **Project** pane and then expand the `Trippville_GIS` geodatabase.
2. Expand the **Base** feature dataset, so you can see the feature classes it contains.
3. Click on **Base_Anno** and drag it into the map view to add it to your map.
4. Click on **Bookmarks** in the **MAP** tab on the ribbon and select **New bookmark**.
5. Name your new bookmark as `Entire City 1:24000` and click on **OK**.
6. Zoom into your map until you are able to see the annotation layer you just added clearly.
7. Click on **Bookmark** again. This time, select the **Entire City 1:24000** to return your map to the original view.
8. Save your project.

You have now successfully created the two maps requested by the Director. You created the first map that showed all the commercially zoned parcels within the City. You then created a new map, which limited the commercially zoned parcels to only those between *1* and *3* acres.

Exercise 4C – using map and layer files

The Director has decided that he wants an additional map added to this project. He needs to know the location of the City's sewer system in relation to the commercial properties that are between *1* and *3* acres. This is because the new business wants to be able to connect to the City's sewer system.

Luckily, you have a map file of the City's sewer system, which was created for another project. This will serve as a great starting point for this new map you need to add to your project. Map files allow you to share previously created maps, so they can be used in other projects without having to start all over with a blank map.

Once you import the map file, you will then need to add the commercial properties that are between *1* and *3* acres. To make this easier, you will create a layer file from the map you just created and then use that file to add the layer to the sewer map.

Step 1 – adding a new map using a map file

In this step, you will add a map of the City of Trippville's Sanitary Sewer system using an existing map file. This will add a new preconfigured map to your project, which you will add too:

1. If you closed ArcGIS Pro after the last exercise, launch the application and open the project `%Your Name%Ex4B`.
2. Right-click on **Maps** in the **Project** pane and select **Import**.
3. In the **Import** window, select **Folder** in the **Project** tree in the left-hand side of the window.
4. Double-click on the `Chapter4` folder.
5. Select the `Sanitary Sewer Map.mapx` file and click on **Select**.

Your project should now have a new map that shows the location of the Trippville Sewer System, which looks as follows:

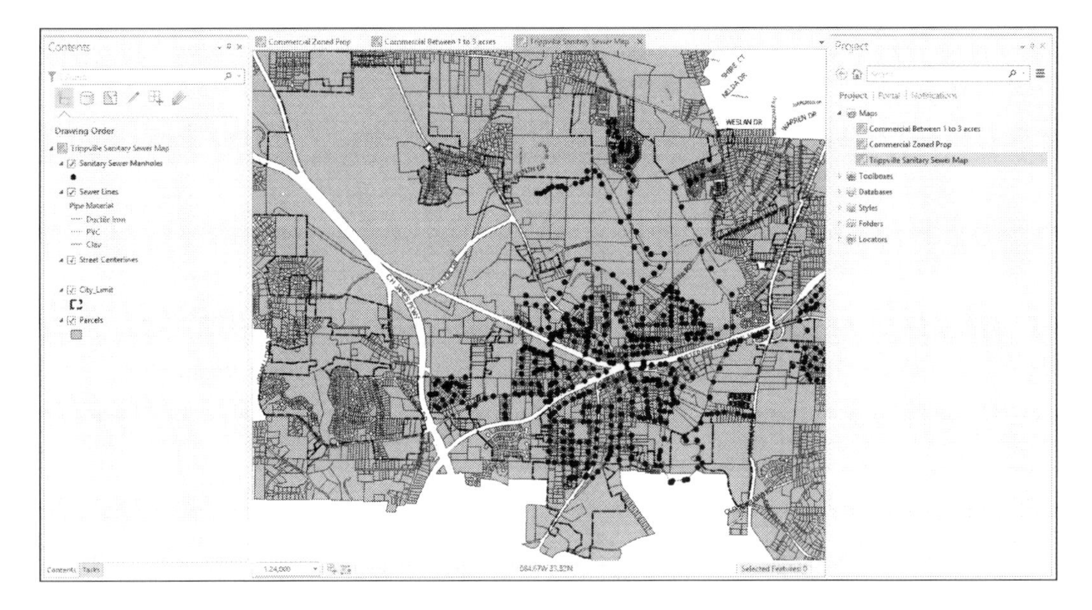

Step 2 – creating and using a layer file

Now you need to add a layer showing the location of the commercial properties that are between *1* and *3* acres. You will create a layer file from the previous map you created and then use the layer file to add the new layer to the sewer map:

1. Select the tab for the `Commercial Between 1 to 3 acres` map in the **View** area. If you closed this map, expand **Maps** in the **Project** pane and right-click on the `Commercial Between 1 to 3 acres` map and select **Open**.
2. Right-click on the `Commercial Properties 1 to 3 AC` layer in the **Contents** pane. Select **Save as Layer File**.
3. Select **Folders** in the left-hand side pane under **Project**. Double-click on `Chapter4`.
4. Name the new layer file `Commercial Prop 1 to 3AC` and click on **Save**.
5. Click back on the **Trippville Sanitary Sewer Map** tab to activate it.
6. Click on the **Add Data** button located in the **MAP** tab on the ribbon.
7. Select **Folders** and then the `Chapter4` folder.
8. Select the `Commercial Prop 1 to 3AC.lyrx` file you created and click on the **Select** button.

You have just added a layer showing the commercial properties that are between *1* to *3* acres that are located in the City to your map. Now the Director will be able to see which of those parcels is near the sewer system and thus be able to have a City Sewer service. Using the layer file made adding this new layer much easier and ensures that the symbology was consistent with the other maps within the project. It is also possible to copy and paste a layer from one map to another within the same project. This method will also keep the symbology consistent between maps and scenes within the project.

Step 3 – labeling

Now the Director wants you to label the pipe sizes for the sewer system, so he can ensure that there is enough capacity to handle the load the new business may add to the system. You need to turn on and configure labeling for the sewer line layer:

1. Select the **Sewer Lines layer** in the **Contents** pane.
2. Now you need to determine which field contains the size of the sewer pipes. Right-click on the **Sewer Lines** layer and select the **Attribute** table. This will open the **Attribute** table in a pane located below the main map view.

3. Scroll through the available attribute fields to determine which field contains the pipe size.

Question: What is the field name or alias that contains the size of the sewer pipes?

____

____

4. Close the **Attribute** table for the sewer lines.
5. Click on the **LABELING** tab in the ribbon.
6. Click on the **Label** button to turn on labels for **Sewer Line layer**.
7. Set the **Field** to the one you identified in the question mentioned earlier.
8. At the current full view of the map, it is hard to read the pipe sizes. So, you will set a scale visibility range for the labels, which will allow them to become visible at a scale that makes them easier to read. In the drop-down list located next to **Out Beyond** on the **LABELING** tab in the **Visibility Range**, select **1:10,000**. The size labels should disappear.
9. Zoom in until the size labels appear.
10. To make the pipe size labels stand out a bit more, make them bold and adjust the color on the **LABELING** tab in the **Text Symbol group** tab.
11. The placement of the labels is not optimum, so you need to adjust the placement. Click on the small arrow with the line above it in the **Label Placement group** tab. You should see several placement options listed that were not shown before.
12. Select the **Water (Line) placement** option.

Step 4 – configuring label conflict and placement options

You have successfully added the pipe size labels to your map. However, there are some additional configuration options, which might improve how they look and perform. You will now make some adjustments to those:

1. Right-click on **Sewer Lines** layer and select **Label Properties**. This will open the **Label Class** pane on the right-hand side of the interface. This pane allows you to make further adjustments to the labeling configuration settings to refine their appearance.

2. Click on the **Position** tab and then select the **Conflict Resolution** icon, which looks like three text boxes stacked on one another.
3. Expand **Remove duplicate labels** in the **Label Class** pane.
4. In the drop-down menu that appears underneath, select **Remove within a fixed distance**.
5. Set the search radius to `500` and the units to **Map Units**. It should look like this:

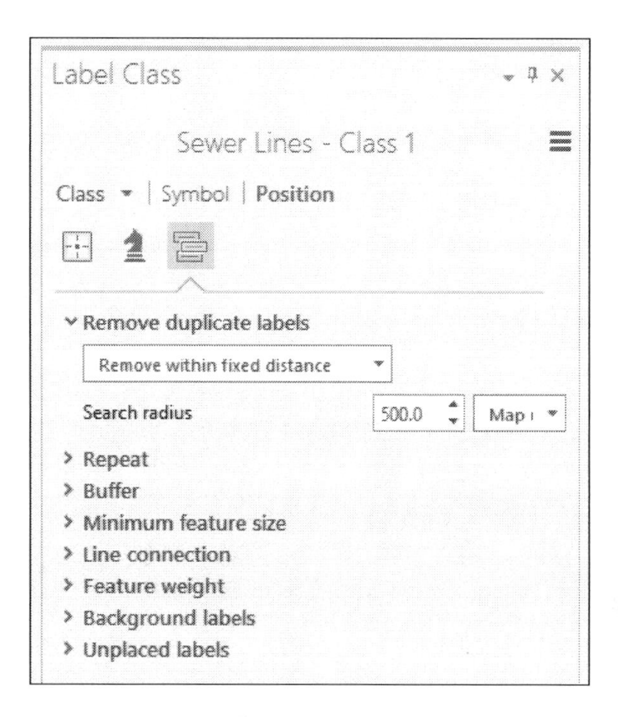

This makes the labels appear a little less cluttered. However, you are not done yet. There are still some adjustments, which will improve the labels.

6. Expand the **Minimum feature size** option in the **Label Class** pane.
7. Set the **Minimum** for labeling size to `100` and the **units** to **Map Units**. This will further reduce the clutter in the map.
8. Save your project.

Continue to try other labeling options to see how they can impact the placement of the labels and the overall appearance of the map. Pan and zoom to other places in the map to see if your settings still work well in different locations.

Challenge

The Community and Economic Development Director was so impressed with the maps you created that he has another project that he would like your help with. A company would like to open a new manufacturing center within the City. So, they are looking for a parcel that is zoned light or heavy industrial, which is near the railroad and can easily be provided with both a water and sewer service. The parcels need to be *5* to *10* acres.

Using the skills you have learned, create a new project that contains the following maps for this scenario:

- A map showing the location of all heavy and light industrial properties within the City (zoning codes HI and LI).
- A map showing the location of all industrial zone parcels between *5* and *10* acres.
- A map showing the location of industrial zoned parcels between *5* and *10* acres along with sewer system. Symbolize and label sewer system the same way you did in `Ex 4C`.
- A map showing the location of the industrial zoned parcels between *5* and *10* acres along with the water lines and fire hydrants. Symbolize water lines based on their size. The symbology for the fire hydrants is up to you.

Summary

In this chapter, you learned how to create a new project and add multiple 2D maps to that project. You were introduced to several methods to add new maps to your project, including using a project template, adding a blank map, and using a preconfigured map file. Once you created the maps, you then used various methods to add new layers.

You learned how to use the **Add Data** button in the ribbon to add one or more layers to a single map. You also saw how you can add new layers directly from the **Project** pane using database and folder connections. The use of layer files to add preconfigured layers was also explored.

Once the layer was added, you were shown different methods to configure the symbology and labels for the layer. You learned how to use the attribute table to define different symbology, which allowed you to highlight specific values associated with the feature. You also learned how you could use symbology and definition queries to filter your layers, so you only displayed features with specific values.

5

Creating 3D Maps

As the world becomes a more crowded place, the need to view data in a 3D environment grows. With the infrastructure of a modern urban area expanding both up and down as well as out, the importance of seeing what is above and below the ground is becoming critical to planning new projects, responding to emergencies, and managing the infrastructure.

ArcGIS Pro allows users to create 3D maps natively. In the past, this required users to have extensions such as 3D Analyst for ArcGIS. This ability to create 3D maps opens up a whole new world and way to view your data quite literally. You can extrude data from above or below the ground to see things such as does an existing natural gas line interfere with the new sewer line that is being planned or will the new office building obstruct the view of the mayor. This opens the door to new levels of analysis that were not always possible with traditional 2D maps.

In this chapter, you will explore the following:

- How to create a new 3D map
- Use a custom elevation model as the surface of your map
- Extrude data above and below the ground level

3D maps

ArcGIS Pro allows you to easily add 3D maps to your projects. 3D maps are also referred to as *scenes* in ArcGIS Pro. When you add a new 3D map or scene to a project, you must first choose which type or view mode you wish to use. ArcGIS Pro supports two: global and local.

A **global scene** is designed to support large areas where the curvature of the earth has a noticeable impact. This might be used to show the track of a hurricane or the path of a gas pipeline, which crosses many states or countries.

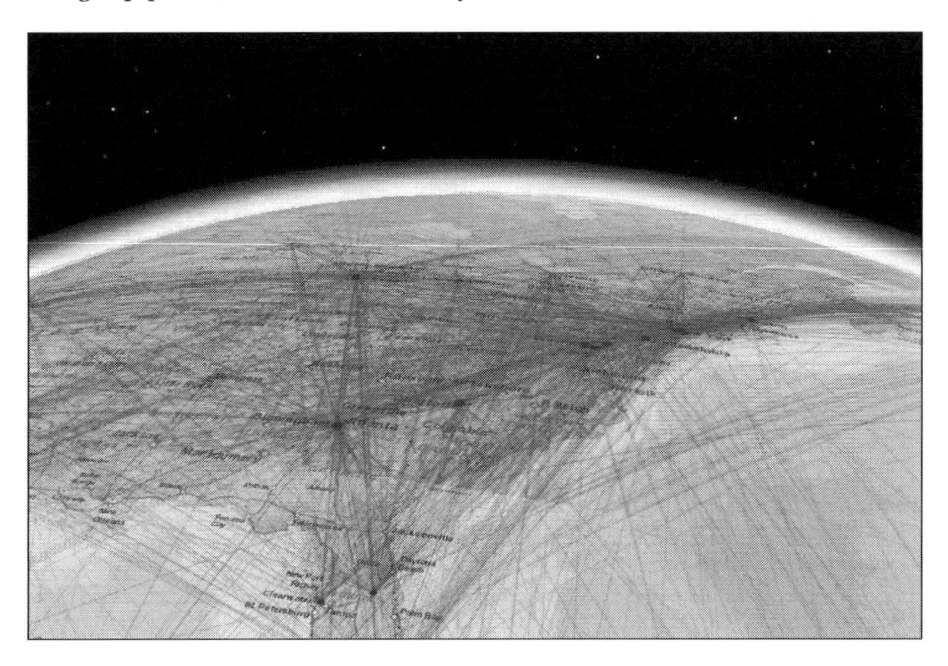

The preceding image is an example of a global scene, which shows the airline flight paths over the east coast of the United States. The curvature of the earth certainly has an impact on these over long distances. Because of the extent of the area covered by a global scene, these typically use a geographic coordinate system.

A **local scene** is designed to display smaller areas where the curvature of the earth has less impact and most often uses a projected coordinate system. A local scene might be used by a city to show its downtown area or by engineers planning a new road extension with a bridge or by a cell phone company to site a new cell tower within a community.

The following image shows an example of a local scene that depicts the buildings in a specific project area. They have been extruded vertically based on their height to create a 3D view showing how each building is related to those surrounding it both horizontally and vertically.

Warning: At the time this book was being written, there was a known issue with displaying 2D layers in a 3D scene, which uses a projected coordinate system and Esri's default world elevation surface. If you set your scene to use a projected coordinate system instead of a geographic coordinate system, your 2D layers will not be visible if you use the default Esri world elevation surface. To overcome this issue, you have two options. The first is to use your own local elevation model. The second is to use a geographic coordinate system in your scene. You may also want to visit `http://my.esri.com` or `http://pro.arcgis.com/en/pro-app/get-started/release-notes.htm` to see if there are any patches for this or other issues.

Exercise 5A – creating a simple 3D scene

In this exercise, you will create a simple 3D scene using an Esri project template. You will add and extrude a layer, so it has height above the ground level.

Step 1 – creating a project and 3D scene

In this step, you will create a new project using one of the templates included with ArcGIS Pro. This template will include a local 3D scene automatically:

1. Open ArcGIS Pro.
2. From the **Create a new project** pane, select the `Local Scene.aptx` template. This template automatically generates a 3D scene in your new project.
3. Name your new project `%your name%Ex5A` and set the location to `C:\Student\IntroArcPro\My Projects`. Click on **OK**.

Your new project should open with a scene already created. In your **Contents** pane, you should see two categories: 2D and 3D layers. The 2D layers are draped across your ground surface. The 3D layers may be extruded above or possibly below the ground. In the next step, you will add a layer and make it 3D.

Step 2 – adding a layer

In this step, you will add several layers to your map. Some will remain 2D layers and will serve as a general backdrop. You will make the building layer 3D and extrude it to show the heights of the building:

1. Make the **MAP** tab active in the ribbon and select **Add Data**.
2. In the tree located in the left-hand side of the **Add Data** window, expand **My Computer**.
3. Select the **C:** drive in that same pane.
4. Double-click on **Student**, **IntroArcPro**, and lastly **Databases**.
5. Double-click on the `Trippville_GIS` geodatabase and the **Base** feature dataset.
6. While holding your *Ctrl* key down, select the following layers:
    - `City_Limit`
    - `Parcels`
    - `Buildings`

7. Click on the **Select** button once you are done.

The new layers are added to your scene and ArcGIS Pro automatically zooms you to the extent of the new layers you added.

8. Adjust the draw order of your layers in the **Contents** pane by dragging them into the following order:
   - `Buildings`
   - `City_Limit`
   - `Parcels`

When completed, your **Contents** pane should look like the following:

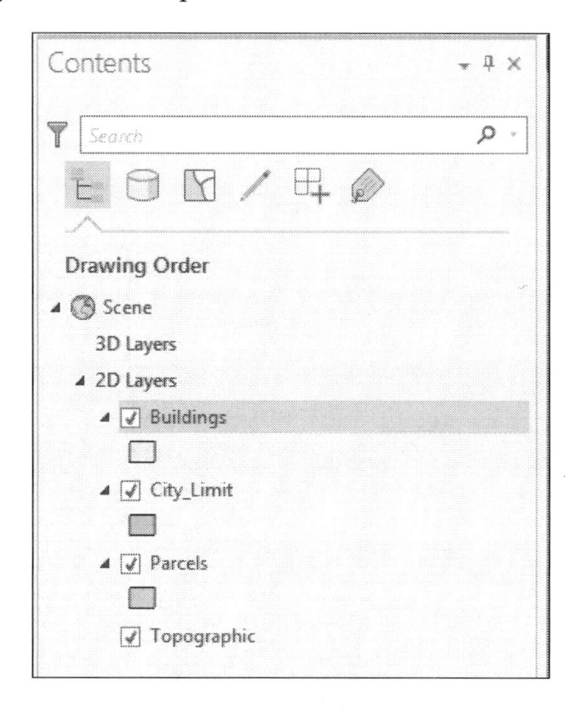

9. Using the skills you have learned so far, set the symbology for the layers you just added to these settings:
   - `City_Limit`

     Color: no color

     Outline Color: black

     Outline width: `3 pt`

- `Parcels`

  Color: 10% Gray

  Outline color: black

  Outline width: `1 pt`

- `Buildings`

  Color: Dark Umber

  Outline color: Black

  Outline width: `1 pt`

Your scene should now look very similar to the following:

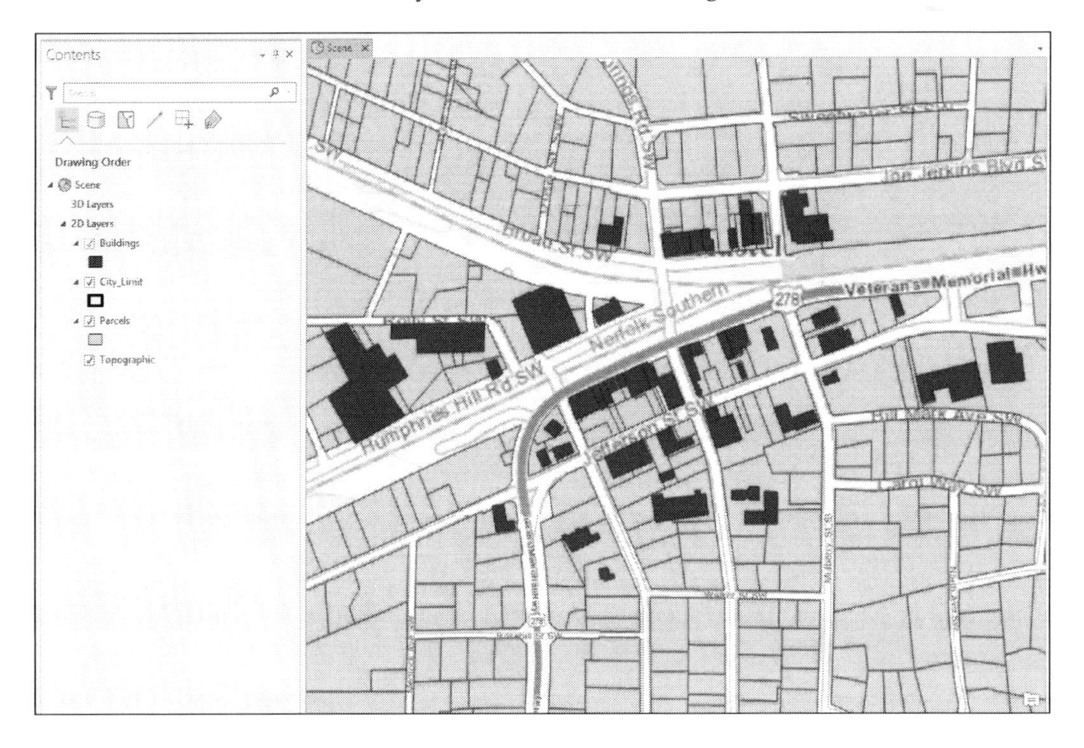

10. Save your project.

Step 3 – making a layer 3D

In this step, you will turn the `Buildings` layer from a 2D layer into a 3D layer. You will extrude the height of each building based on an attribute field to show how tall each building is:

1. Drag and drop the `Buildings` layer from the 2D category to the 3D category.

When you do this, you may note that parts of the building outlines disappear. That is because as a 3D layer parts of them are now below the elevation surface.

2. Make sure that you still have the `Buildings` layer selected and click on the **APPEARANCE** tab on the **FEATURE LAYER** contextual tab.
3. In the **Extrusion** group tab, click on the small drop-down arrow below the type and select **Base Height**.

> Extruding a layer allows it to be displayed in 3D. It determines the method ArcGIS Pro will use to determine the height of features and how to display the height. The **Base Height** option will add the value of a specified field to each vertex of a feather to create the 3D view. There are four types of extrusion you can choose from in ArcGIS Pro. You will learn more about those later.

4. In the cell located next to the type, set the field or expression that will be used to extrude the buildings, so their true height is shown. Click on the small drop-down arrow and select **Est_HGT**.

Your building layer should now appear to have some depth.

5. Right-click on the `Buildings` layer and select **Zoom to Layer**.
6. Select the **Explore** tool on the **MAP** tab in the ribbon.
7. Move your mouse pointer near the center of the cluster of buildings.
8. Hold your scroll wheel down and then move your mouse toward the top of the scene to make it rotate along the *z*-axis. This allows you to see the extruded buildings displaying their height. Continue to rotate the map until you can clearly see the building's height.

Your scene should now look similar to this:

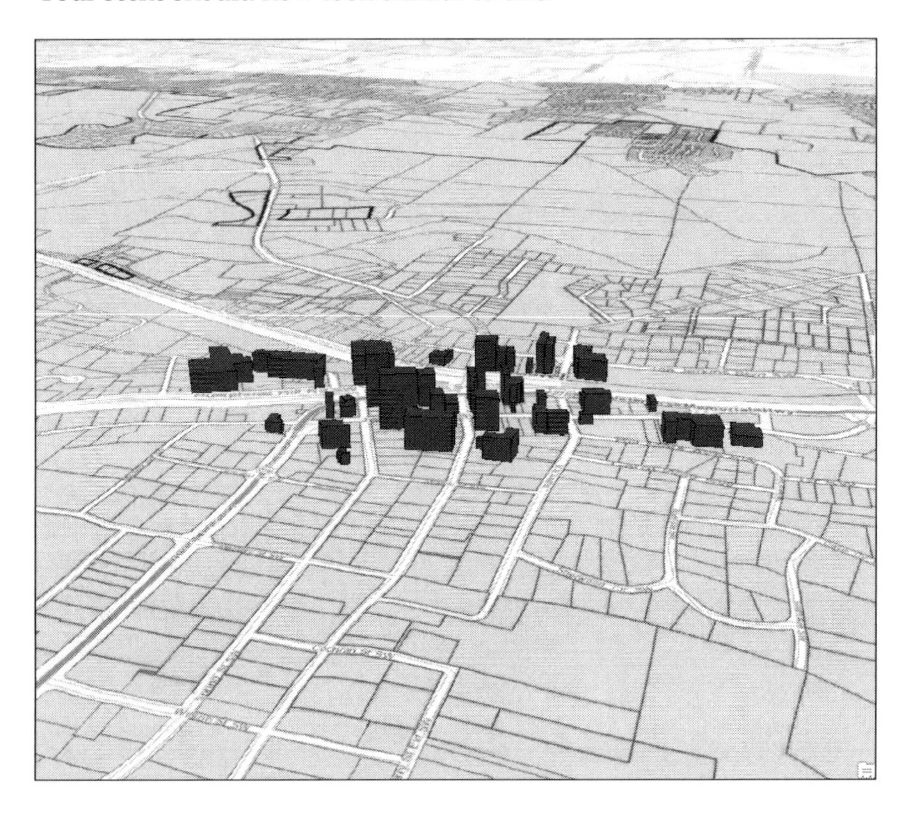

9. Continue to use your mouse to zoom, rotate, and pan within your new 3D scene. Experiment with how the navigation works in a 3D environment.
10. Save your project and close ArcGIS Pro.

Congratulations! You have just created your first 3D scene in ArcGIS Pro. For many of you, this may be the first 3D map you have ever created.

Elevations

When you create a 3D map within ArcGIS Pro, there are several elevations you need to be concerned about. You need to define a ground elevation. The ground elevation provides a baseline for all of your 3D data. It is the canvas to which all other layers are applied within your map.

Once you have identified your ground elevation, then you need to determine the elevations of your 3D features. These may be directly related to the ground or at an absolute elevation.

Ground elevations

The ground elevation is used by ArcGIS Pro to represent the surface of the earth within the 3D scene you are creating. Each scene must have a ground surface specified. The ground surface cannot be deleted. It can be changed and does have various properties, which can be adjusted.

By default, ArcGIS Pro will use the Esri world elevation surface, which is published through ArcGIS Online. You can choose to use your own elevation surface if desired. Your own local surface is often much more accurate, so it will provide better results. Things you can use as a ground surface include a Digital Elevation Model, Triangulated Irregular Network, and web elevations surfaces.

Digital Elevation Models

Digital Elevation Model (DEM) is a raster dataset that represents the elevations over a defined extent. Raster data is made up of a series of equal-sized cells with each cell containing a numeric value. What the value represents will depend on the purpose of the raster. In the case of a DEM, the values raster cells represent are the average elevation for the area covered by the cell. The following is an example of a DEM:

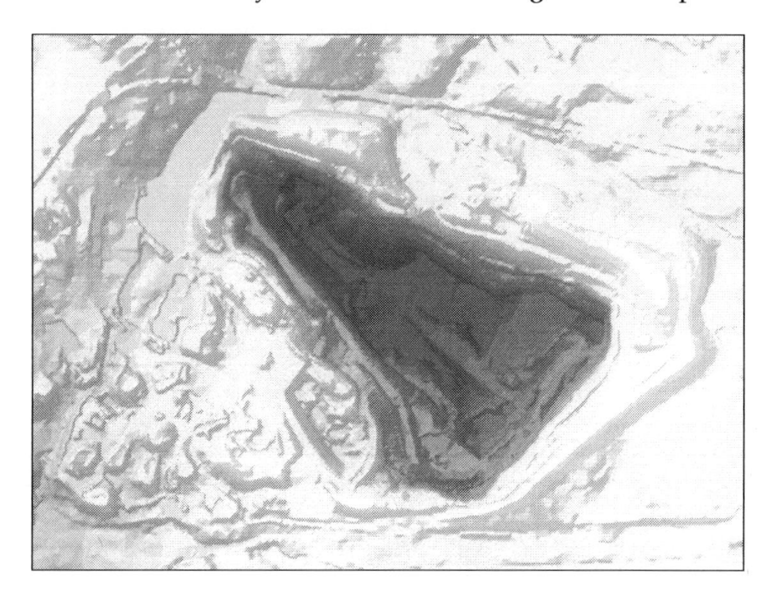

If you were to zoom in on this DEM at some point, you would start to see individual squares, which are the cells that make up the raster data. The squares you see are now commonly referred to as a pixel even though that is not correct. Each cell is assigned the value of the averaged elevation it covers. The cell might be 1 x 1 foot or 2 x 2 m. It is the size of the cells that determines the resolution and plays a part in its accuracy.

How does the DEM know the average elevation for the areas covered by its cells? Like all GIS data, it can be derived from many different methods. It might be created based on elevation information acquired by a land surveyor. It might have been compiled by a **photogrammetrist** using stereo photography methods. It may have been interpolated using contour lines or some combination of all these methods.

Common raster formats are as follows:

- JPG, JPEG, JP2
- Tiff
- PNG
- SID (commonly called Mr. SID)
- ECW
- IMG
- GRID (Esri Native Raster Format)
- BMP (commonly called a bitmap)

Creating a DEM in ArcGIS Pro requires you to have the Spatial Analyst extension. However, if you already have access to a DEM, you can create a 3D scene and use it as your ground surface without the extension.

Triangulated Irregular Networks

Triangulated Irregular Network (**TIN**) is a vector-based representation of a surface. It is constructed from a series of nodes and lines, which form a network of adjacent triangles. The triangles form facets of a 3D surface, which include areas of the same elevation.

Here is an example of a TIN as it is being drawn. You can see the triangles that make up the TIN as it is being drawn:

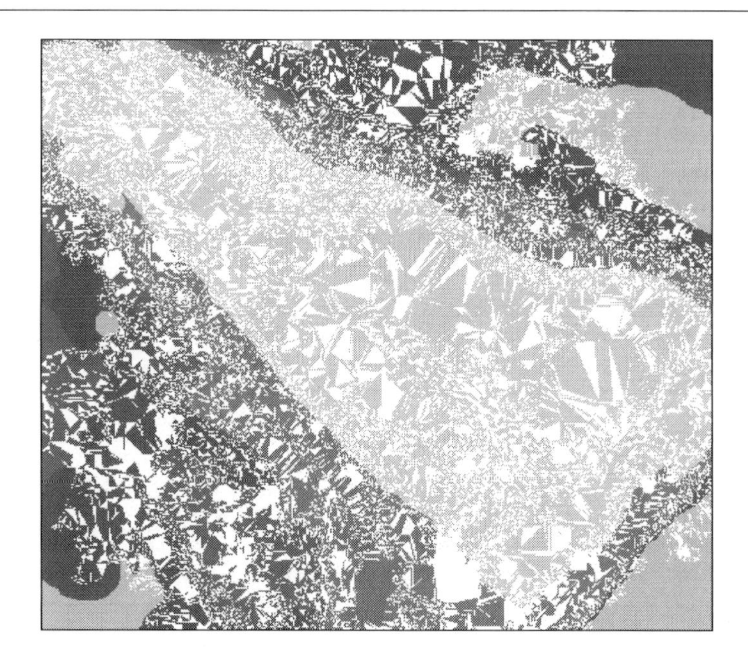

Once all the triangles are drawn, you are presented with what appears to be a solid surface. It is easy to see the elevation changes within the TIN, as shown in the following image:

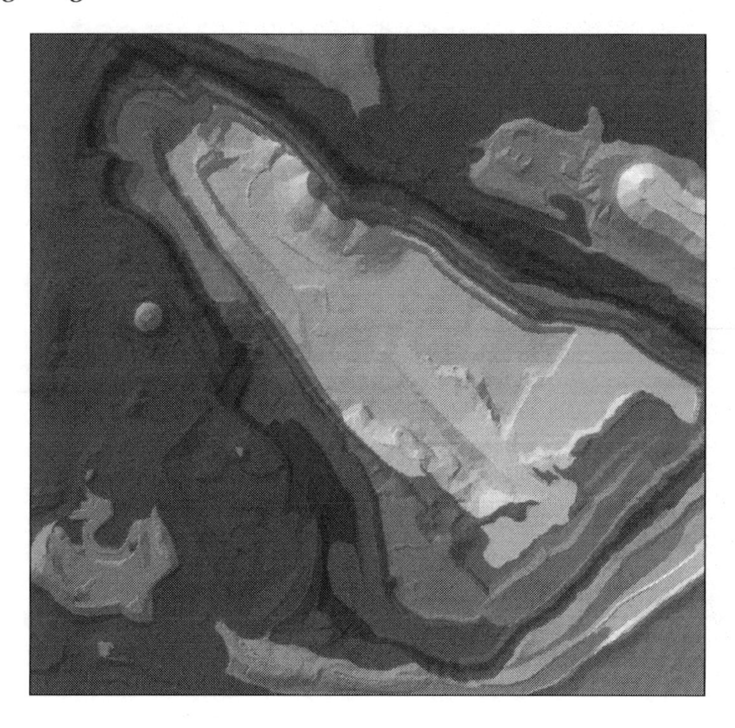

Because a TIN is vector-based data it tends to be smaller in size to an equivalent DEM. However, due to their complexity, they can take much longer to draw or process when used in analysis.

Being vector-based TINs can be created from a range of input data, such as streams, ridge lines, roads, and so on. When the TIN is generated, these input features will remain in the same location and be matched with nodes or edges within the TIN. This can allow the TIN to achieve a high degree of accuracy and detail that is not normally possible with a DEM.

Creating a TIN does require you to have the 3D Analyst extension. However, if a TIN already exists, you may use ArcGIS Pro without an extension as a ground surface or a layer.

Extruding features

Extruding features is how you turn them from flat 2D shapes into 3D objects. It provides them with height. For example, you can extrude power poles or trees so that you can see how tall they are. Both of these are normally stored at point features. When you extrude them so they have height, they become vertical lines. You can see an example of buildings and power poles that have been extruded to create 3D features as follows:

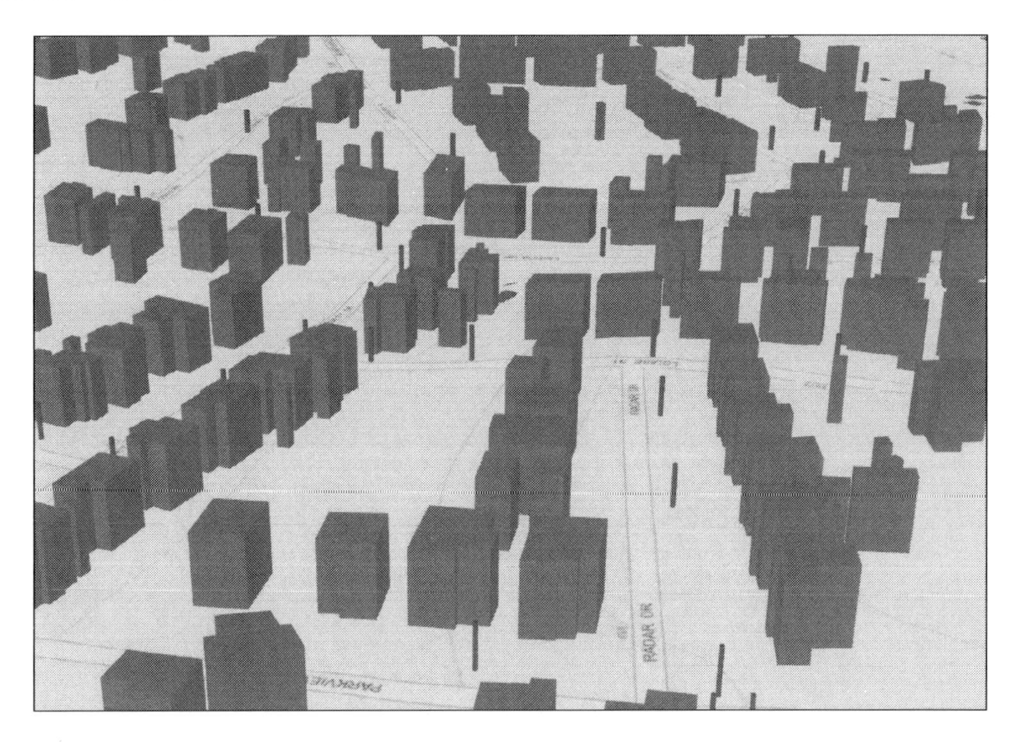

The buildings in this scene started as a 2D polygon layer, which just shows the footprint of the buildings on the ground. By extruding them, you can now see their height above the ground gaining a greater appreciation of the spatial relationships between them and other features. The power poles were a 2D point layer, which were extruded to show their height. Now, you can visualize the relationship between the buildings and the power poles.

ArcGIS Pro provides four methods for extruding features. The one you use will depend on what information you have for your features. The following table explains these four methods:

Extrusion type	Description	Requirements	Data types
Minimum height	This adds the extrusion height to the minimum *z* value producing a flat top feature.	This requires an attribute field that contains an elevation value for the lowest point of the feature along the ground surface.	Lines or polygons
Maximum height	This adds the extrusion height to the maximum *z* value producing a flat top feature.	This requires an attribute field that contains the elevation for a feature at its highest point along the ground surface.	Lines or polygons
Base height	This adds the extrusion to the base elevation of each feature vertex. The result may not be a flat top depending on the base heights of feature vertices.	No attribute fields are required. However, the extrusion can be based on a field such as measured height if available.	Points, lines, and polygons
Absolute height	Features are extruded to a specified height regardless of base elevations or other *z* values.	This requires an attribute field that contains the top elevation of each feature.	Points, lines, and polygons

Other methods to display 3D data

Extrusion is not the only method to display data in 3D. ArcGIS Pro includes many 3D symbol styles, which can add a level of realism to your scenes. The 3D symbols are designed to display features using commonly expected textures, materials, and details. For example, they can provide buildings with a brick appearance or show a fire hydrant with a real looking hydrant like you would expect to see driving down the road. Often these symbols have built in 3D settings, which do not require extruding the features.

You can also use the **CityEngine** symbology rules to apply even more realistic symbology to your scenes. **CityEngine** is another application for Esri, which allows users to create advanced 3D renderings of their data.

The following is an example of what can be accomplished using the 3D symbology styles in ArcGIS Pro. Here, you can see a real looking streetscape, which includes buildings, fire hydrants, and power poles:

Exercise 5B – creating a local scene

The Community and Economic Development Director is working on a presentation for a group of concerned citizens. He has asked you to create a 3D scene of the central downtown business district, which he can use in the presentation.

In this exercise, you will create a local scene to fulfill the Director's request. This scene will use a locally developed DEM, which is more accurate than the Esri world elevation model. You will then add and symbolize layers using the 3D symbology to create a more realistic scene.

Step 1 – open a project and add a local scene

In this step, you will open the last project you created and add a new local scene:

1. Start ArcGIS Pro.
2. Open the project you created in `Exercise 5A`. It should be in your list of recently opened projects.
3. In the **Project** pane, right-click on **Maps** and select **New Scene**.
4. Once the new scene is generated, right-click on it and select **Rename**.
5. Type `Local Scene` as the new name and press your *Enter* key.

You have just created a new scene in the project you created in the last exercise. Remember that a single project can contain multiple maps and scenes. This makes them easier to manage and access. By default, new scenes are created as global scenes. Now you will convert it to a local scene:

6. Click on the **VIEW** tab in the ribbon.
7. In the **View** group tab, select **Local**. This converts your scene to a local scene.

Step 2 – setting the ground surface

In this step, you will assign the DEM that the city engineer has provided as the ground surface. This DEM was created by the city engineer based on survey data he had collected by a professional surveyor:

1. In the **Contents** pane, right-click on `Local Scene` and select **Properties**.
2. In the **Map Properties** window, select **Elevation Surface** from the pane on the left-hand side.
3. Expand the ground surface by clicking on the small arrow located to the left. This will allow you to see the currently assigned ground surface.

> **Question**: What is the name of the currently assigned ground surface?
>
> ____
>
> ____
>
> **Question**: What is the location of the currently assigned ground surface?
>
> ____
>
> ____

4. Click on the *add elevation source* button. It looks like **+** on a yellow-filled polygon.
5. Expand **My Computer** and then select your **C:** drive.
6. Now navigate to `C:\Student\IntroArcPro\Databases\Trippville_GIS.gdb`.
7. Select **DEM**.

Your **Map Properties: Local Scene** window should now look like this:

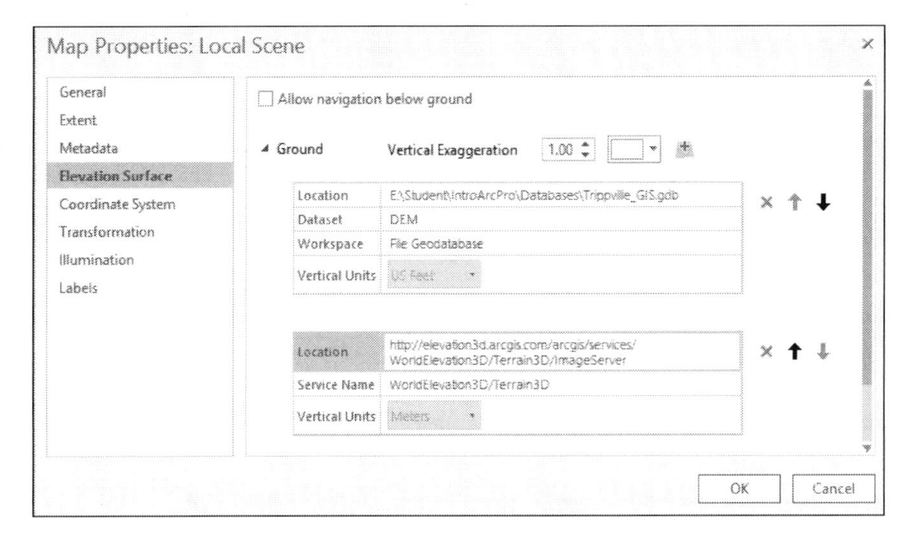

8. Click on the small red **X** located next to the **WorldElevation3D/Terrain 3D** surface to remove it. Since you are using your own DEM, which covers your area of interest you will not need it.
9. Click on **OK** to close the **Map Properties** window.

You have just set the ground surface for your scene to the DEM provided by the city engineer. Don't be concerned if your scene currently appears to be blank. This will be fixed as you add more layers to the scene.

Step 3 – adding layers and setting symbology

Now that you have the ground surface defined, it is time to start adding the layers requested by the Director:

1. Using the skills you have learned, add the following layers to your scene.
    - `Buildings`
    - `Fire_hyd`
    - `Power_Poles`
    - `Light_Poles`
2. Drag the `Buildings` layer to the 3D category. The layer may disappear from the map display. You will fix this issue in a moment.
3. Double-click on the symbol patch located below the layer to open the **Symbology** pane on the right-hand side of the interface.
4. Ensure that the **Gallery** tab is selected. In the search area, type `Building` and then set the search to **All styles**.
5. Select **International Building** under **Procedural Symbols**.
6. Right-click on the `Buildings` layer in the **Contents** pane and select **Zoom to Layer**.
7. Make sure that the **Explore** tool is selected on the **MAP** tab in the ribbon.
8. Place your mouse pointer near the center of the buildings and depress the scroll wheel on your mouse. While holding your scroll wheel down, move your mouse-pointer toward the top of your scene to rotate your scene, so you can view the height of the buildings layer, which is now using the 3D symbology style you just applied.
9. Once you have rotated your scene, zoom in to get a closer look. You will note your buildings now have a much more realistic appearance than they had when you just extruded them in the last exercise.
10. Now drag the `Light Poles` layer to the 3D category. Again, they may also disappear from the map.
11. Click on the point symbol located below the layer name to open the **Symbology** pane.
12. In the **Symbol** pane, ensure that the **Gallery** tab is active.
13. In the search area type `Light`. You may also need to ensure that **All styles** is still set as the search option.
14. Select the *Street Light 6* symbol as this most closely matches the ones used by the City of Trippville. (You can choose a different one if you wish.)

15. Right-click on the `Light Poles` layer in the **Contents** pane and select **Properties**.
16. Click on the **Display** option in the left-hand side pane.
17. Check the box that says **Display 3D symbols in real-world units** and click on **OK**. This will allow you to set the correct height on the light poles. The standard height of the light poles used in the City is 18 feet.

Before you can set the height for the light poles, you will need to make one more change to the project settings. By default, the 3D symbols use meters as the units for specifying height. You need to change this in the **Project Options** to **feet**:

18. Click on the **PROJECT** tab in the ribbon and select **Options**.
19. Click on **Units** under **Project**.
20. Expand the **3D Symbol Display Units** and click on the radial button located at the end of the row for **Foot_US** in the grid as follows:

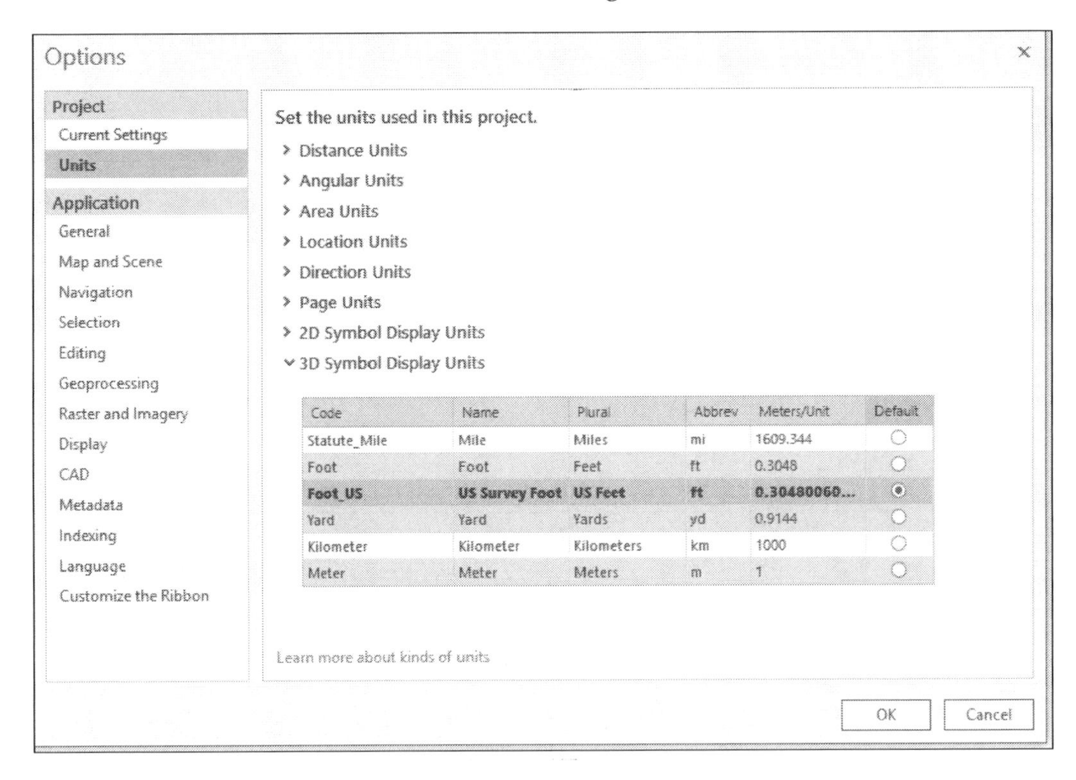

21. Once you verify the setting, click on **OK**.
22. Click on the back arrow located on the upper-left-hand side of the **Project** window to return to your scene.
23. Save your project to ensure that the project settings changes are not lost if ArcGIS Pro encounters issues.
24. Click on the **Properties** in the **Symbology** tab ensuring that the `Light Poles` layer is still selected. If you do not see the **Properties** option in the **Symbology** tab, try clicking on the point symbol below the `Light Poles` layer in the **Contents** pane.
25. Click on the **Layer** button which looks like three stacked squares. It is located between the **Symbol** button, which resembles a paint brush, and the **Structure** button, which resembles a wrench.
26. Set the **Height** to `18 Ft` and then click **Apply**.
27. Repeat this process for the `Fire_Hyd` and `Power Pole` layers. Use the following settings. Remember to set the 3D display settings to use real-world units in the layer properties for each layer:
    - `Fire_Hyd`
        - **Symbol Search**: Fire Hydrant
        - **Symbol**: Fire Hydrant Red from 3D Street Furniture
        - **Height**: `3 Ft`
    - `Power Poles`
        - **Symbol Search**: Pole
        - **Symbol**: Telephone Pole 1
        - **Height**: `35 Ft`

Your scene should now look similar to this depending on your rotation and zoom scale if you have correctly applied all the settings.

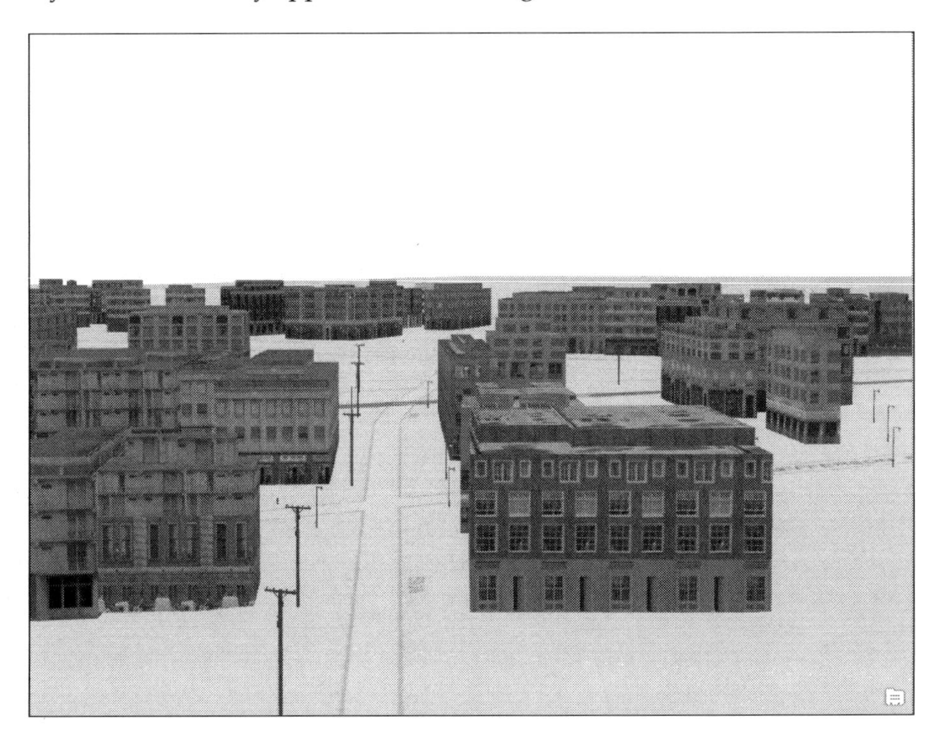

You are almost done. Now you need to apply a basemap that adds a more realistic back drop and apply a sky to the scene:

28. Click on the **Basemap** button on the **MAP** tab in the ribbon.
29. Select the **Imagery** basemap. This will allow you to use the aerial photography as a basemap, which will add a touch of realism to the scene.
30. Now to add the sky, you will apply a fill to the entire scene. Right-click on **Local Scene** in the **Contents** pane and select **Properties**.
31. Select the **General** option in the left-hand side pane of the **Map Properties** window.
32. Set the **Background** color to **Sodalite Blue** (you can choose a different color if you think another will work better) and click on **OK**.
33. Save your project.
34. Take some time to pan, rotate, and zoom within the scene you just created. It has a very realistic appearance, which is exactly what the Director was hoping to accomplish.

Your final scene should look similar to this. Yours may be slightly different depending on your rotation and scale.

Congratulations, you have created a realistic looking 3D visualization of a downtown area.

Summary

As you saw, ArcGIS Pro supports some very powerful tools to visualize data in 3D. This opens new levels of understanding using your GIS data. You learned how you can create different scenes to support different spatial extents.

You also learned how you can use different methods to turn 2D data into 3D features using extrusion or 3D symbol styles. When using extrusion, you can illustrate the true height of features using one of the four methods supported in ArcGIS Pro. Using 3D symbols can add a level of visual realism to your scenes.

The key element to create a 3D scene is to define a ground surface. You learned that you can use the default Esri-provided world elevation surface or one of your own. If you decided to use your own, it can be a DEM or a TIN. Using your own ground surface can increase the accuracy of the elevation data within your scene and the relationships between features along the *z*-axis.

6
Creating a Layout

You have seen how you can create informative and interesting 2D and 3D maps using ArcGIS Pro. While these are impressive, they are not complete. The last step to producing any map is to put it in a frame with other information, which helps the viewer comprehend the data you are presenting. This frame is called a layout.

A single layout can contain one or many 2D or 3D maps. A single project may contain many layouts. It is usual to have many different layouts within a project to meet various needs and requirements. You might have a large layout to use during presentations and a smaller one to include in a report.

Each layout you create will include several elements, such as a north arrow, scale, and title, in addition to one or more map frames. What you need to include in your layout will depend on the story you are trying to impart to the viewers. Before you create your layout, you need to take some time to figure out:

- What is the purpose of this layout?
- How will it be used?
- Who will be using it?

The answers to these questions will impact how you design your layout. They will help you determine the size, scale, and details you need to include. So, some thought should be given before you even start building the layout in ArcGIS Pro.

While working within an organization, there is often a desire to use common styles when creating your layout so that each one is branded and identifiable as coming from your organization. This means that you want to ensure that everyone within your organization uses the same style north arrow, scale bar, and/or logo within the layouts they create. What you need is a template layout with these elements predefined. Not only will this help give all layouts created by your organization the same look and feel, but it can greatly increase the speed of printing and sharing maps or posters.

The one thing you always need to keep in mind is that the layout represents your final product. It will form the basis on which all your efforts are judged. It will not matter how good your data or analysis happens to be; if it is not presented in a clear and professional manner, it will be of little use. Designing a well-thought out layout will lend credibility to all your hard work. It allows you and all your efforts to shine. It is also an opportunity to let your creative side show. The one thing I always tell those who are new to GIS when they reach this point is "make it pretty." While this seems simple, there are plenty of examples of GIS maps, which fail to meet that simple guideline.

In this chapter, you will learn the following:

- What to consider when designing a new layout
- How to add a new layout to a project
- How to design a layout that contains one or more map frames
- How to create and use a layout template

Things you need to consider when designing a layout

Many ArcGIS users believe that once they complete their analysis or create their maps, they are done. They think that they will be able to quickly pull it all together in a quick layout and print it out. This is not always the case. As mentioned previously, the Layout frames all your hard work. As a result, it is important to put some thought and planning into the creation of a layout.

Generally speaking, there are three basic factors you need to address when creating a new layout. First is the purpose, why does this map exist. Second is the audience, who will be using the map. Third is the situation, how will it be used or presented.

All these three factors will impact how you design your layout. These will guide the size of the map, the orientation, content, and more.

Purpose

Each map or poster exists for a reason. It has some purpose. It may be trying to show the location of City Hall or what parcels are in the flood zone or how to get from one location to another. The purpose defines the overall theme of the map. It also helps determine what content needs to be included. Some common map purposes include the following:

- To show location of features

- Highlight specific attributes associated with features, such as zoning classification, population size, pipe material, or road type
- Show spatial relationships between features in one or more layers
- Present the results of the analysis
- Meet legal requirements, such as the Official Land Use Map for a city or Tax Map for a county

It is now easy to see how the purpose will be a factor in determining the content of your layout. If you are creating a zoning map, you will certainly need a map frame that shows the zoning classification of each parcel. You will also need a legend that allows the map viewers to understand the different classifications. If this is the official zoning map for the city, you may also need to add places for city officials to sign the map and a date of adoption.

Based on the five purposes listed previously, try to determine the primary purpose for each of the following maps:

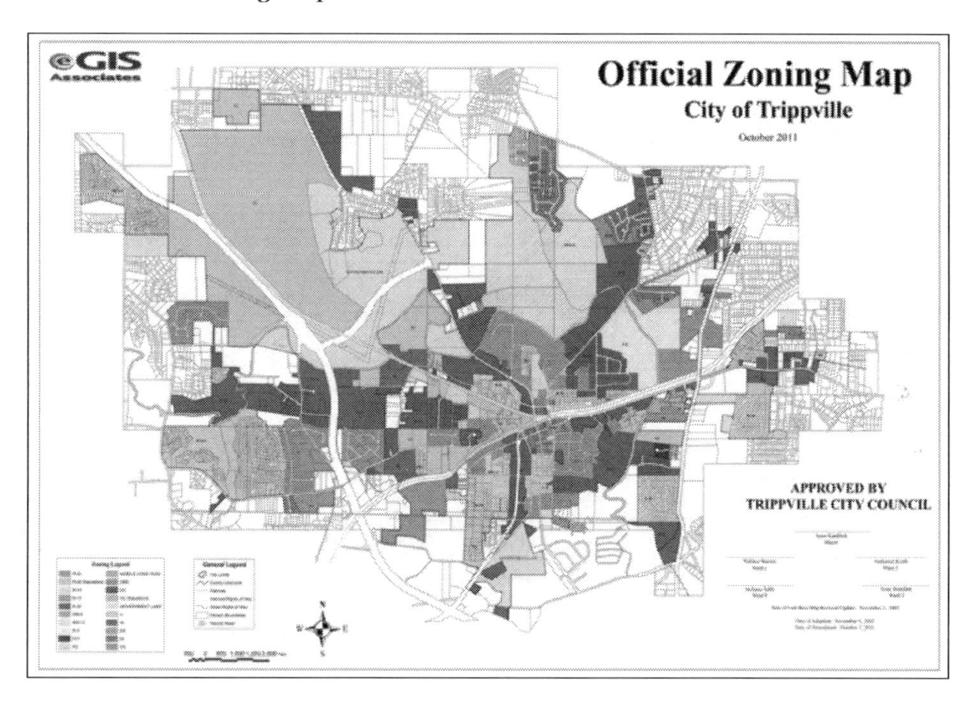

Purpose of Map 1: ______________________________

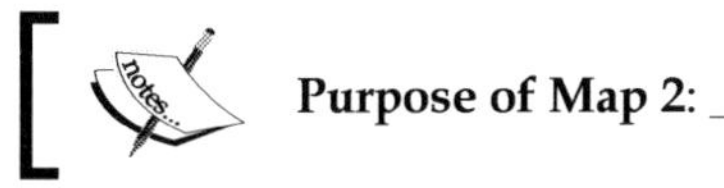

Purpose of Map 2: ____________________

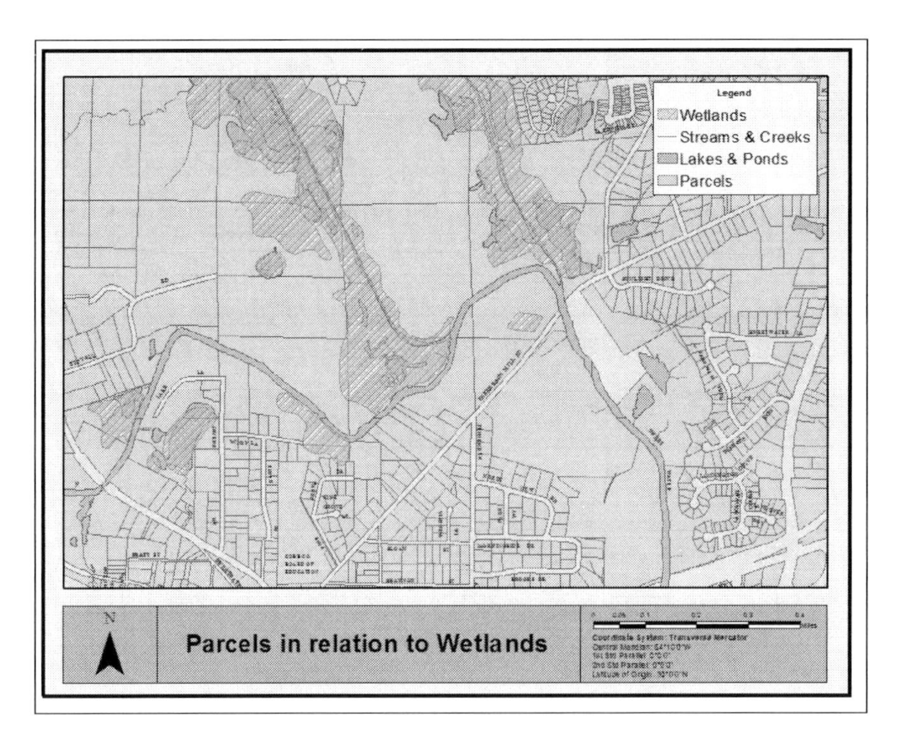

Purpose of Map 3: ____________________________________

The audience

The audience that will use and interpret your map will also impact its design. You should consider the following factors about your intended audience:

- Age
- Education or knowledge level
- Physical abilities or disabilities

If you are preparing a map for an older audience, you may need to make fonts bigger to make it easier to see. This may mean that you cannot put as much detail into a single map to avoid over-cluttering your map, which might mean that you need to create multiple maps for this audience. Age may also impact the choice of symbology styles you use. If you are making a map for a very young audience, you might want simple brightly colored symbols for example.

Age can also impact the methods you use to present the maps you create to ensure the greatest impact. Younger audiences tend to prefer digital media to printed maps. While older audiences tend to want printed maps rather than digital media.

The author's story: A

This true story helps to illustrate my point. A couple of years ago, I was assisting a County Property Appraiser's office when an older gentleman walked into the office. He was not happy about the appraisal of his property. The Chief Appraiser for this county walked him over to a computer and pulled up a map of the property in question and symbolized it based on appraised value along with some other values. The screen showed that this gentleman's parcel was appraised at the same value as equivalent neighbors nearby. The older gentleman would not however accept what the computer was showing him. While this was happening, I went to another computer in the back and printed a copy of the same map the Chief Appraiser was showing to the gentleman on the computer. I then took it out to the gentleman and reviewed it with him. Then, he was happy. The minute he could physically hold the map in his hands, it became real and had meaning. I told him that we would make sure that the computer had the same information as the printed map (which it already did) and the gentleman left happily.

The education and knowledge level of your intended audience will also have an impact on your map and layout design. For example, if you are creating a map of your sewer system for a group of civil engineers, they are going to need a lot of detail. They will want to know pipe sizes, flow direction, treatment capacities, and more. However, if you are creating a sewer map for the general public, that level of detail will confuse many because they do not have the engineering education or knowledge. The public is more likely to be just interested in knowing whether they have a sewer service.

Physical abilities or disabilities are something you should also take into account. Will the audience include a lot of people who wear glasses? Will some of them be color blind? These are factors to consider when designing your layout and maps.

Question: If you know your audience might include someone who is color blind, what can you do or change so that they could successfully use your map as well?

____

____

The author's story: B

Here is another true story to help illustrate this point.

I was working with an engineer on a project to generate a street improvement plan for a city. During this project, each road in the city was inspected and scored. I then consolidated all this data and developed a simple ranking system of good, fair, and poor to determine which roads were in the greatest need of repair or upgrading. The engineer asked me to produce a map showing the location of the roads and their ranking. Since these were roads, I thought that I would use standard traffic light colors to identify the ranking of each road. I made all the good roads display green, the fair roads display yellow, and the poor roads display red. Then I presented the map to the project engineer. When I did this, he asked me which roads were ranked good and which ones were poor. I told him the good roads were green and the poor ones were red as the legend showed. He then informed me that he was color blind. He could not distinguish red and green. So, I had to go back and make adjustments to my design. I applied a line pattern to each rank in addition to the color so that the engineer could tell the ranking of each road segment.

A situation

A situation is all about how your map will be presented and used. Will it be presented in a digital format or will it be printed? Will it be hung on a wall, used during a presentation or taken out into the field? Is this a legal document? If so, are there any defined requirements? All of these considerations will impact your design.

You may wonder why you would design a map differently if you are going to print it versus publish it digitally. Well, the simple answer is that each has its own limitations. When printing a map, you are limited by the capabilities of your printer. Is it color or just black and white? What sizes will it print? How much memory does it have? It does little good to design a color 36 inch by 48 inch map with aerial photography if all you have is a small desktop color printer that has only a couple megabytes of memory. Your map will overwhelm such a printer. Another thing you need to consider when printing maps, especially if you use an ink jet style printer, is paper quality. The quality of the paper you use in an ink jet printer has a big impact on the quality of your final output and even what you can print. Low-grade 20 lb bond paper will not produce a high-quality map. It will also not do well with a map that has a large amount of fill or aerial photos. The paper just cannot absorb the amount of ink that is applied, so it will become wrinkled and rip very easily. It might even damage your printer.

There are several options you have for publishing a map digitally. You can publish to a PDF file. This allows people without GIS software to view your creation on a wealth of devices and even add the map to a website without the need for special GIS web servers. PDFs can also be secured by setting a password to open them. Current versions of the PDF format even support layers and attribute data. This allows you to create an interactive map even for those without GIS software. When creating a PDF, you must consider the file's size. The more you include in a map and the higher the resolution, the larger the PDF becomes.

You can also publish your maps to ArcGIS Online or ArcGIS for Server. These allow even greater levels of access and functionality. However, these are web-based technologies, so you must always be concerned about performance when you are designing a map, which will be published in this manner. Simpler is always better. You will need to avoid complex symbology and labels whenever possible. Reducing the number of layers also helps.

Another popular way to present a map digitally is by using a projector. This is especially common when presenting the map at a meeting. Projectors tend to wash or fade colors. So, you may need to choose a more intense color palette for your map if you intend to project it. Also remember that projectors will impact the scale of your map. Though you set the map to one scale in ArcGIS Pro, the projected image will not always be at that same scale. Factors such as the projector's angle to the projection surface or the shape of the projection surface or settings on the projector itself, such as the Keystone, can cause the map scale to be skewed.

Those are just a few of the ways situation that might impact your map design. Let's see what other things you can think of, where a given situation might impact your design. Answer the following questions:

Question: You are preparing a map of your water system, which will be given to the field crews to help them locate the system in the field. The maps will be stored in their trucks and used in all kinds of weather. How might this impact your design?

____

____

Question: You are preparing the official zoning map for a city. This will be the legal zoning map as required by the City's zoning ordinance and will be hanging in City Hall for city officials and the citizens to use. What factors should you consider that might impact your design?

____

____

Creating a layout in ArcGIS Pro

Now that you know what factors can impact the design of your layout, it is time to learn how to actually create a new layout in ArcGIS Pro. ArcGIS Pro supports several methods to create new layouts:

- You can start from scratch by adding a new blank layout to your project
- You can import an existing map document file that was created with ArcMap
- You can copy an existing layout within your project
- You can import a layout file as a template.

You will now explore a couple of these methods using the project you created in Exercise 4B and 4C. The Director has asked you to print a few of the maps you created in that exercise. So, you will create a layout for each map the Director wants printed.

Exercise 6A – creating a simple layout

The Director wants you to print a copy of the map you created previously that identified commercial properties that were between 1 and 3 acres. He needs to use the map in a meeting he will have with the executives of the company looking to locate into the City along with other city officials.

Step 1 – open ArcGIS Pro and your project

The first step is to open the project in which you will be creating the layout. See the following:

1. Start ArcGIS Pro.
2. If you see the `project %your name% Ex4B` listed in the **Open a recent project** list, select it and proceed to step 2. Otherwise, click on **Open another project**.
3. Click on the computer located under **Open**.
4. Select the **Browse** button.
5. In the **Open Project** window, select your `C:` drive from the tree in the left-hand side pane.
6. Navigate to `C:\Student\IntroArcPro\My Projects\Tripp Ex4A` and click on `%your name% Ex4B.aprx`. Then, click on the **Select** button.

Your project should open to a familiar map. You should see up to three tabs across the top of the map view area for each of the three maps you created in the exercises from *Chapter 4, Creating 2D Maps*.

Step 2 – adding a new blank layout

In this step, you will add a new blank layout to your project and then add a map frame, which will display the map that shows the commercial properties that are between 1 and 3 acres in size.

Since this map will be used in a meeting with multiple people, you will create a large layout. This will allow you to create a map that is easy for a group to view and use in a meeting. For this meeting, the Director only wants to focus on those parcels that are within the size limit specified. He does not need to see the entire City:

1. Activate the **INSERT** tab in the ribbon.
2. Click on the **New Layout** button in the **Project** group tab.

3. From the list of available layout options, select **Architectural Landscape E 36" x 48"**.

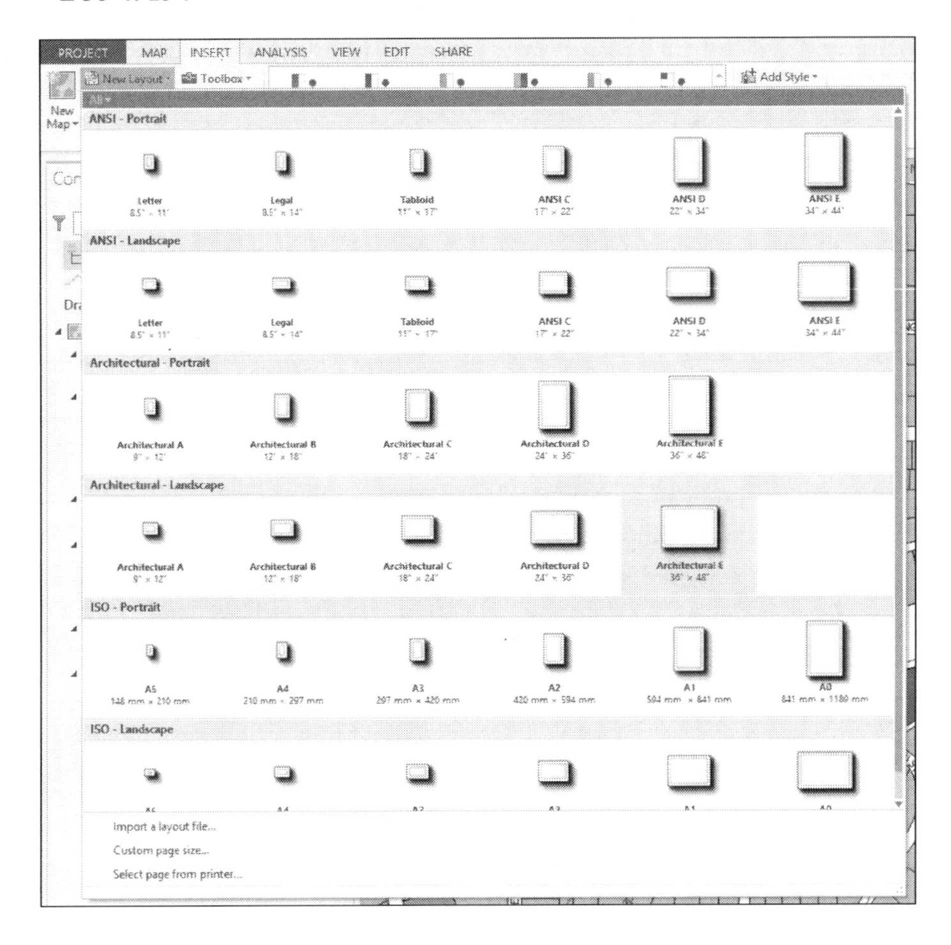

You should now see a new blank layout present in your project. In addition, a new folder has been added to your **Project** pane named `Layouts`.

1. Click on the small drop-down arrow located below the **Map Frame** button located in the **Map Frames** group on the **INSERT** tab.
2. Select the **Commercial Between 1 to 3 acres** map as illustrated here:

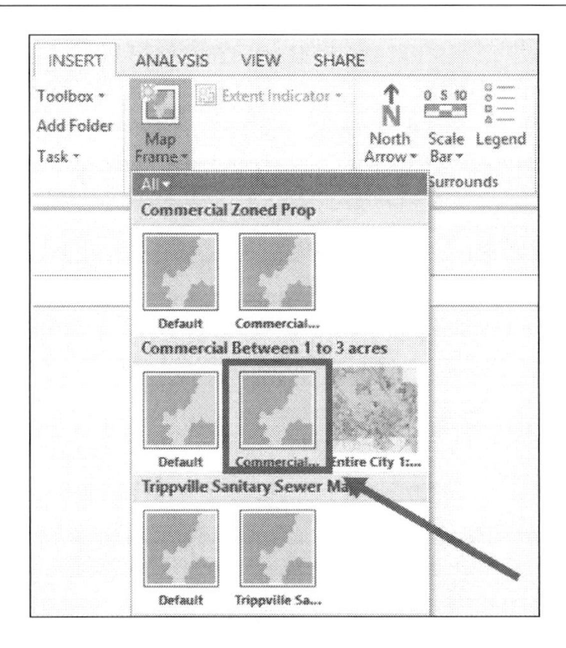

3. Click on the **Full Extent** button located in the **Map** group of the **Layout** tab. It looks like a small globe. This will zoom the map frame to the full extents.
4. Save your project.

   Now that you have added the map frame to your layout, you need to size and position it, so you have room to add a title and other elements.

5. Right-click on the newly added map frame in the **Contents** pane and select **Properties**.
6. Click on the **Placement** button in the **Element** pane located to the right of the interface.
7. Make the following adjustments to the settings:
   - **Size**:

     Width = 46 in

     Height = 30 in

   - **Position**:

     X = 1 in

     Y = 4.5 in

8. Close the **Elements** pane.

Step 3 – displaying the desired area

Now that you have added the desired map to your layout and sized it appropriately, you need to focus on the area that contains the parcels of interest to the Director for his meeting. You should note that these parcels are concentrated on the east side of the City. So, you will need to zoom into that area:

1. In the **Layout** tab, select the **Activate** button located in the **Map** group.
2. The **MAP** tab should automatically appear with the **Explore** tool active. Using the **Explore** tool, click on a point that is located near the center of the commercial parcels, which are between 1 and 3 acres in size.
3. Using your scroll wheel and mouse, zoom your map until it looks similar to this:

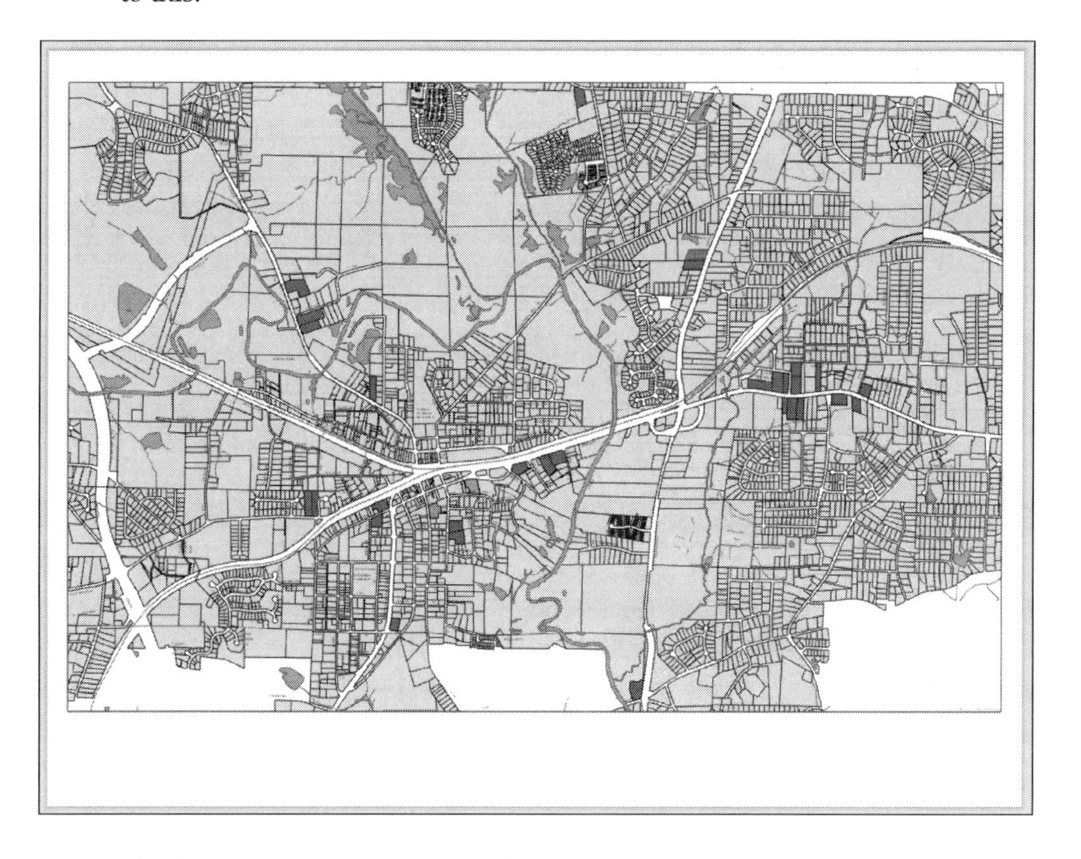

This has gotten your map very close to where it needs to be, but the Director wants the map printed at a scale where one inch will equal an even number of feet, such as 400 or 500 feet. This will allow him to use a ruler to check distances between features in the map easily.

4. Check the current scale of your map frame by looking at the scale window located in the lower left-hand corner of the map view.

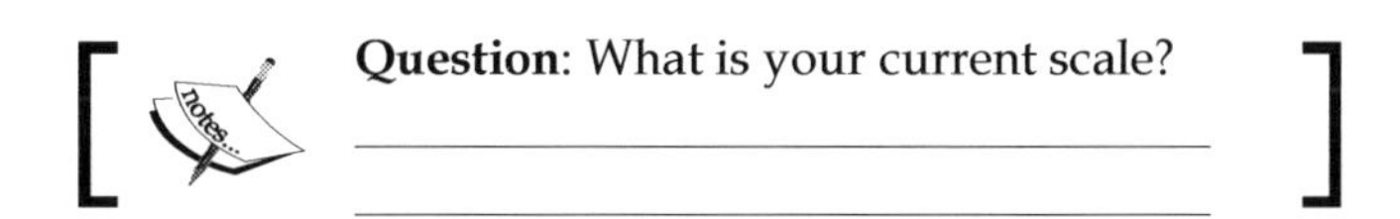

Question: What is your current scale?

____

____

Your scale should be somewhere between 1:4600 and 1:5000. This is very close to a scale that would make 1 inch equal 400 feet. The 1:4800 scale is equal to 1 inch equaling 400 feet if your map is set to use feet as its units:

5. Type 1:4800 in the small-scale window.
6. If required, pan your map some more until all the parcels of interest are visible.
7. Once you have verified that all parcels of interest are visible, click on the **Layout** tab in the ribbon. Click on the **Close Activation** button.
8. Save your project.

Step 4 – adding other elements

You are very close to completing your layout. You just need to add a few more elements to your layout, such as a title, north arrow, legend, and scale:

1. Click on the **INSERT** tab on the ribbon. From here, you can insert various elements into your layout.
2. Using your scroll wheel, zoom into the lower center of your layout. This is where you will place the title.
3. Click on the small drop-down arrow located to the right of **Dynamic Text** and select **Name of Map**. You will use the name of the map you added to the layout as the title:

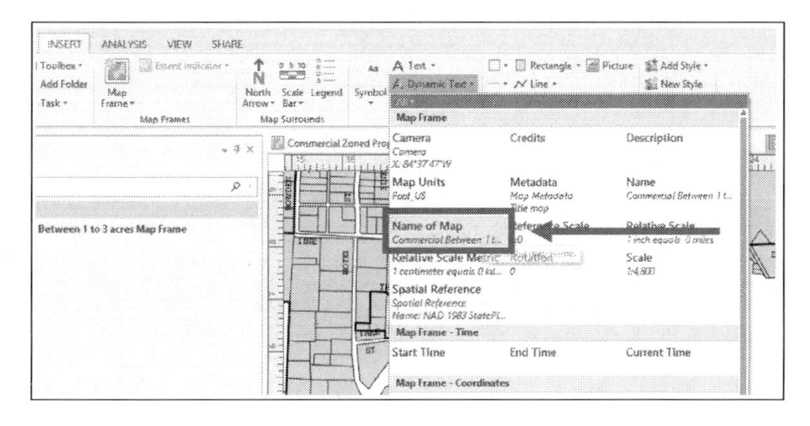

Dynamic text is text that references a specific property of the **Map Frame**, **Project**, **Layout**, or your computer system. Dynamic text will automatically update if the value for the specific property changes.

1. The **Elements** pane should automatically appear. Click on the **Options** button located in the **Elements** tab.
2. Expand the **General** option and rename this element `Title`.
3. In the **Text** window, delete **Name of Map** from the window and click on **Apply**.
4. Click on the **Placement** button in the **Elements** pane. Make the following adjustments to the **Placement Properties**:
    - Width = 18.5 in
    - Height = 1.5 in
    - X = 14.75 in
    - Y = 2 in
5. Now you need to insert a north arrow. Make sure that the **INSERT** tab is active and click on the drop-down arrow below **North Arrow**. Select a north arrow style you like, such as ArcGIS North 11. (I generally like to keep it simple since the north arrow is only one small element in a larger picture.)
6. In the **Elements** pane, adjust the size and position as follows:
    - Width = 2.0 in
    - Height = 2.5 in
    - X = 2.8 in
    - Y = 2.75 in

This will place a north arrow similar to ArcGIS North 11 in the lower left-hand corner of the layout. If you choose a longer north arrow style such as ArcGIS North 4, you may need to make additional adjustments to the size and position to get the arrow to fit appropriately. Use your judgment as to what looks best. Remember that this is a chance for you to exercise some artistic flare.

1. Zoom in to the area between the north arrow and map title.
2. Insert a scale bar using the same basic method you use to insert the north arrow.
3. Select the **Design** tab in the **Scale Bar Tools** contextual tab. This tab allows you to control various design setting for the scale bar. Since the Director wants the scale to be based on inches you must make some adjustments to the scale bar.

4. The scale 1:4800 is the same as 1 inch = 400 feet. So, you need to adjust your scale bar to show that. Change the **Resize Behavior** to **Adjust number of divisions**. This will keep the division spacing from changing allowing you to set it, so the divisions are shown in inch increments.
5. Set the Units to Feet.
6. Change the Division Value to 400. This sets the division increments so that they are 1 inch.
7. Now make the following changes to the **Placement Properties** in the Elements pane similar to the method you used to change the north arrow settings:
    - Width = 4 in
    - Height = 0.30 in (this can vary depending on which scale bar style you choose)
    - X = 7.0 in
    - Y = 2.5 in
8. Click on the **Layout tab** and click on the **Full Extent** button.

   Your layout should now look similar to this. Your north arrow and scale bar may look different depending on which style you selected.

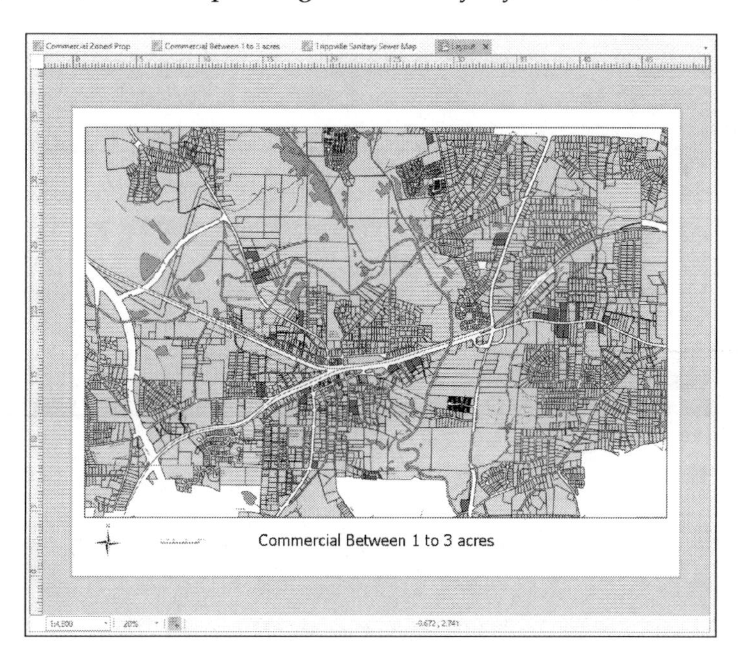

9. Save your project.

Challenge

Using the skills you have learned to insert a north arrow and scale bar, add a legend to your layout. Position the legend in the lower right corner of the layout to help balance its overall appearance. A legend has a few more components than a north arrow, so this may be a bit more complicated. You may want to go to `http://pro.arcgis.com/en/pro-app/help/layouts/add-a-legend.htm` for hints or assistance.

Exercise 6B – creating and using a layout template

The Director was very happy with the map you made for his meeting and now wants another. He would like to have a map printed at the same size, but that shows all the commercially zoned parcels in the City.

In this exercise, you will use the layout you created in the last exercise to create a template file. You will then use that template to create a new layout.

Step 1 – saving a template file

In this step, you will create a template layout based on the layout you created in the last exercise:

1. Click on the **Sharing** tab on the ribbon.
2. Click on the **Layout File** button in the Save As group.
3. In the **Save Layout** as `PAGX File` window, select your `C:` drive and then navigate to `C:\Student\IntroArcPro\My Projects`.
4. Name the new layout file **Layout Template** and click on **Save**.

You have created a layout file that can be used as a template for other layouts. New layouts based on this template will contain all the same standardized elements.

Step 2 – creating a new layout using a layout file

You will now use the file you created in the previous step to add a new layout to your project:

1. Click on the **INSERT** tab in the ribbon.
2. Click on the small drop-down arrow located next to **New Layout**.
3. Select **Import a layout file**, which is located near the bottom of the displayed window.

4. In the **Import** window, navigate to `C:\Student\IntroArcPro\My Projects` and select the **Layout Template.pagx** file you created in Step 1.

A new layout has been added to your project that looks exactly like the layout you created in the last exercise. Now you will need to configure the layout to display the correct map and make adjustments to a few of the elements.

Step 3 – configuring a new layout

In this step, you will configure the layout to display the map that shows all commercially zoned parcels in the city:

1. Select the **Commercial Between 1 to 3 acres** map frame in the **Contents** pane to open the **Elements** pane, so you can adjust properties.
2. Click on the **Options** button in the **Elements** pane.
3. Expand the **Map Frame** option if needed.
4. Click on the drop-down arrow for the **Map** option and select **Commercially Zoned Prop**. This assigns the correct map to display in the layout.
5. Change the element name to **Commercially Zoned Property**.

Therefore, the title appears with the word property spelled out fully; you need to change the name of the map that is being displayed since the title is dynamic text.

6. Select the **Commercially Zoned Prop** tab at the top of the map view. This will make that map visible.
7. In the **Contents** pane, right-click on **Commercial Zoned Prop** and select **Properties**.
8. In the **General** properties, rename the map **Commercially Zoned Property** and click on **OK**.
9. Click back on the new layout you have created (Hint: it may be named `Layout2`).

The title should now be changed to reflect the new name of the map displayed in the map frame. Now you need to adjust the scale of the map, so it displays all the commercially zoned property and the legend so it shows your layers:

10. Using the skills you have learned, change your scale to 1: 7200, so you can see all the commercially zoned properties in the City as desired by the Director.
11. Select the legend in the **Contents** pane.
12. Select the **Design** tab in the **Legend Tools** contextual tab.

13. Click on the **Layers** drop-down menu and turn on all layers by clicking in the box located next to each layer name.

You have just created a new layout using a layout file as a template. It automatically included the same style north arrow and scale bar. All the layout elements were properly positioned. After making some simple adjustments, you quickly created a new layout, which is ready for printing.

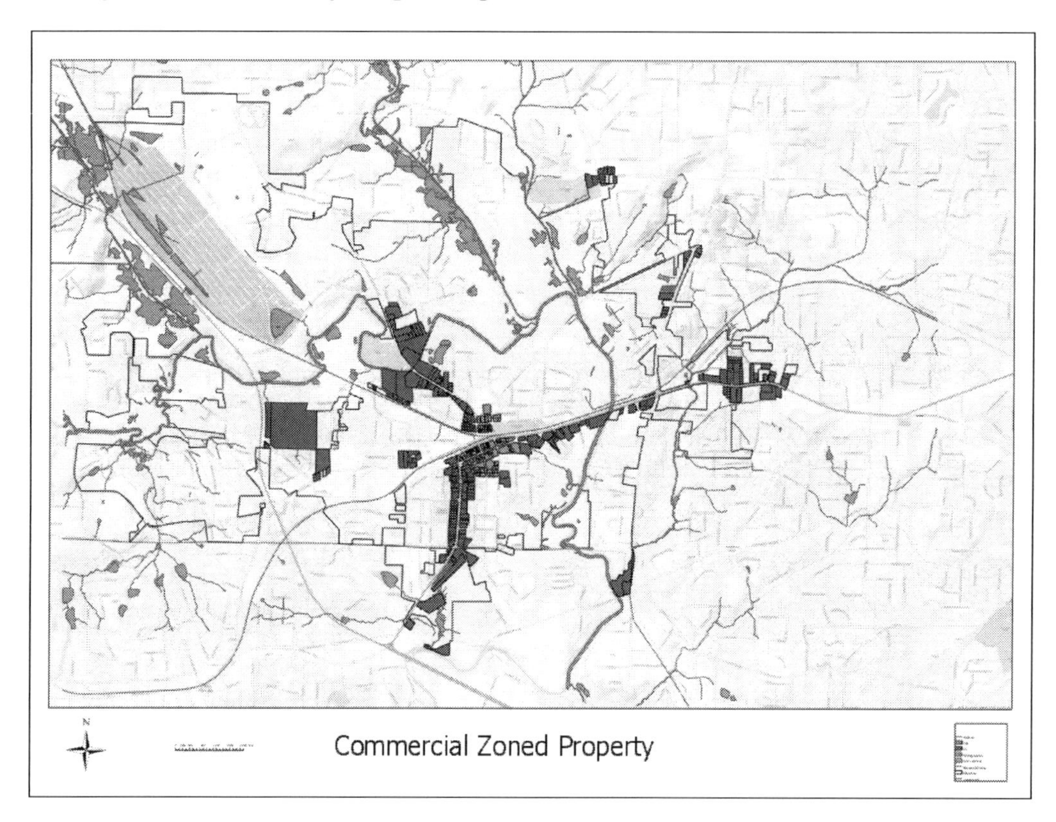

The challenge

You had to change the scale in the new layout you created so that the scale bar no longer has divisions, which are 1 inch in size. See if you can adjust the scale bar so that the divisions are once again 1 inch in size. It just happens that 1:7200 equals 1 inch = 600 feet. If you need help or assistance, you can go to `http://pro.arcgis.com/en/pro-app/help/layouts/scale-bars.htm`.

Summary

You learned how to create simple but effective layouts to frame the 2D and 3D maps you create using ArcGIS Pro. Layouts allow you to present your efforts in a professional manner, allowing others to appreciate your efforts. This makes your analysis, maps, and data more useful to others and helps put all your hard work into perspective.

As you learned, before you create your layout, there are several factors you need to consider. These include purpose, audience, and situation. Each of these will impact the overall design of your layout. So, it is important to think these through before you start building your layout.

Each layout you create will contain several elements, which will need to be configured. You will need to choose which map or maps from your project to include, what north arrow to use, determine the scale, and more. The factors you considered will help to drive the inclusion or exclusion of some elements. To help standardize elements between layouts, you learned how to create a layout template. This will help brand your maps and speed their production.

7
Editing Spatial and Tabular Data

One of the favorite questions I am often asked from administrators, directors, and elected officials is "when will our GIS be done?" The honest answer is never. All the layers of information we store in a GIS are changing. New roads are built. Parcels are split and sold. New sewer lines are installed and more. All of these changes need to be incorporated into our GIS databases. To do this, you must be able edit your GIS data.

ArcGIS Pro allows you to edit your GIS data. You can add new features or modify existing features. You can also edit the structure of your GIS database. With ArcGIS Pro, you can create new feature classes, add new attribute fields, create new tables, add domains, and so on. It also supports editing of both 2D and 3D data.

ArcGIS Pro is still very much a work in progress. There are some data structures, such as topologies, geometric networks, and the Parcel Fabric, which cannot be edited with the current version of ArcGIS Pro, which is 1.1. It is expected that over time, as ArcGIS Pro continues to mature, that these limitation will be removed.

In this chapter, you will learn the following topics:

- How to create and manage feature templates
- How to create update spatial and tabular data
- Know which data types can be edited
- How to make changes to your geodatabase schema

Editable data formats

ArcGIS Pro allows users to access and use a lot of data formats to make maps, perform queries, and so on. But using and accessing is far different from being able to edit. It is important to know the limits of ArcGIS Pro with certain common data formats.

Data formats – editable or not

The more you work in GIS, the more data formats you will encounter. Esri's ArcGIS Platform supports many of the most commonly used formats. However, as we said earlier, there is a big difference between being able to view and query data and being able to edit it.

The following table outlines some of the limits ArcGIS Pro has with various data storage formats you are likely to encounter on a regular basis:

Data format	Display	Edit	Comments
Personal geodatabase	No	No	
File, workgroup, and enterprise (SDE) geodatabase	Yes	Yes	Workgroup and enterprise geodatabases require Standard or above license. Topologies, Geometric Network and Parcel Fabric editing not currently supported.
Shapefiles	Yes	Yes	
Coverage	No	No	
CAD files (DWG, DGN, and DXF)	Yes	No	ArcGIS Pro will display DWG files created with AutoCAD 2016 or earlier.
ArcGIS feature service	Yes	Yes	Published with ArcGIS Server.
ArcGIS feature layer	Yes	Yes	Published from ArcGIS Online or Portal with ArcGIS. Editing must be enabled.
Web map services	Yes	No	ArcGIS Pro can access web map services, including ArcGIS Server, ArcGIS Online, WMS, and WMTS.
Excel Spreadsheet	Yes	No	
DBF file	Yes	Yes	
Text files (TXT or CSV)	Yes	No	

This list is just those data types that are most frequently encountered with a focus on vector or tabular data. There are many other GIS data formats. Shapefiles and geodatabases are the primary spatial formats that ArcGIS Pro is designed to interact with completely. Even the geodatabase currently has some limits on what can be edited using ArcGIS Pro.

The personal geodatabase format is not supported in ArcGIS Pro. Esri has been slowly reducing support for this type of geodatabase, which is built on top of Microsoft Access Database technology. This is largely due to the limitations of Microsoft Access, which is restricted to 2 Gigabytes in size and whose performance slows the larger the database gets. This is the reason Esri developed the file geodatabase.

It is strongly recommended that if you are still using personal geodatabases, you migrate them to a file workgroup or enterprise geodatabase, especially if you wish to begin using ArcGIS Pro or ArcGIS Online. These geodatabase types also offer much better performance and storage capacity. The following table shows a general comparison between the three types of geodatabase, which can be created with the ArcGIS Platform:

	Personal	**File**	**Workgroup/ Enterprise**
Supporting database	Microsoft Access	Individual files designed by Esri	SQL Server Express, SQL Server, Oracle, DB2, Informix, and PostgreSQL
Storage size	2 GB (Performance degrades as size increases)	1 TB at the base of the database plus 1 TB per feature dataset	10 GB plus depending on supporting database
Number of editors	1	1 per feature dataset (if map references layers from multiple feature datasets, each dataset will be locked when editing)	10 or more depending on supporting database

How to know what format data is in

Now that you know that ArcGIS Pro allows you to work with multiple data storage formats, but you can only edit a few. How do you tell what format the data you are using is stored as? That is a good question.

Just as ArcGIS Pro allows you to visualize your data, it also provides visual clues about the data you are working with. Different icons are displayed next to various types of data to help you easily identify the type of data you are working with. The following image illustrates some of the icons used by ArcGIS Pro to identify different types of data.

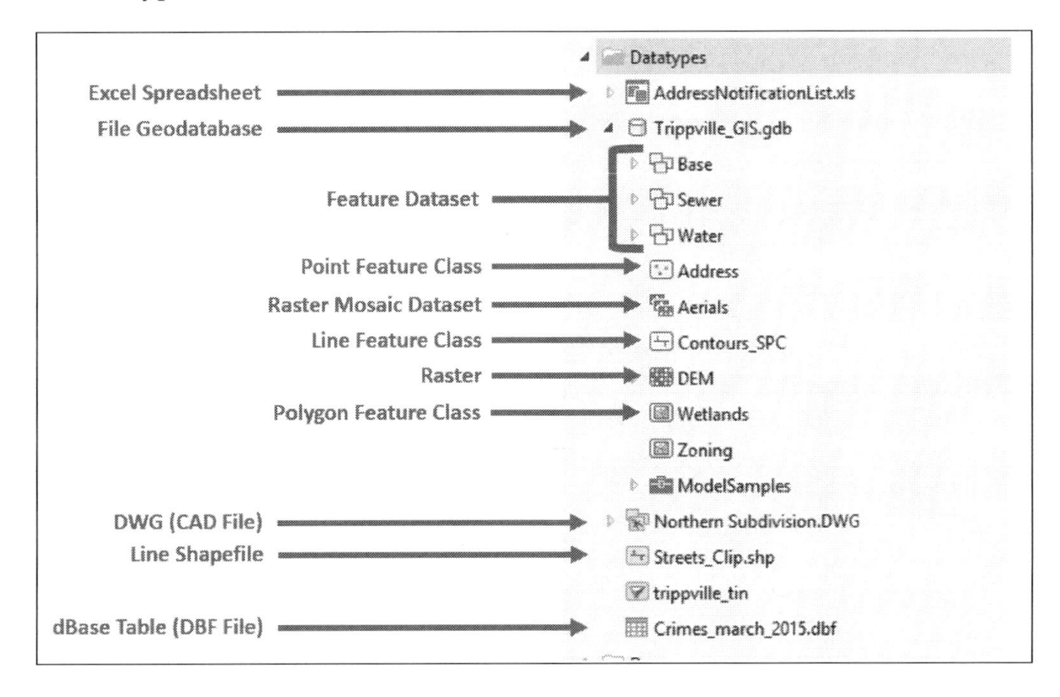

As you can see, ArcGIS Pro uses different icons to differentiate data types. For example, Shapefiles use green icons with a graphic to tell whether the Shapefile contains points, lines, or polygons. CAD files are identified with blue icons. DWG, DXF, and DGN files all use the same blue icons. ArcGIS does not distinguish between files created with AutoCAD, MicroStation, or one of the other many drafting and design software packages used by engineers and surveyors.

Some data formats support the storage of multiple data types, whereas others only allow users to store a single data type. For example, the geodatabase allows you to store points, lines, polygons, raster, and more within a single database, whereas a Shapefile will only allow you to store a single data type. A Shapefile will be a point or a line, or a polygon Shapefile. It cannot contain more than one data type in a single Shapefile.

How to edit data using ArcGIS Pro

The world is ever changing, so your GIS needs to keep up with those changes. Whether it is splitting a parcel, adding a road, adding a new attribute field, or creating a new layer of data, it is important that your GIS data reflects the most current conditions of the real-world features it represents and meets the needs of your organization.

ArcGIS Pro contains various tools that allow you to do all of this. You can add new features to an existing layer. You can modify existing features to show changes. You can create new layers and tables. In other words, ArcGIS Pro allows your GIS to grow, change, and flourish as reality changes.

How to start editing features and attributes

To begin editing data in ArcGIS Pro is fairly easy. The first step is to open a project that contains layers, which reference data that is stored in an editable format, such as a Shapefile or geodatabase. From there, all you need to do is click on the **EDIT** tab in the ribbon. It is that easy.

For those that have been using ArcGIS for Desktop, this may seem too easy. What happened to start editing? In ArcGIS Pro, you no longer need to start editing. You can immediately start editing once your project is open. ArcGIS Pro also does not limit you to editing data in one workspace at a time. If your map contains layers that point to data stored in a geodatabase and as Shapefiles, you can edit them all at the same time. You no longer need to start and stop editing each time, you need to switch between workspaces.

What is a workspace? A workspace is the location in which your data is stored. It can be a database or a folder. So, a geodatabase is considered a single workspace. A folder that contains Shapefiles or other data files would be another workspace.

Preparing to edit

Before you actually start editing your data, you need to take some time preparing both your data and ArcGIS Pro. Generally speaking, you should take the time to do the following before you start editing:

1. Ensure that all the spatial data you plan to edit is in the same coordinate system. This avoids errors that can be caused due to transformation issues.

2. Add and symbolize all layers you wish to edit to your map. The simpler you can keep the symbology for each layer in the map, the faster it will redraw as you pan and zoom during editing. Save the complex symbology for printing. Since ArcGIS Pro supports multiple maps in a single project, you may want to have one map you use to edit data and another to include in a layout for printing.
3. Simplify your attribute fields so that only those you need are visible. This will increase your efficiency and reduce the chance you mistakenly edit an attribute value that should not be changed.
4. Adjust **Project Option** settings for editing from the **Project** pane by:
    - Ensuring that proper units are set for distance, angle, direction, and area
    - Ensuring that edit settings are set as desired such as how and when to save.
5. Set which layers you wish to edit in the **List by Edits** within the **Contents** pane.
6. Set snapping options.

Taking the time to go through these steps before editing for the first time in a project will make editing easier and reduce the chance of errors.

The EDIT tab

The **EDIT** tab on the ribbon as shown in the following screenshot, is where you go when you want to begin editing your data. Here, you will find many of your most commonly used editing tools. The tools on this tab allow you to modify existing features or to help create new features. This tab not only allows you to edit spatial data but also attributes and standalone tables as well:

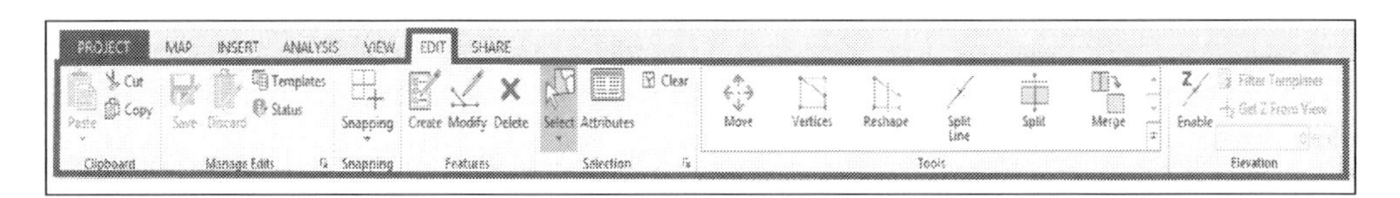

As you can see that the **EDIT** tab contains seven groups: **Clipboard**, **Manage Edits**, **Snapping**, **Features**, **Selection**, **Tools**, and **Elevation**. Now we will explore these groups and some of the tools.

The following screenshot shows some tools in **Clipboard**:

The **Clipboard** group contains tools, which will copy or paste to and from your computer's clipboard. Using these tools, you can copy features in one layer to another or cut them from one layer and paste them into another. You can also use these tools to duplicate features in the same layer.

The following screenshot shows tools present in **Manage Edits**:

The **Manage Edits** group helps you control your edits. Here, you can save your edits or discard them. If you discard edits, ArcGIS Pro will revert all your data back to the way it was before the last time you saved. This means that it will undo all edits you have performed since you last saved your edits. Once you save your edits, you cannot discard or undo them.

ArcGIS Pro does allow you to set up an automatic save under your **Project Edit** option. You can choose to automatically save at specific time intervals or after a number of operations. Setting up an automatic save can help ensure that you don't lose edits you have performed if your system or ArcGIS Pro were to crash. The drawback to auto-saving is that while it is saving you cannot work. You must wait for it to finish saving, so you want to make sure that you set a sensible interval if you choose to enable auto saves. Also, remember that you cannot undo any edit after it has been saved. So, if you mistakenly edited the wrong feature and ArcGIS Pro saves, you will need to manually reverse your edit or restore a backup of some sort.

The **Templates** button will open the **Manage Templates** pane on the right-hand side of the ArcGIS Pro interface. From this pane, you can manage your feature templates, which are used to create new features. Those will be discussed in more detail later.

Finally, you can check the editing status of the maps in your project. It will tell you what layers are editable, which are not, and any warnings you should be aware of when editing the data referenced in that map.

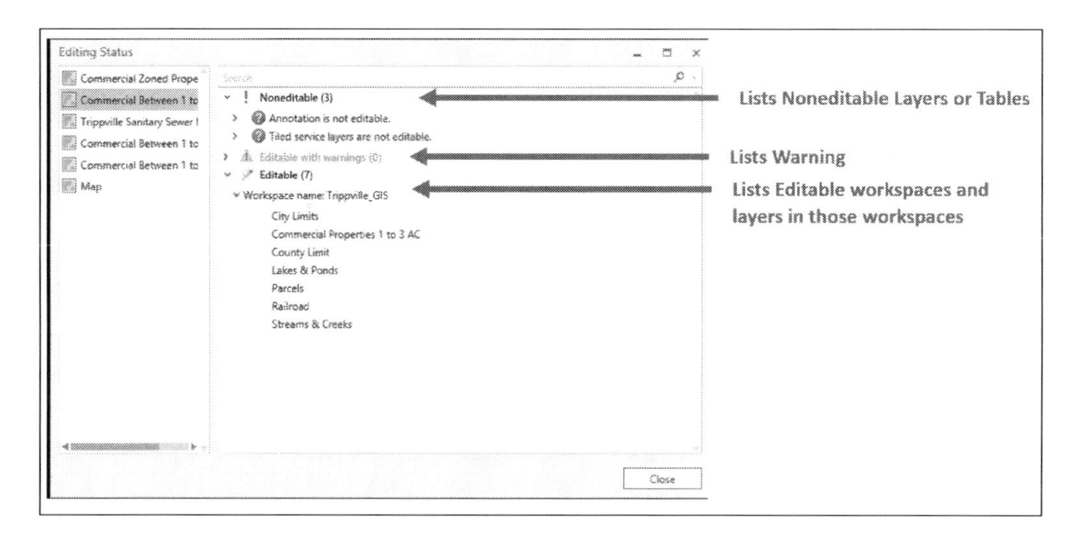

The next group is **Snapping**. It includes a single button with a drop-down menu, which controls snapping and snapping options. **Snapping** allows you to easily draw features so that they maintain connections to other features on the same or a different layer while editing. From the **Snapping** group, you can set where you want to snap to new features or sketches to existing features.

A sketch is something you create while editing. It can represent a new feature such as a new street centerline or parcel polygon. Sketches can also be shapes you draw to modify or reshape existing features such as a line drawn to split a polygon. Sketches are temporary and only exist in the memory of your computer. If the sketch represents an update or change to one of your layers, it is not committed back to your GIS data until you save your edits. Once you save your sketch becomes a true feature. This is why you should save often.

With **Snapping** enabled, you can snap to the end, edge, midpoint, intersections, and vertex of other features as shown in the following screenshot:

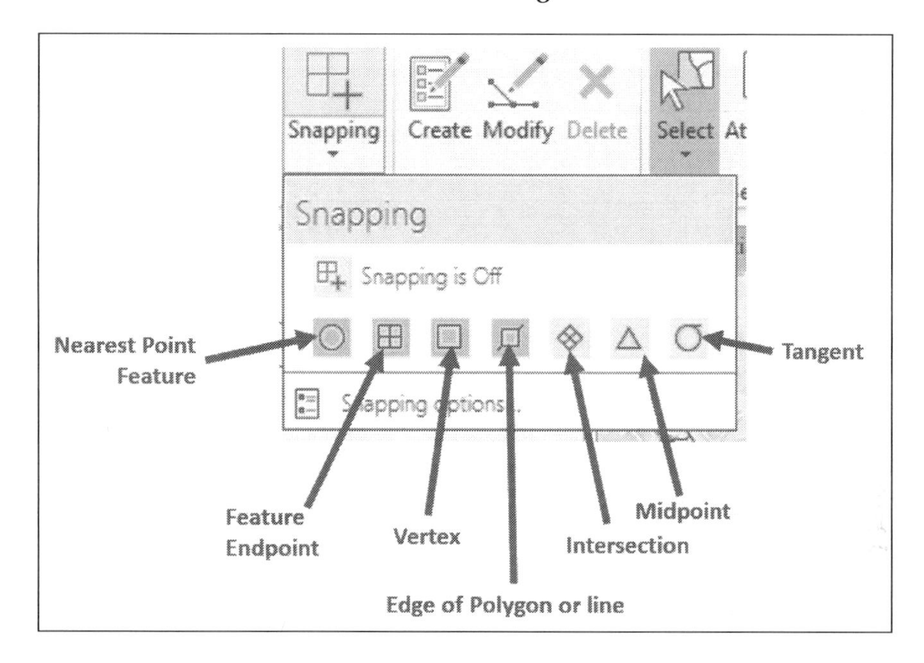

From the **Snapping** drop-down, you can also enable and disable **Snapping**. Holding down your space bar will also temporarily disable **Snapping** as long as it is held down.

The following screenshot shows tools present in the **Features** tab:

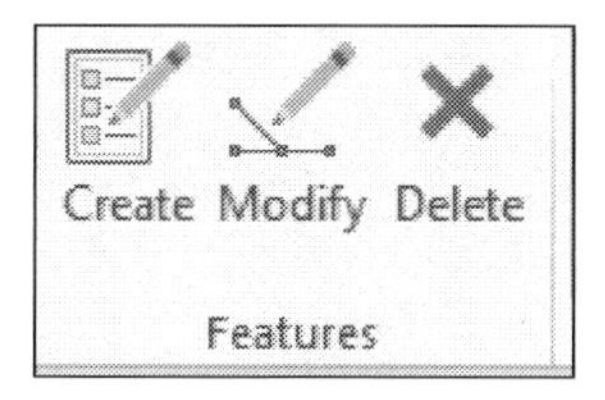

The **Features** group contains tools for editing features. You access the **Create Features** pane by clicking on the **Create** button. The **Create Features** pane contains the feature templates used to create new features in your layers. The **Modify** button opens the **Modify** pane, which includes tools to change existing features, such as **Move**, **Rotate**, **Split**, and many more. Finally, it is the **Delete** button, which erases selected features or records.

The **Selection** group includes the tools to select features interactively within your map, viewing attributes and clearing your selection. The **Select** tool includes five different shapes, which can be drawn to select features for editing as shown in the following screenshot:

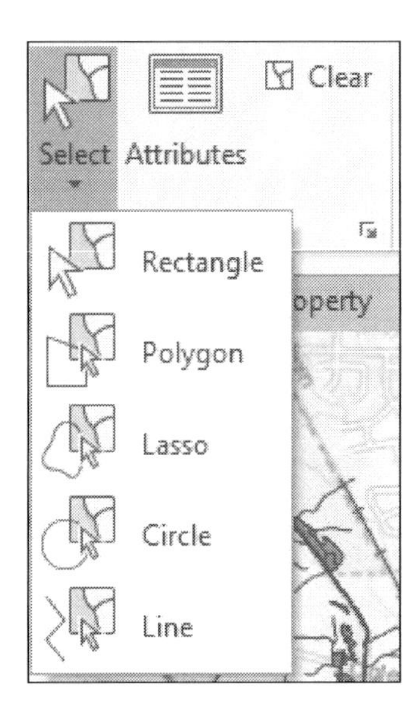

Most of these are self-explanatory with the exception of **Lasso**. The **Lasso** select tool allows you to draw a freeform polygon shape. The drawing follows the movement of your mouse pointer as long as you hold the left mouse button down. Releasing the mouse button completes the shape.

The **Tools** group contains tools for modifying features. You will use them when you need to perform tasks such as splitting a parcel which has been subdivided into more than one parcel. There are also tools for splitting a line, editing vertices, moving a feature, and merging features.

The last group is **Elevation**. This contains tools for editing 3D features.

Creating new features

Creating a new feature is a very common editing task. It might add a new road or a new sewer manhole or a new stormwater detention pond. As new things are built, we need to add them to our GIS. Creating new features require a feature template.

Feature templates

Feature templates define the properties required to create a new feature. This includes the target layer, default construction tool, default attribute values, and symbology.

These templates are tied to your map's contents. So for each layer in your map, you will have a matching feature template. If you symbolize a layer with unique values, graduated colors, or graduated symbols, you will have a feature template for each unique symbol associated with that layer. For example, if you symbolize a parcel layer based on its zoning classification and there are three zoning classes: residential, commercial, and industrial, you will have three feature templates for that layer.

When you add a layer to your map or adjust the symbology, your available feature templates are adjusted to match. You can also manually create feature templates, adjust the properties of existing templates and delete templates. This is all accomplished in the **Manage Templates** pane.

There are two types of feature templates within ArcGIS Pro, feature templates, and group templates. You have already been learning about feature templates, so now let's take a quick look at group templates. Although a feature template will create a new feature on a single layer, a group template will create multiple features on multiple layers based on the rules you define. Within the group template, you will define a primary feature template. The primary feature template drives the creation of other features.

Exercise 7A – creating new features

In this exercise, you will update several layers based on a plat you have been given for a new subdivision, which has been built in the City of Trippville. This is part of your normal duties as the city's GIS specialist. The plat is from a local surveyor and has been scanned.

Using the information shown in the plat, you will need to update the sewer system layers, road centerlines, and parcels.

Step 1 – opening your project and preparing to edit

In this step, you take some time to make sure that everything is ready to begin editing. You will open your project to verify that the layers you need to update are editable. You will ensure that there are not any warnings or messages that might cause problems while you update the data. You will also verify other editing option settings:

1. Open ArcGIS Pro.

2. Using skills you have learned, open the `Creating new features.aprx` located in `C:\Student\IntroArcPro\Chapter7`.

When the project opens, you should see a single map that contains the layers you will be updating. It should look very similar to this:

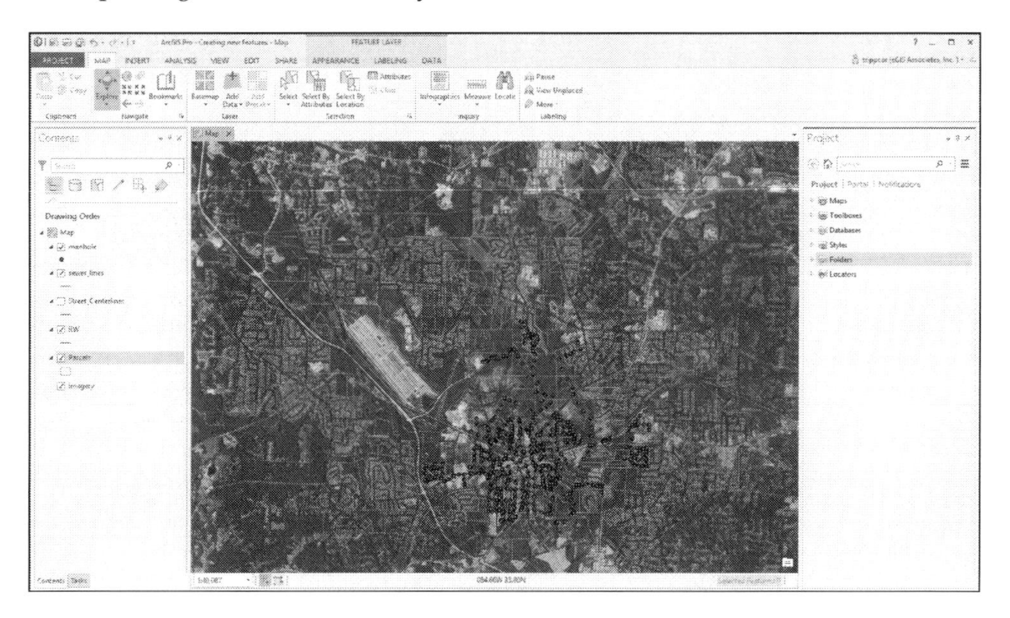

Now that your project is open, you need to verify a few settings:

3. Click on the **List By Data Source** button in the **Contents** pane to verify the location of the source data for the layers in your map.

Question: What geodatabase is being referenced by the layers and where is it located?

____

____

4. Right-click on the **Parcels** layer and select **Properties**.
5. Select **Source** from the pane on the left-hand side of the **Layer** properties window.
6. Scroll down in the right-hand side pane until you see **Spatial Reference** and expand it by clicking on the open arrow head located next to it. This will show you what coordinate system this layer is in.

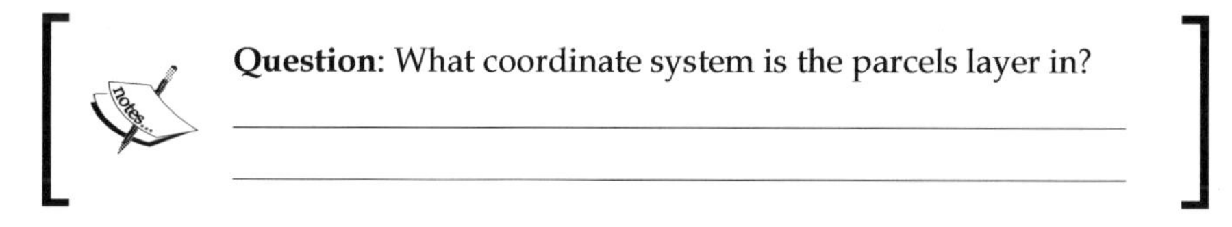

Question: What coordinate system is the parcels layer in?

7. Using that same method check the remainder of the layers within the map. See what coordinate system they are in.

Question: Are all your layers within the same coordinate system?

ArcGIS Pro will allow you to edit data that is in different coordinate systems. However, the recommended best practice is to have all data being edited within the same coordinate system. This helps avoid errors caused by using different transformations.

Now it is time to set up the symbology for the layers you will be editing. For the most part, the current symbology will work. However, the Public Works Director wants the new sewer lines entered with the correct size and material. To make that process more efficient, it would be good to have a feature template with those default values defined ahead of time. Since feature templates are linked to the layers in your **Contents** pane, you will change the symbology for the sewer lines to be based on the pipe's size and material.

Luckily, there is already a layer file, which has the symbology settings already defined. You will be able to import those settings from the layer file without having to configure the symbology for the `sewer_lines` layer from the beginning:

8. Select the `sewer_lines` layer in the **Contents** pane.
9. Select the **APPEARANCE** tab.
10. Click on the **Import** tool located in the **Drawing** group. This will open the **Geoprocessing** pane to the right side of the interface.
11. Click on the **Browse** button located at the end of **Symbology Layer**.
12. In the **Symbology Layer** window, click on **Folders** located in the left pane of the window.
13. Double-click on the `Chapter7` folder.

14. Click on the `Sewer Lines.lyrx` file and the **Select** button.

    Your **Geoprocessing** pane should now look like this:

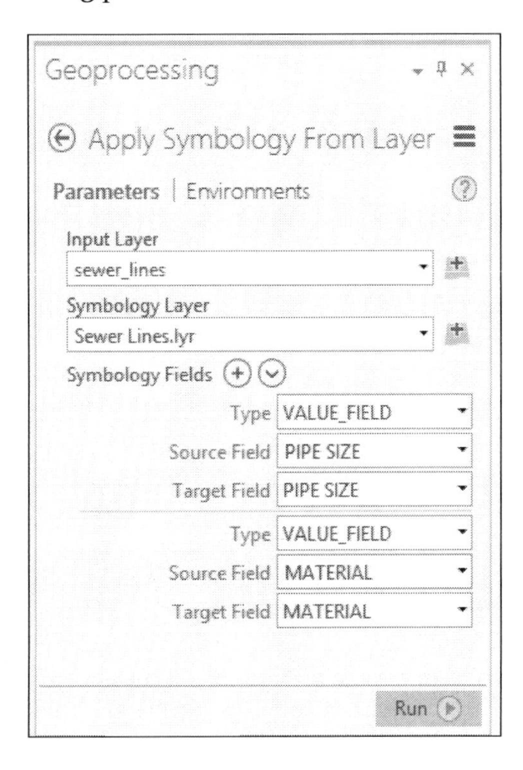

15. Once you have verified that everything is filled out correctly, click on **Run**.
16. When the process has completed, close the **Geoprocessing** pane.

You have just used a layer file to import pre-set symbology for your **Sewer** layer, which allows you to distinguish each sewer pipe's size and material. If for some reason this did not work, you can right-click on the existing **Sewer** layer and select **Remove**. Then, go to the **Project** pane and the **Folders** connection. In the `Chapter7` folder, you can right-click on the `Sewer Lines.lyrx` file and choose **Add to Current Map**.

ArcGIS Pro 1.1 seems to have an intermittent issue importing symbology settings based on two or more fields. If you happen to experience this, you may remove the existing `sewer_lines` layer and use the **Add Data** button on the **MAP** tab to add the `sewer_lines.lyrx` as a new layer with the correct symbology. Hopefully, Esri will address this in future versions.

You are almost ready to start editing. There are a couple of other settings you need to check:

17. Click on the **PROJECT** tab in the ribbon and select **Options**.
18. Select **Units** and verify that the following settings are chosen. If they are not, then select the correct units:
    - **Distance Units: Foot_US**
    - **Angular Units: Degrees Minutes Seconds**
    - **Area Units: Square_Foot_US**
    - **Location Units: Foot**
    - **Direction Units: Quadrant Bearing**
    - **All Others: Accept assigned values**
19. Select **Editing** in the left-hand side pane of the **Options** window.
20. If needed, expand the **Session** section in the right-hand side pane.
21. Ensure that **Automatically Save Edits** is not enabled. Since you are new to ArcGIS Pro, you don't want edits to be saved until you have verified them to be correct.
22. Also make sure to save edits when saving a project is not enabled.
23. Feel free to examine the other **Editing** options that are available. Once you are done exploring, click on **OK**.
24. Click on the **Back** arrow located in the top-left corner of the **Project** window to return to the main ArcGIS Pro interface.

You have one last thing to verify. You need to make sure that **Snapping** is enabled and what will be snapped too. You also need to verify the snapping tolerance:

25. Click on the **EDIT** tab in the ribbon.
26. Click on the small drop-down arrow located below **Snapping**.
27. Select **Snapping** options.
28. Set your **XY** tolerance to **10 Map Units**.
29. Set your **Snap** tip color to **Mars Red** and click on **OK**.
30. Click on the arrow below **Snapping** once again. Verify what snapping position options are enabled.

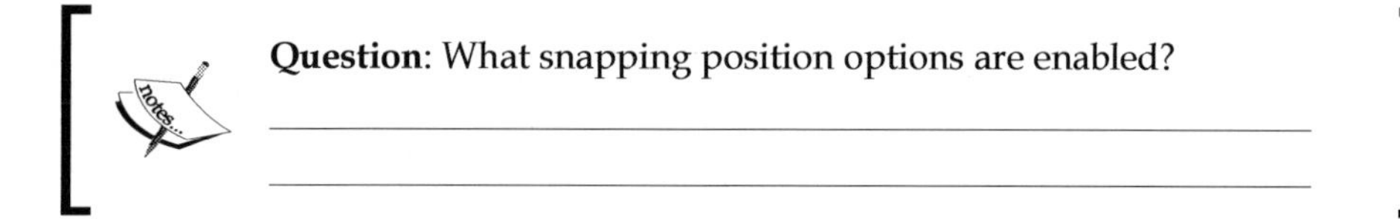

Question: What snapping position options are enabled?

____

____

You can change the snapping tolerance and snapping positions as needed while you edit. As a new user, you will need to change those settings frequently as you try to find a happy medium that will work in most cases. Once you figure out what works best for you, you will not need to change the tolerance as often. The snapping positions will be changed much more frequently because they depend largely on what you are editing:

31. Ensure that **Snapping** is enabled by clicking on the **Snapping** button on the **EDIT** tab.
32. Set your snapping positions to endpoint and edge by clicking on the small arrow below **Snapping** and selecting those options.
33. Save your project.

You are now ready to begin editing. You have taken the time to ensure that your editing environment has been properly set up.

Step 2 – adding your source data

The surveyor had provided you with a paper copy of the plat for the new subdivision. The plat shows the layout of the parcels, streets, sewer, and water features in the new subdivision. Luckily, one of your other staff members scanned and georeferenced the scanned plat, so you can easily add it to your map.

ArcGIS Pro 1.1 does not support **georeferencing** like ArcGIS for Desktop's ArcMap does with the georeferencing toolbar. This functionality is planned for a future release. For those not familiar with the term georeferencing, it simply means identifying the location of a feature in a real-world coordinate system such as WGS 84, which is a type of latitude and longitude used by GPS, or **Universe Trans Mercator** (**UTM**), or one of the many other real-world coordinate systems.

The following steps will help you to add a scanned georeferenced plat to your map:

1. Click on the **MAP** tab in the ribbon. Then, select **Bookmarks**.
2. Choose the `New Subdivision 1` bookmark to zoom you the location of the new subdivision.
3. In the **Project** pane, expand **Folders** and the `Chapter7` connection.
4. Right-click on the `Forrest Park Subdivision.jpg` and choose **Add to Current Map**.

Your map should now look like this. Your zoom scale and display area may be slightly different depending on the size of your monitor and its resolution.

There is no need to have the imagery base-map visible at this time. It will only slow you down as you pan and zoom during editing. So, turn off the imagery layer by unchecking the box next to the layer.

You have just added the scanned plat as a layer to your map. This will allow you to use it as a guide to add new features and update your GIS database. This is just one example of a data source you can use. There are many other ways you may acquire new information to use in your GIS.

Step 3 – drawing a new sewer line

Now that you have the plat added to your map to use as a guide, you will start drawing new features. You will start with simple lines and points that make up the sewer system:

1. Zoom into the northeast corner of the new subdivision, so you can see where the new sewer line connects to the existing sewer line.
2. Click on the **EDIT** tab in the ribbon.
3. Click on the **Create** button to open the **Create Features** pane on the right-hand side of the interface.

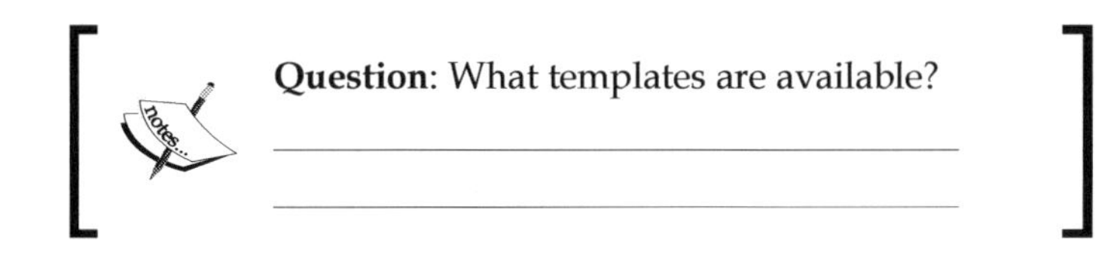

Question: What templates are available?

You will now add a new sewer line and manhole by tracing the features shown in the plat. Before you do that though, you will examine a feature template.

4. Right-click on the **8 inch PVC** template in the **Create Features** pane and select **Properties**. This should open the **Template Properties** window for that template.

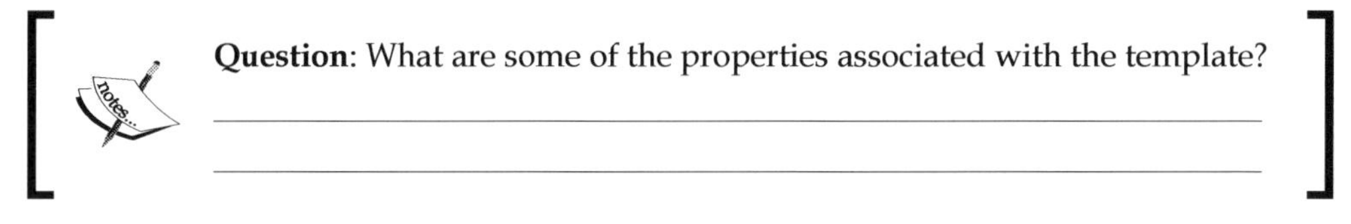

Question: What are some of the properties associated with the template?

5. Under **Attributes** look at the values next to **Pipe Size** and **Material**.

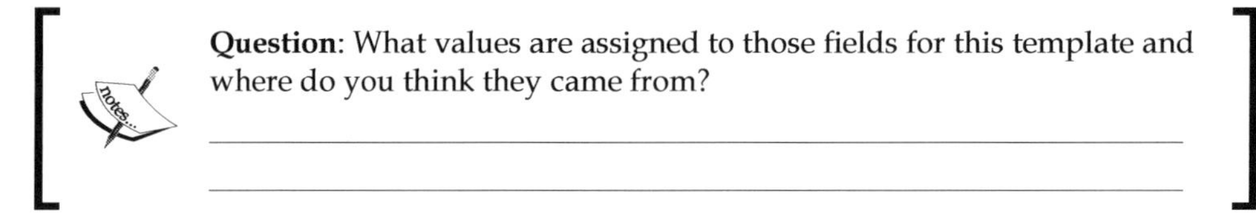

Question: What values are assigned to those fields for this template and where do you think they came from?

6. Since you are adding new pipes that were recently constructed, you will set the value for the **Condition** field to good for this template. Click in the cell next to condition and select **Good** from the drop-down list.
7. Click on **OK** to close the **Template Properties** window.

The reason you were presented with a drop-down list for the condition field is that it had a coded value domain assigned to that field. A coded value domain is a list of predefined accepted values, which can be entered into a field. When editing, you can only select a value that is included in the domain. This helps increase editing efficiency and reduces errors.

Now that you have configured your feature template for the sewer lines, you are ready to begin drawing new features:

8. Select the **8 inch PVC** template in the **Create Features** pane. You know to use this template because your development ordinance requires this size and pipe material for all new residential subdivisions.
9. Click on the end of the existing 8 inch ductile iron pipe near the intersection of the existing **GA HWY 50** and the new **Oak Place**. You may need to verify that snapping is enabled if your cursor does not automatically snap to the end of the existing pipe.
10. Move your mouse to the manhole located to the west at the intersection of the new **Oak Place** and **Pine Drive**, as illustrated in the following screenshot and double-click:

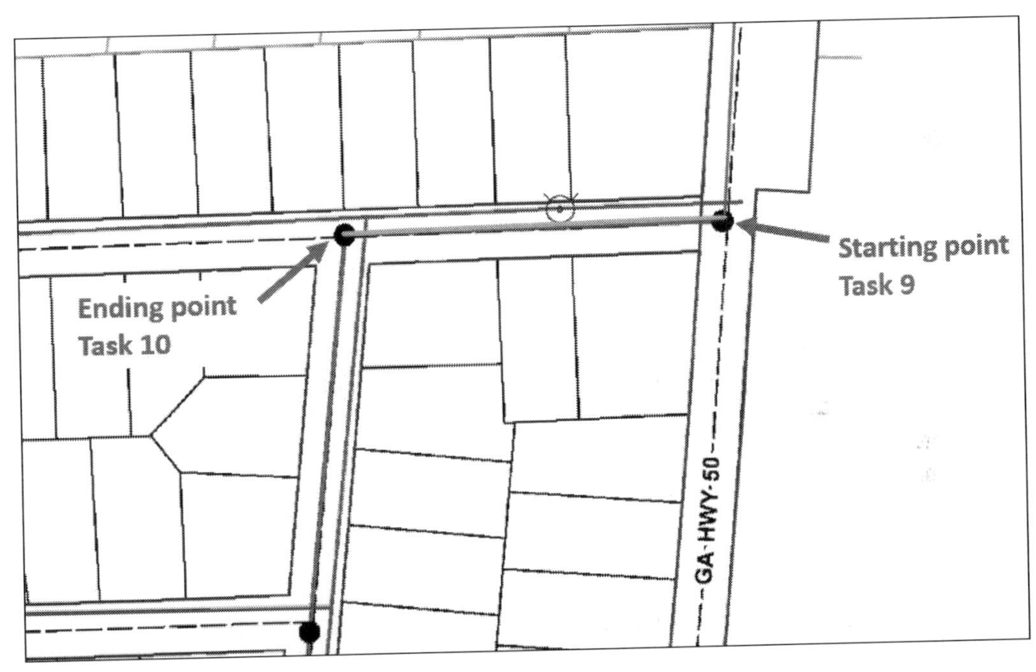

If you forget to double-click or move your mouse away too fast and only single-click, you can press your *F2* key to finish the sketch.

You have just drawn one of the new sewer lines that were constructed for the new subdivision. Drawing the new features is only part of creating a new feature. Next, you need to update the attributes associated with the feature. The power of GIS comes from the combination of both spatial and attribute data, which allows you and others to ask your map questions and get answers. Those answers are only as good as the data. So, it is just as important to keep your features attributes up to date as it is to draw them correctly.

Step 4 – update attributes

Now you will update the attributes for the new feature you just created. Attributes are the information about that feature. They will vary from one feature to another and from one layer to another.

The following steps will guide you through how to update attributes:

1. Click on the **Attributes** button located in the **Selection** group of the **EDIT** tab in the ribbon. This will open the **Attributes** pane on the right-hand side of the interface.

The values for **Pipe Size**, **Material**, and **Condition** have already been assigned by the template you used to draw the line.

2. Click in the cell to the right of `LINEID` and type `1200`. This field is used to identify each sewer line in the system, which can then be linked to a work order management system to track the maintenance history of each sewer line segment.

Challenge

Using the skills you just learned, draw the remaining sewer lines shown on the plat for the new subdivision. Assign the new pipes you draw sequential `LINEID` numbers. You will need to create individual pipe segments between each of the manholes shown on the source plat you are using to digitize the locations of the sewer lines.

Step 5 – drawing the manholes

Now that you have drawn the sewer lines, you need to add the new manholes as well. A manhole is generally located at the end of each pipe segment.

To draw the manhole follow these steps:

1. Select the **Create Features** pane using the tab at the bottom of the right-hand side pane in the ArcGIS Pro interface.
2. Click on the **Manhole** template and ensure that the **Point** tool is select underneath the template name.
3. Click on the west end of the first pipe you drew in step 3 to add a new manhole.
4. Click on the **Attributes** tab located next to the **Create Features** tab.

You should now see all the attribute fields associated with the manhole layer. You do not have all the information to fill out all these now. However, you can update a couple of them by:

5. Clicking in the cell next to **Condition** and select **Good** from the drop-down list.
6. If you completed the earlier-mentioned challenge, continue to add the other manholes shown on the plat using the same process, setting the condition for each new manhole to **Good**.
7. After you are done with adding new manholes, turn off the **Forrest Park Subdivision** plat in the **Contents** pane to view your handy work.

If you completed the challenge and added all the new manholes, your map should now look like this:

8. If you are happy with the new sewer features you have added, click on the **EDIT** tab and the **Save** button in the **Manage Edits** group. This will save your edits back to the `Trippville_GIS` geodatabase.

When editing, your edits are only shown on your computer and stored in the computer's memory. They are not committed to the source for the layer, so others can see them until you save your edits. Until they are saved, all edits are considered a sketch, which no one else can see. This also means that if your computer crashes or ArcGIS Pro fails for any reason before you save, all your edits will be lost and unrecoverable. So if you do not have the auto-save function enabled within the ArcGIS Pro options, make sure to save often.

Step 6 – adding the roads

Now that you have learned how to add simple new features, it is time to do something a bit more challenging. You will add the road centerlines and rights-of-way. The first step will be to digitize the street centerlines and then use those to construct the rights-of-way.

The following steps will guide you through how to add roads to your map:

1. Turn the **Forrest Park Subdivision** plat back on, so it is visible once again.
2. Turn off the manhole and `sewer_lines` layers.
3. Turn on the `Street_Centerlines` layer.
4. Zoom into the northeast corner of the new subdivision where you first started drawing the new sewer lines in the last step.
5. Click on the **Create Features** tab in the pane on the right-hand side of the interface.
6. Select the **Street_Centerlines** template and ensure that the line tool underneath is active.
7. To draw your first centerline segment, start at the point where the plat shows **Oak Place** intersecting **GA HWY 50**. Click at that intersection to start drawing your first street segment.

8. Move your mouse pointer to the location the plat shows **Oak Place** intersecting with **Pine Drive**. It should be in the same place as one of the new manholes you digitized earlier. Double-click at this location to draw the end of your first segment. Remember that if you only single-click, you can use the *F2* key to finish the sketch.
9. Now edit the attributes for this segment, as shown here:
    - `ST_NAME` = **Oak Place**
    - `RD_Class` = **City**
10. Now you will draw your next segment by once again selecting the **Street_ Centerlines** template and then clicking on the end point for the segment you just completed.
11. Move your mouse pointer along the road centerline shown on the plat and single-click at a location just before it starts to curve. This should be a location near the **O** in **Oak**.
12. To draw the curve, you will use the Arc Segment tool. Look toward the bottom of the ArcGIS Pro interface for this tool on the toolbar that appeared when you started drawing the new street centerline segment as shown in the following screenshot:

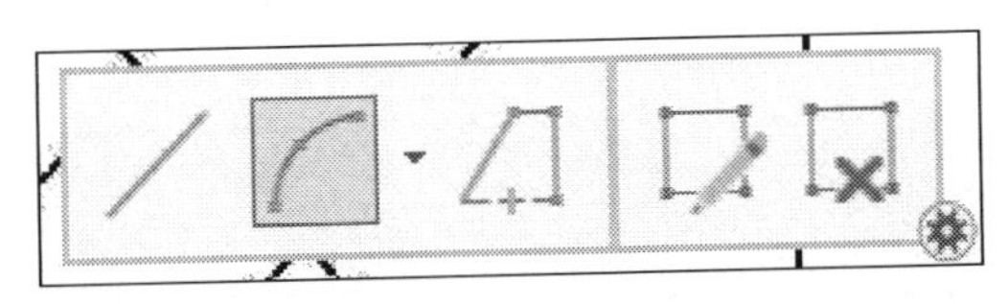

13. This tool allows you to trace three points to define the arc in the centerline. Single-click just once past the **O** in **Oak** where the curve begins. Then, click near the estimated middle of the curve to define the center of the arc. Finally, click near the estimated end of the curve where the road straightens out once more to define the end of the arc. Reference the image after the following task 14 if you need help.

You have just drawn a curve that is embedded in the centerline for the road. This provides a truer representation of the road's location and geometry than digitizing a series of small straight segments would have. You still need to finish this segment:

14. Click on the **Line** tool located on the small toolbar where you found the Arc Segment tool. This will allow you to continue drawing the segment for this portion of **Oak Place**.

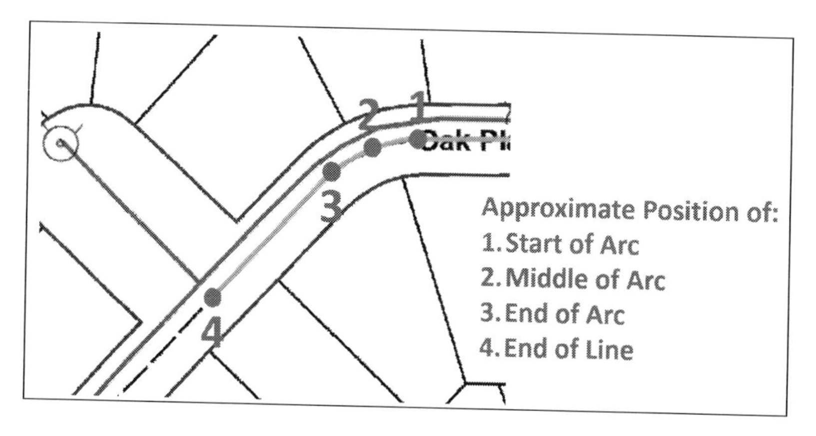

15. The segment should automatically start from the end of the arc you just drew. Move your mouse pointer to the estimated intersection of **Oak Place** and the small unnamed road that runs northwest and double-click to end the sketch.
16. Edit the attributes for this segment the same way you did the first.
17. Save your edits.

[If you do not get the road to draw correctly the first time, do not be too concerned. It takes practice to become proficient using the drawing tools within ArcGIS Pro. As the saying goes, practice makes perfect.]

18. Using the skills you have just learned, continue drawing the remainder of the street centerlines, as shown on the plat. Once completed your map should look very similar to the following screenshot if you turn off the **Forrest Park Subdivision** plat:

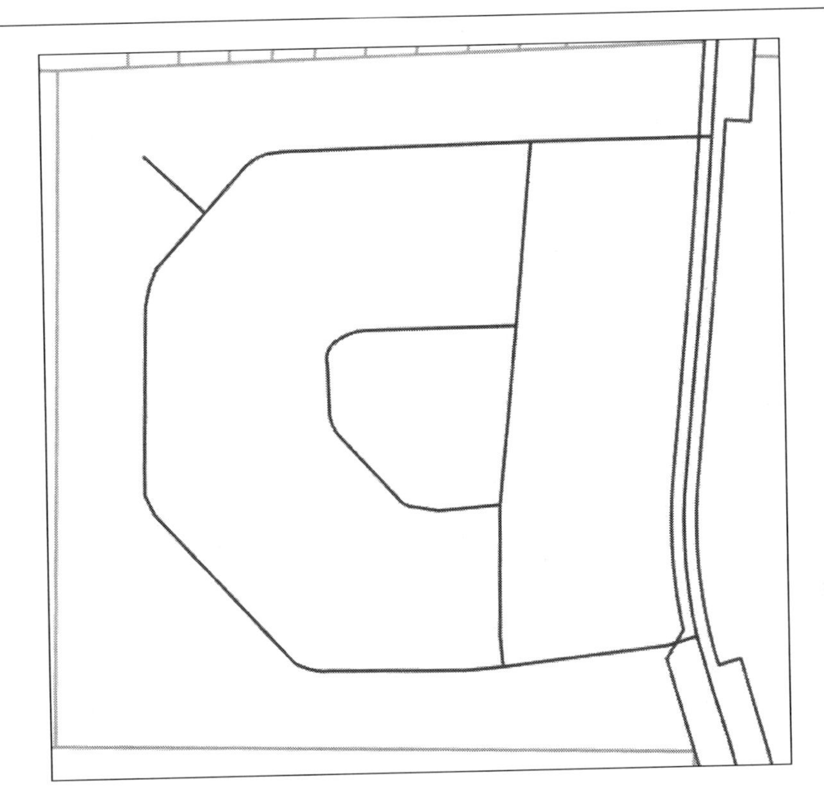

Now to add the rights of way for the roads. You could do this the same way you drew the roads. However, since the rights of way are based on the road centerlines, it is often easier to use the centerlines as a framework to construct your rights of way.

If you were using Esri's older ArcMap application, you could simply use the **Copy Parallel** or **Buffer Edit** tools to accomplish this. However, ArcGIS Pro 1.1 does not currently support those functions. So, instead you can use the **Trace** tool, which is located on the same toolbar you used to draw the arcs:

19. Click on the **Snapping** dropdown and enable **Intersection Snapping** since your right of way will start at the intersection of the existing right of way for **GA HWY 50**.
20. Click on the **RW** template. This is the template for the rights of way.
21. Select the **Trace** tool located underneath the template. It should be the last button in the list of tools.

22. Click on where the centerline of **Oak Place** intersects the right of way for **GA HWY 50**.
23. After you have clicked on that intersection, press the letter **O** for offset. The **Trace Options** window should appear in the middle of your screen.
24. Set the options as shown in the following screenshot and click on **OK**:

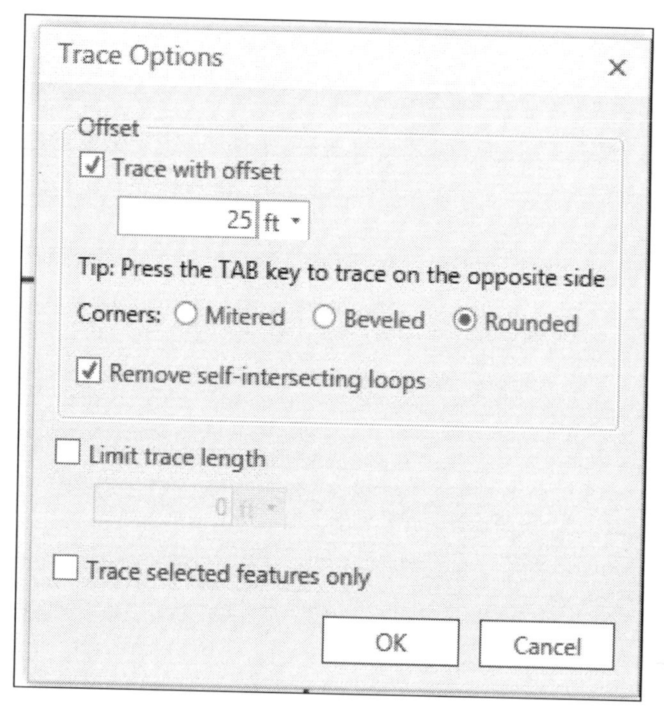

25. Move your mouse pointer along the centerline of **Oak Place** until the right of way you are drawing intersects with the right of way for the small unnamed road shown on the plat. Double-click once you reach that point.

You have just drawn the northern right of way boundary for **Oak Place**. Now you will continue to draw the right of way for the small unnamed road using a different method. You have been provided with information on the specific measurements for the right of way for the small unnamed road, so you will use that to draw the right of way:

26. Ensure that the **RW** template is still selected with the **Line** tool active.
27. Click on the end of the northern right of way you just completed for **Oak Place** to start your sketch.

28. Move your mouse to the northwest along the right of way shown on the **Forrest Park Subdivision** plat and right before you get to the start of the curve right-click to expose the menu as shown in the following screenshot:

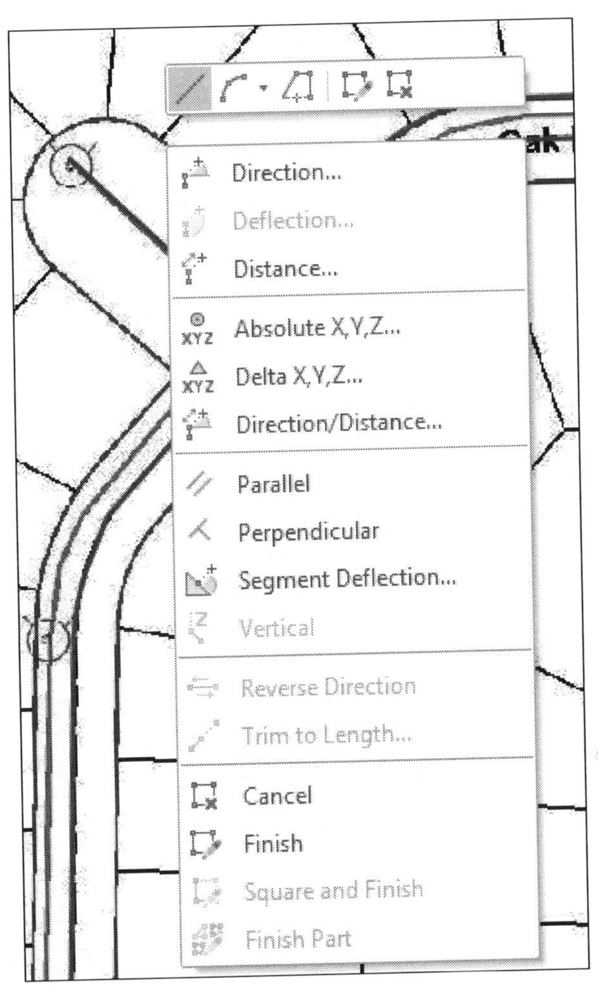

29. Select **Direction/Distance,** so you can enter the specific bearing and distance for the right of way segment.
30. Enter a **Direction** of **N49-52-25W** and a **Distance** of **105 ft** and press your *Enter* key.
31. Locate the **End Point Arc Segment** tool on the toolbar, which contained the other arc and line tool you have used in the past. You may need to click on the small drop-down arrow located next to the **Arc Segment** tool you used in the past.

The **End Point Arc Segment** tool allows you to specify a starting and ending point for the arc and then a radius. In this exercise, you will just match it to the scanned plat. However, if you press the *R* key after identifying the starting and ending point, you can specify a specific radius:

32. Click on the end of the last segment you drew. Then, move to the end of the curve on the southwest side of the unnamed street.
33. Move your mouse pointer to the approximate center of the curve. The following screenshot illustration will help you:

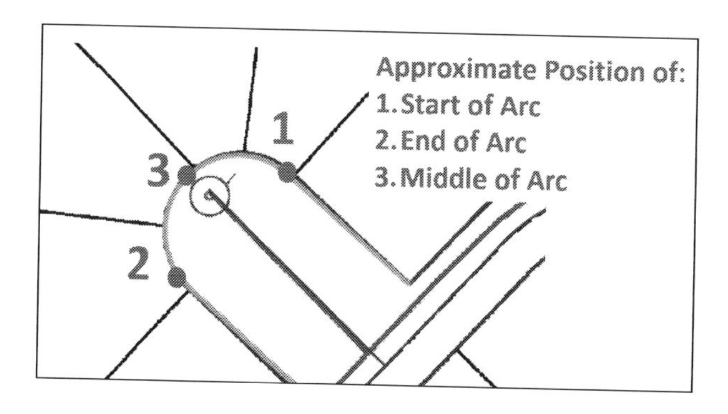

34. Activate the **Line** tool once again and move your mouse pointer along the southwestern right of way for the unnamed road until you get near the intersection with the continuing right of way for **Oak Place** and then right-click again to display the context menu.

[A context menu is any menu that displays when you right-click within the applications interface. These menus will vary depending on the location of your mouse pointer when you right-click.]

35. Select **Direction/Distance** from the context menu and set the **Direction** to `S49-52-25E` and a **Distance** of `105 ft` and press your *Enter* key. You have now completed drawing the right of way for the small unnamed road. If it does not match to the scanned plat perfectly that is OK.

36. Continue drawing the remainder of the northern right of way for Oak Place using the same method you did for the first section. You need to go all the way around till you intersect back to the right of way with GA HWY 50. It should look like this when you are done:

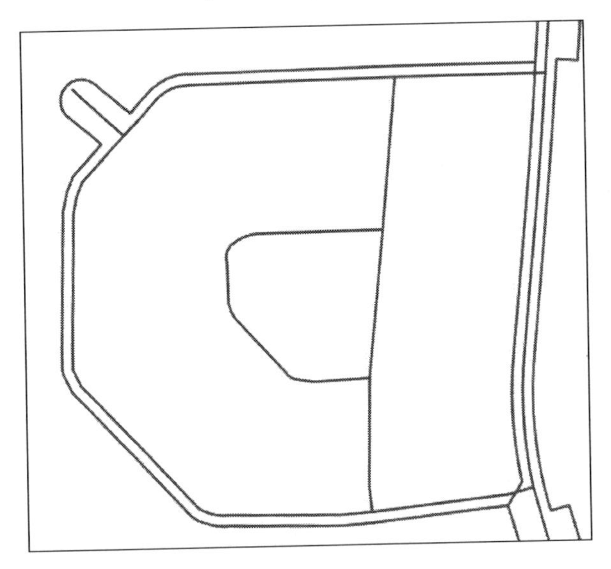

Now that you have drawn the outside right of way for Oak Place, you need to draw the inside right of way to complete the process. This is what you will do next:

37. Zoom back to the area you were in when you started drawing the northern outer right of way for Oak Place.
38. Select the `RW` template in the **Create Features** pane. Activate the **Trace** tool.
39. Click on the intersection of the Oak Place centerline and the right of way for GA HWY 50.
40. Move your mouse pointer a short distance along the centerline of Oak Place and right-click.
41. Select **Trace Options** from the context menu that appears. This does the same thing as pressing the *O* button.
42. Set the trace options to the same settings you used previously.
43. Start moving your mouse pointer along the centerline of Oak Place. You should note that ArcGIS Pro is drawing the northern right of way again. Press your *Tab* key to shift it to the southern right of way.
44. Move your mouse pointer along the centerline of Oak Place until the new right of way line reaches the intersection of the southern right of way for Oak Place and the eastern right of way for Pine Drive.

45. Using the skills you have now learned, draw the remaining rights of way lines shown on the Forrest Park Subdivision plat. Make sure to save your edits often.
46. Save your project. If you are not continuing to the Challenge, close ArcGIS Pro.

You have now seen how to add new features to layers within ArcGIS Pro and how to update the attributes for those features.

Challenge

Using the skills you have learned throughout the book, update the water lines and fire hydrants using the information shown on the Forrest Park Subdivision plat. Note that these layers are not in your map, so they will need to be added. The water lines in the new subdivision are all 6 inch PVC pipes.

Editing your schema

As the needs of your organization grow and change, it is important for your GIS to keep up. This means you will need to make changes to the database schema.

So what does the word schema mean? Simply put, it means the structure of a database. So schema is not just a GIS term, but it is also a term used for other databases as well. In the case of a GIS database, schema refers to things such as what feature classes are stored in the database, what attribute fields are linked to those feature classes, what domains are included in the database, are any tables or feature classes related, and so on.

ArcGIS Pro allows you to make some changes to your GIS data schema. It will allow you to add new feature classes, add fields and create domains. It does have some limitations currently such as you cannot create a topology or geometric network. Hopefully, as ArcGIS Pro continues to mature, those limitations will disappear.

Adding a field

Sometimes, changing your schema is as simple as adding a new field to an attribute table. A field is a column within a database table. The actual act of adding a field within ArcGIS Pro is not overly complicated, especially when compared to editing the spatial data. However, it does require some thought.

When adding a new field, you need to think through the properties that will be associated with that field such as name and field type.

Field name

In a database table each field must have a unique name. That name cannot include spaces or special characters. Underscores are allowed. The allowed length of the field name will depend on the type of database. Is it a dBase, Access, or SQL Server table? Each of these have their own limitations.

As a rule of thumb, I have learned that keeping my field names limited to seven to eight characters works best. This will prevent field names from getting shortened if you export the data to a different format, which does not support the name length of your native database. This often happens when your GIS data is stored as a geodatabase, but you export to a Shapefile. Shapefiles store attributes in a dBase format, which does not support long field names. So, a field that is named `parcel_indentification_number` in your geodatabase may get renamed to `parcel_in` when exported to a Shapefile.

Alias

An alias is a more descriptive name for a field. An alias can contain special characters such as spaces and does not have the length restrictions associated with the field name. Aliases allow users to better understand the purpose of a field. By default, the alias is what is displayed when a table is opened in ArcGIS Pro.

Field data types

When you add fields to database, you must decide what type of data will be stored in that field. This is the field data type. There are several data types depending on the type of database you are working with. Here is a list of some of the most common data types you can use in ArcGIS Pro:

Name	Description	Comments
Text or String	Stores alphanumeric data. Fields can be up to 254 characters long. The default length in ArcGIS Pro is 50 characters.	Does not provide the best database performance compared to other field types. Make sure to set the size as small as possible to conserve storage space.

Name	Description	Comments
Integer (Long and Short)	Stores whole numbers meaning no decimal places. Difference between long and short integers varies somewhat depending on the database but generally short integers can store values between -33,000 to 33,000. Long integers will store values between -2.1 Billion to 2.1 Billion approximately.	Provides the best performance of all field types. This makes them the most optimum type to use if overall database performance is a concern.
Float or Single	Stores decimal values out to approximately six to eight decimal places depending on the database.	Provides median database performance.
Double	Stores decimal values with 15 or more decimal places depending on the database.	Performance similar to Float.
Date	Stores date and time. Format will depend on database.	
BLOB	**Binary Large Object** (**BLOB**) field. These are used to store data which does not fit one of the other field types. Not all databases support BLOBS.	Provides the worst performance of any field type. Can cause issues if exported to a database that does not support this type.
Raster	Stores images or pictures directly in the database. It is a specialized BLOB field.	Same issues as BLOB.
GUID	**Global Unique Identifier** (**GUID**) Provides a unique identifying value to all records and tables within the database that have this field type.	GUIDS are required if you plan to allow mobile or offline editing or use database replication.

Exercise 7B – adding a field and populating values

The Public Works Director has asked you to determine how long each road segment is in miles for a project he is working on. While you do have the road centerline data, it is not attributed with length in miles. So you will need to add a new field to store the length in miles and then calculate that value for each segment.

Step 1 – adding the field

In this step, you will open your project and then add the new field, which will store the length of each road segment in miles.

1. Open ArcGIS Pro and the **Create New Features** project you used in the last exercise.
2. In the **Project** pane, expand **Databases** so you can see the two databases connected to this project.
3. Expand the `Trippville_GIS` database and the `Base` feature dataset.
4. Right-click on the `Street_Centerlines` feature class to expose the context menu.
5. Click on **Design** and select **Fields** as shown in the following screenshot:

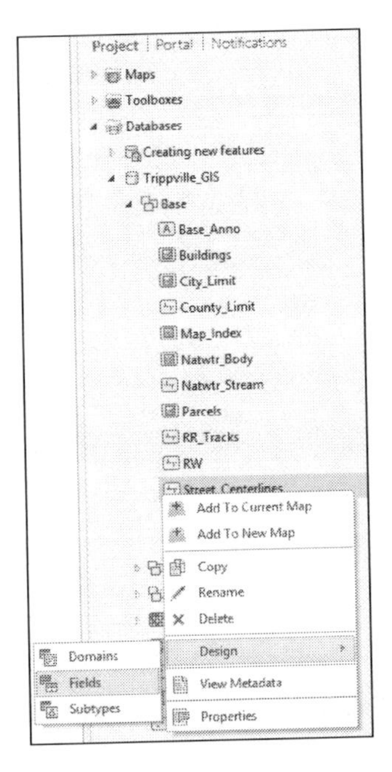

6. A new tab should open in the main view area of the interface where the map was before. This new view shows the current attribute fields and their properties for the `Street_Centerline` feature class. At the bottom of the existing fields, you should see a row that says **Click here to add a field**.

Question: What attribute fields are associated with the `Street_Centerline` feature class and what field types are they?

____

____

7. Click on the **Click here to add a field**.
8. Name the field `Len_Mi`.
9. Then, click on the **Alias** cell and give the new field an alias of `Length` in miles.
10. Then, click on the **Data Type** cell and set it to a **Float**. You are using a single cell because the Director is looking for length in miles only out to two decimals.
11. Leave all other settings the same. The table should now look like this:

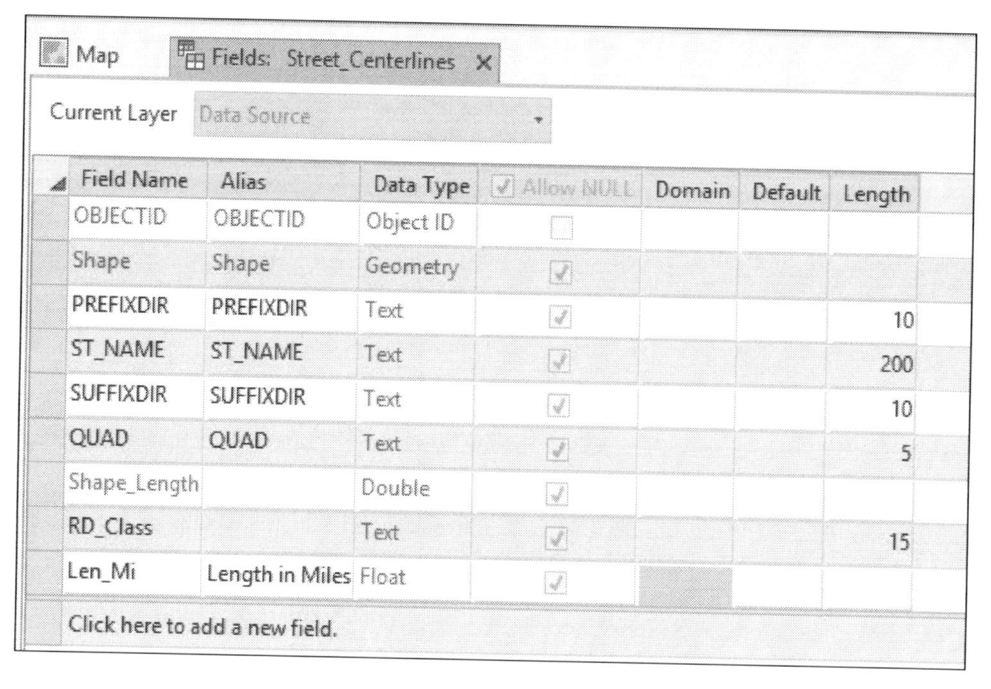

Field Name	Alias	Data Type	Allow NULL	Domain	Default	Length
OBJECTID	OBJECTID	Object ID				
Shape	Shape	Geometry	✓			
PREFIXDIR	PREFIXDIR	Text	✓			10
ST_NAME	ST_NAME	Text	✓			200
SUFFIXDIR	SUFFIXDIR	Text	✓			10
QUAD	QUAD	Text	✓			5
Shape_Length		Double	✓			
RD_Class		Text	✓			15
Len_Mi	Length in Miles	Float	✓			

12. Once you have verified your new field has been added correctly, click on the **Save** button located on the **Fields** tab in the ribbon.
13. Close the **Fields** view.

You have just added a new field to the `Street_Centerlines` feature class, which you can use to store the length of each road segment in miles. However, that field is empty. You need to populate it.

Step 2 – populating the field using the field calculator

In this step, you will populate the length of each road segment in miles into the field you just created. You could do this manually for each segment using some of the editing skills you have already learned. That would be very time consuming. There is a much more efficient way.

One of the fields you identified as being associated with the street centerlines was `Shape_Length`. So, you already have values showing the length of each road segment. It is in feet though so to populate your new field with the correct values in miles you need to convert feet into miles. ArcGIS Pro includes a tool called the **field calculator**, which will allow you to convert feet into miles, while at the same time, populate that value into the table for all features in the feature class:

1. If needed, activate the **Map** view in ArcGIS Pro by clicking on the tab at the top of the **View** area.
2. Right-click on the `Street_Centerlines` layer in the **Content** pane.
3. Select **Attribute Table** to open the **Table** window at the bottom of the interface.
4. Right-click on the new field you created. It should be shown using the alias you specified when you created the new field, **Length** in miles. Also note that the field is `Null`. `Null` which means it is empty. It has no values stored.
5. Select **Calculate Field**. This will open the **Geoprocessing** pane in the right-hand side of the interface.
6. Within the **Fields** box, double-click in `Shape_Length`. This will insert the name of the field with the correct syntax into the expression box located below the **Fields** and **Helpers**.
7. Click on the / symbol located below the **Helpers** box to add that to the expression. That symbol will divide the value stored in the `Shape_Length` field for each record by a value you will specify next.
8. In the expression box, type `5280` after the / symbol. That is the number of feet in a mile. Your expression should look like this:

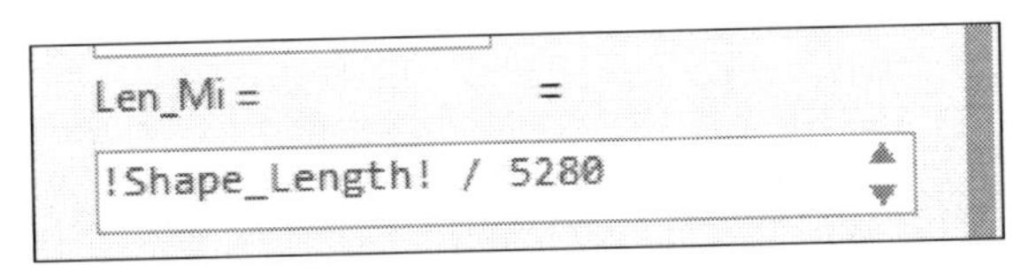

9. Once you verify your expression, click on the **Run** button located at the bottom of the **Geoprocessing** pane.
10. When the process is complete, close the **Geoprocessing** pane.

When the **Calculate Field** is complete, you should now see the field you created filled with values. If you scroll through the table, you will see that it calculated a length in miles for all the records in the table. The **Calculate Field** tool is very powerful and efficient way to populate a field.

If you are not proceeding to the next exercise, save your project and close ArcGIS Pro.

Importing a new feature class

Adding a field is just one change you can make to the schema of your GIS database. You can also add entirely new feature classes by creating a new one from scratch or importing it from another data source.

It is not uncommon to have data stored in other formats and locations, which would be beneficial in your GIS. Moving all those different data sources into a single GIS database makes them easier to use, find, and manage. ArcGIS Pro contains tools for importing, exporting, and converting data. This allows you to build a comprehensive GIS database, which can be integrated with other solutions.

Exercise 7C – importing a Shapefile

A local consultant was hired to locate all the water valves within the City of Trippville. They provided the Public Works Director with a Shapefile containing the water valves they located and attributes collected.

You now need to import this Shapefile into the City's geodatabase, so it is stored in the same location as the other feature classes that make up the City's water system.

Step 1 – open the project and add a layer

In this step, you will open the project you have been working with throughout this chapter. Then, you will add the Shapefile, which was provided to the Director from the consultant that contains the water valve data:

1. Open ArcGIS Pro and the **Creating New Features** project.
2. On the **MAP** tab in the ribbon click on the **Add Data** button.
3. Click on **Folders** in the left-hand side pane.
4. Double-click on `Chapter7`.
5. Select `Water_Valves.shp` and then click on the **Select** button.

You have now added the data collected by the consultant to your map. Now you need to export that to your geodatabase.

Step 2 – export to geodatabase

Now you will export the Shapefile to a feature class in the City of Trippville's geodatabase. You will store the newly imported feature class in the `Water` feature dataset along with the other water system related feature classes:

1. Right-click on the `Water_Valves` layer you just added to your map.
2. In the displayed context menu, go down to data and select **Export Features**, as shown in the following image:

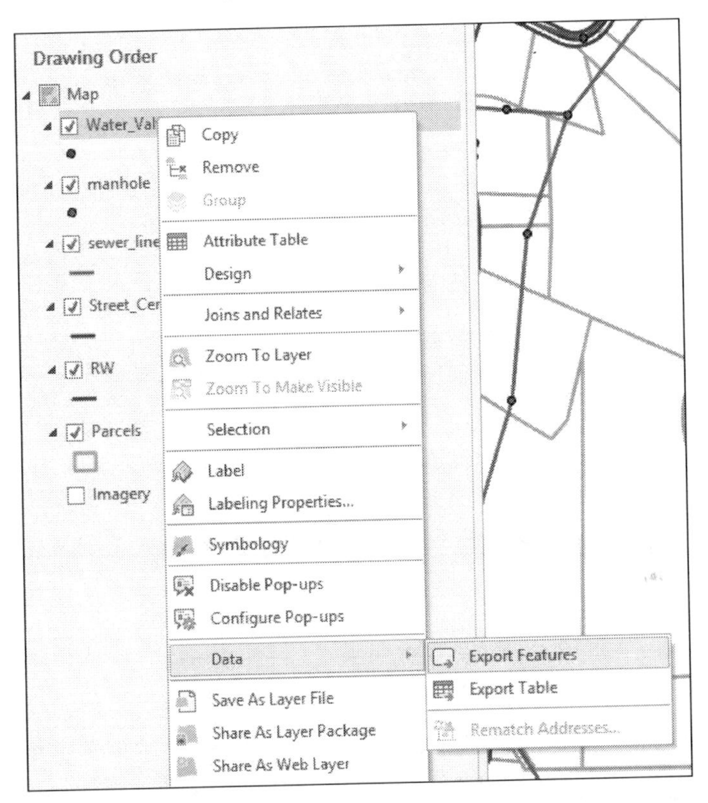

3. The **Geoprocessing** pane should open with the **Copy Features** tool. Click on the browse button located to the right of the **Output Feature Class**.
4. Click on **Databases** in the left pane of the **Output Feature Class** window.
5. Double-click on the `Trippville_GIS` geodatabase.
6. Double-click on the `Water` feature dataset.
7. In the cell next to **Name** type `Water_Valves` and click on **Save**.
8. Then, click on the **Run** button at the bottom of the **Geoprocessing** pane.

When the **Copy Features** tool completes its run, you should see a second `Water_Valves` layer added to your map. This new layer is now using a new feature class that was created in your geodatabase as its source. You will now verify that:

1. Click on the **List By Data Source** button located at the top of the **Contents** pane.
2. Verify that you see a `Water_Valves` layer listed under the `Trippville_GIS.gdb` workspace and one in the `Chapter7` folder.
3. Right-click on the `Water_Valves` layer in the `Chapter7` folder and click on **Remove**. This removes the layer, which was based on the original Shapefile from the map, so only the one which references the City's primary geodatabase remains.
4. Save your project and close ArcGIS Pro.

You have just converted a Shapefile to a geodatabase feature class and verified that the conversion was successful. Importing the data that was in the Shapefile into your geodatabase created a new feature class, which is also a change to the schema.

Summary

The world is not a stagnant place so neither should your GIS. It must be able to keep up with the changes that happen all around us. In this chapter, you learned that ArcGIS Pro has many powerful tools, which allow you to keep your GIS current based on those changing needs and features.

You learned how to add new features using various editing tools within ArcGIS Pro. You were able to create new features and update their attributes to reflect their true nature. You experienced how feature templates can be used to create new features more efficiently by redefining values.

You also learned how to make changes to your database schema. You added a new field to an existing feature class and used the **Calculate Field** tool to populate that field with values. You also learned how you can import data into your geodatabase, so it is easier to access and manage.

8
Geoprocessing

You have learned that ArcGIS Pro has some powerful tools for visualizing and maintaining GIS data. But what about analyzing that information? Can it help you identify the land owners you need to contact along a road that will be repaired or get a count of customers within a given area? ArcGIS Pro has the tools to help you answer these types of questions and a whole lot more.

Answering questions such as these are done using geoprocessing tools within ArcGIS Pro. So what is geoprocessing? Put simply, it is the manipulation of data used inside ArcGIS. Geoprocessing tools can analyze data or convert data from one format to another or add an attribute field to a table or project data from one coordinate system to another. The sky is the limit on what you can do with geoprocessing tools.

In addition to the tools included in ArcGIS Pro, you can purchase extensions that add more capability to ArcGIS Pro. The geoprocessing framework also provides avenues to create your own custom tools using ModelBuilder, Python, and ArcObjects. You will learn more about ModelBuilder and Python in later chapters.

In this chapter, you will learn:

- What determines which geoprocessing tools you have
- How to access geoprocessing tools
- How to understand the analysis process
- How to use some of the most common geoprocessing tools for analysis

One thing to keep in mind as you begin to use geoprocessing tools is that most of them create new data. This means that even if you choose the wrong tool or use a bad setting in the tool, your original data is protected in most cases. This provides a level of protection and helps to put your mind at ease knowing the chance of damaging your data by doing the wrong thing is greatly decreased.

What determines which tools you can use?

There are two things that determine what geoprocessing tools will be available to you in ArcGIS Pro. The first is your licensing level. The second is what extensions you might have. Let's take a look at these two items and their impact on the geoprocessing tools that will be available for your use.

Licensing levels

If you remember from *Chapter 1, Introducing ArcGIS Pro*, ArcGIS Pro has three different licensing levels: Basic, Standard, and Advanced. The license level you have will directly impact the geoprocessing tools you will have access too.

The Basic level is the most limiting license level with the least amount of geoprocessing tools. With this level, you will be able to access simple analysis tools such as Buffer, Union, and Intersect. You will not be able to use tools that allow you to find the closest feature in another layer or that erase overlapping areas between two or more layers.

The Standard level includes a few more geoprocessing tools in addition to all the ones available at the Basic level. Many of these are focused on data management and maintenance. For example, the Standard level includes geoprocessing tools for creating relationship classes within a geodatabase. Standard also includes tools for managing an SDE geodatabase, such as tools that allow you to create a new database user or reconcile a version. More tools for creating and managing topologies and geometric networks will be added to future releases.

The Advanced license level has the greatest number of geoprocessing tools. It includes all the tools found in the Basic and Standard licenses plus additional analysis tools. The Advanced license will allow you to locate the nearest feature to another feature or calculate the distances between points. It also has tools that allow you to erase areas of overlap between two or more layer. These are just a couple of examples of analysis tools you will only find in the Advanced licensing level.

Here is a comparison of the number of geoprocessing tools found in the Basic and Advanced licensing levels of ArcGIS Pro. The Standard license's level will fall somewhere in between these two:

	Basic	**Advanced**
Analysis	15	25
Cartography	0	7
Conversion	28	32
Data Management	206	265

	Basic	Advanced
Editing	0	12
Geocoding	7	7
Linear Referencing	7	7
Multidimension	8	8
Server	8	8
Space time pattern mining	4	4
Spatial Statistics	23	24
Total	306	399

Each licensing level has other limitations besides just which geoprocessing tools are available. You should check Esri's ArcGIS for Desktop Functionality Matrix to see a complete listings of the differences between the licensing levels. ArcGIS Pro will be slightly different as not all the functionality found in ArcGIS for Desktop has been ported into ArcGIS Pro. You can view the functionality matrix by going to `http://www.esri.com/~/media/Files/Pdfs/library/brochures/pdfs/arcgis1021-desktop-functionality-matrix.pdf`.

Extensions for ArcGIS Pro

Esri also has several extensions for ArcGIS Pro. These are included when you purchase the associated ArcGIS for Desktop extension. **Extensions** are add-ons for the core ArcGIS Pro product that provide extended functionality to all licensing levels. Each extension has a focused area of increased functionality that includes additional geoprocessing tools.

There are currently six different extensions that have been developed for ArcGIS Pro. Eventually, all the extensions that have been developed for ArcGIS for Desktop will also be ported to ArcGIS Pro. The current ArcGIS Pro extensions include:

- Spatial Analyst
- 3D Analyst
- Network Analyst
- Geostatistical Analyst
- Data Reviewer
- Workflow Manager

The name of each extension helps to identify its purpose. To use these extensions, not only do you need to have purchased them but a license must also be assigned to the user in ArcGIS Online or Portal for ArcGIS. We will take a quick look at the first three extensions, which are the most commonly used.

Spatial Analyst

The **Spatial Analyst** extension is primarily used to perform spatial analysis using raster-based data, though it does have some vector analysis capability. Operations you perform with the Spatial Analyst extension include: terrain analysis with a **digital elevation model (DEM)**, calculate slopes, determine watersheds, perform hydrological analysis, classify images, and more.

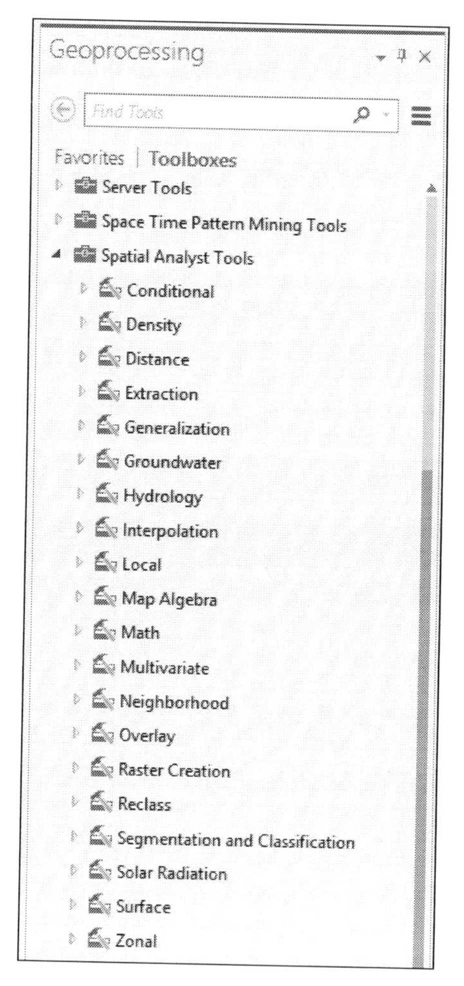

The Spatial Analyst extension includes more than 170 geoprocessing tools, which are found in a toolbox with the same name as the extension. These geoprocessing tools are organized into 20 different toolsets within the toolbox.

3D Analyst

The **3D Analyst** extension allows you to work with and analyze 3D data within ArcGIS Pro. It has many tools in common with Spatial Analyst. The big difference is 3D Analyst is designed to work primarily with 3D vector data as opposed to raster data. It does have some ability to create and analyze raster data but that is not its strong suit. It is not uncommon to use 3D Analyst and Spatial Analyst together. For example, you might take 3D vector data such as elevation contours and use 3D Analyst to create a DEM for additional analysis with the Spatial Analyst extension.

3D Analyst allows you to work with many 3D datasets including **triangulated irregular networks (TIN)**, lidar LAS datasets, and standard data formats. If your data does not have an elevation or height associated with it, 3D Analyst can drape your 2D data over a surface and then use the surface to calculate an elevation for your features in relation to the draped surface.

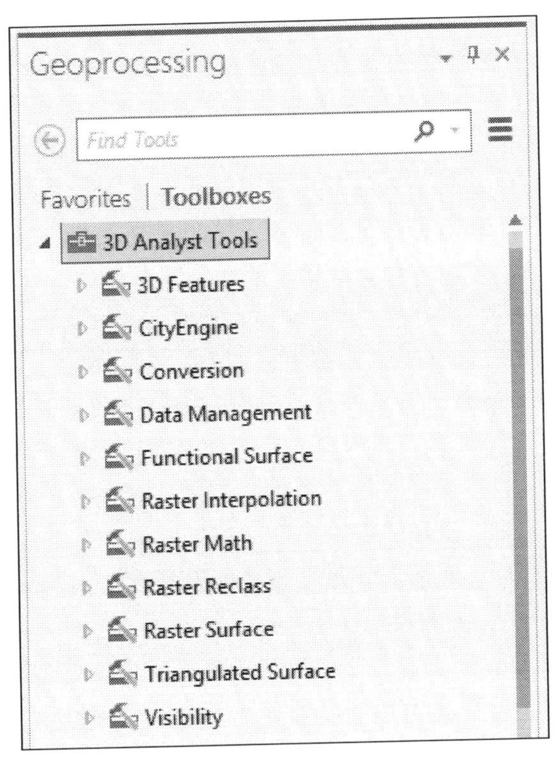

The 3D Analyst extension includes just over 100 geoprocessing tools grouped in 11 toolsets. The 3D Analyst extension for ArcGIS Pro 1.1 does not support the display of a terrain dataset; however, it does allow for the creation of those datasets by the included geoprocessing tools. It is believed that future versions of ArcGIS Pro will fully support terrain datasets.

Network Analyst

The **Network Analyst** extension has tools for creating and analyzing network datasets. A network dataset is a collection of linear features that are connected by nodes which allow for bi-directional flow. These are typically associated with transportation-related networks such as roads or railroads or sidewalks or bike paths. These typically are not used for utilities as those normally are single-direction flow networks.

With Network Analyst, you can calculate the best routes for vehicles, determine service areas based on drive time requirements, find the nearest features along the network, and more. So you might use this extension to help site a new fire station based on the drive time coverage of existing fire stations.

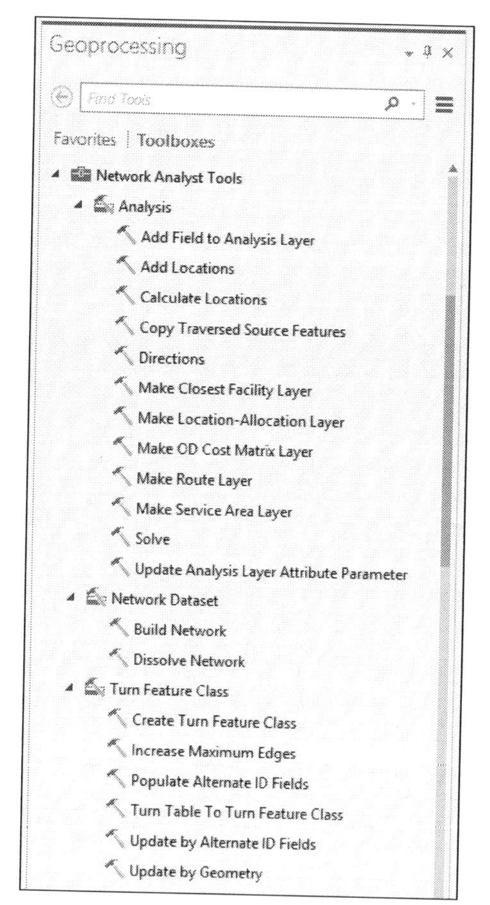

The Network Analyst extension for ArcGIS Pro currently includes 20 geoprocessing tools organized in three toolsets.

Exercise 8A – determining the license level and extension

As you have just learned, the license level of ArcGIS Pro and extensions you have been assigned will impact what you have the ability to do in ArcGIS Pro. So it is important to know what license level and extensions you have to work with.

In this short exercise, you will determine what license level you have and what extensions, if any, you have been assigned from within ArcGIS Pro. If you happen to be the administrator for your organization, you can also log in to ArcGIS Online or Portal for ArcGIS to determine what licenses have been assigned to each user. However, not everyone has administrative rights so it is important to know how you can determine this from ArcGIS Pro.

Step 1 – open ArcGIS Pro

The first step is to open ArcGIS Pro and then determine what license level you have available:

1. Open ArcGIS Pro as you have done in past exercises.
2. Click on **About ArcGIS Pro** located in the lower-left corner of the **Open a recent project** starting window.

The **About ArcGIS Pro** window shows you what version of ArcGIS Pro you are using. It also allows you to check to see if there are any software updates for ArcGIS Pro.

> **Question**: What version of ArcGIS Pro are you using?
> ______________________________
> ______________________________

Step 2 – determining the license level and extension

Now you will see what license level you have been assigned and if any extensions have also been assigned to you:

1. Click on **Licensing** in the left side pane.
2. Review what licenses are available to you. The section will tell you what license level is available and the middle section what extensions have been assigned to you.

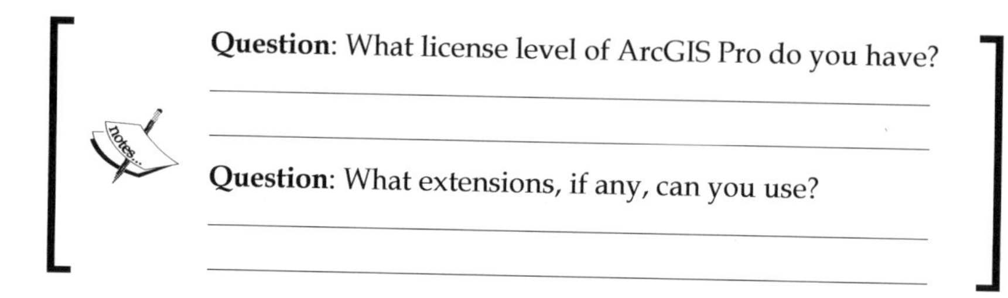

Question: What license level of ArcGIS Pro do you have?

Question: What extensions, if any, can you use?

3. Close ArcGIS Pro once you have answered the preceding questions.

Now you know what you have to work with when trying to perform analysis within ArcGIS Pro.

The analysis process

Analysis normally starts with a question. The question can be a simple one, such as: what is the total length of roads within the city? They can also be very complex, such as: I need to know where the best place is to locate my new business within the city so that it has a water and sewer service, is near major roads, and is in a location that will get business customers during the day and families in the evening.

These questions start you on the analysis process. This process is most often not linear. You will find that once you answer the initial question, it leads to other questions that start the process all over again. So the general analysis process looks like this:

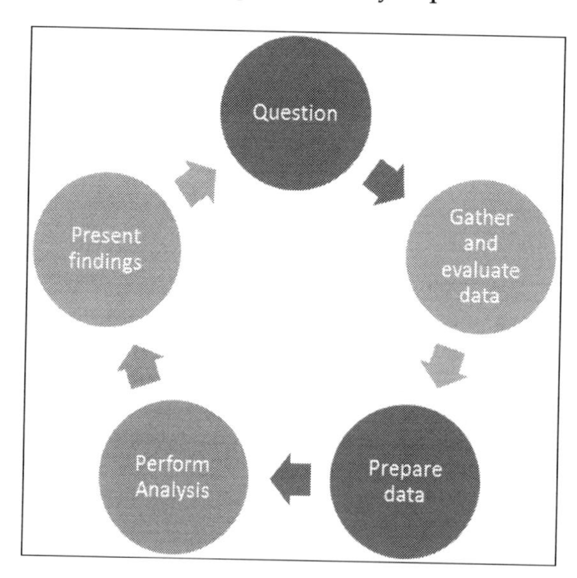

The initial question should cause you to ask other questions as well. The question period establishes the specifications of what needs to be answered and what data you will need to answer the question.

Once you know exactly what question you are trying to answer, you need to begin gathering the data you will need to perform the analysis. As you gather the data, you need to evaluate it. Does it have all the information you need? Is it in the format you need? Are the units the correct ones to answer the questions? These are all examples of things you need to consider as you gather the data you will use for your analysis.

Next, you will need to prepare your data for analysis. This may include simplifying it by clipping it to a specific area, projecting it to a different coordinate system, converting data formats, merging layers, generalizing information, or updating it.

Once you have prepared your data, you are then ready to begin your analysis. This can often require the use of multiple geoprocessing tools along with other tools such as selecting features based on attributes or locations.

After you have completed your analysis, you will need to present your findings. You can do that by creating a map and layout as you have already learned. ArcGIS Pro 1.2 is also going to allow you to create charts and graphs to display your results. This functionality is not supported in the current ArcGIS Pro 1.1 version.

Preparing data for analysis

As you gather and evaluate data for analysis, it is not unusual for the data to need some preparation work to get it into a state you can use for your analysis. For example, you may download data from ArcGIS Online that is in a different coordinate system than the primary one you use for your data. So you would need to project the downloaded data to the coordinate system you use for the rest of your data.

Common data preparation tasks include simplifying data, standardizing units, merging layers, and updating the data. So some of the most widely used geoprocessing tools to perform these tasks are:

- **Clip**
- **Dissolve**
- **Project**
- **Append**
- **Merge**

These tools are available at all licensing levels.

The Clip tool

The Clip tool is used to extract data based on the boundary of other data. For example, if you wanted to determine which portions of streets were located in the city limits, you could use the Clip tool to cut out the parts of the streets that are inside the city limits to their own layer. The Clip tool acts like a cookie cutter.

The Clip tool is found in the **Analysis** toolbox and the **Extract** toolset. It can be used to clip points, lines, or polygons. However, the clipping layer must be a polygon. Here is an example of the Clip tool in action:

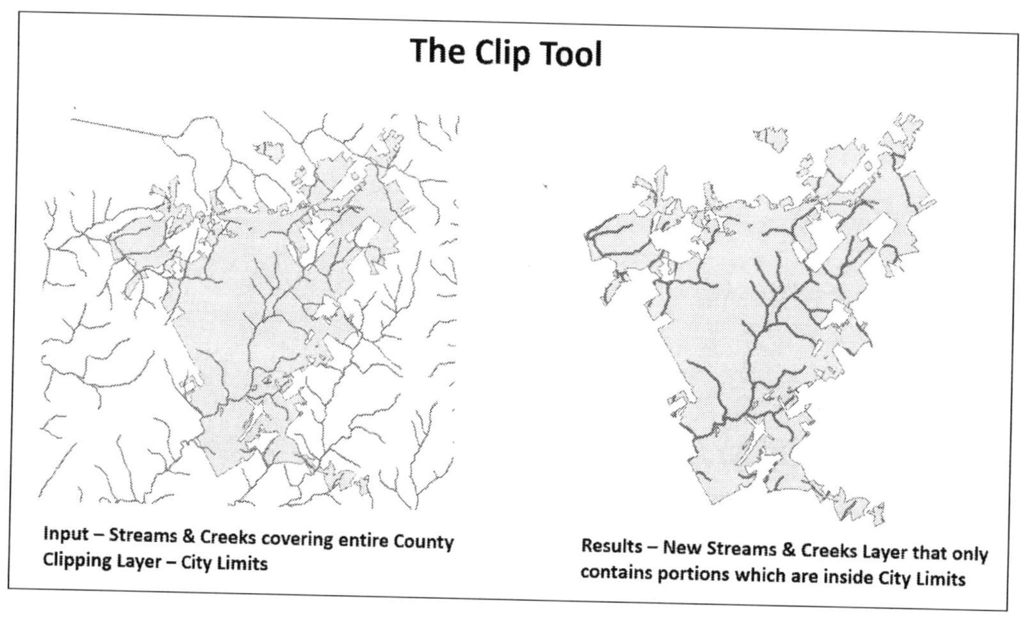

In this example, you are trying to isolate the portions of the creeks and streams that are located inside the city limits. The creek and stream layer contains the location of all of them in the entire county. So you have clipped them using the city limits layer, which is a polygon. The result is a new layer that contains only those portions of the streams and creeks which are located inside the city limits.

The Dissolve tool

The Dissolve tool is used to simplify or generalize a data layer based on a common attribute value. So, for example, if you had a parcel layer that showed each parcel and each parcel was coded with its designated zoning classification but you wanted to know the total amount of area for each zoning classification within the city, you could use the Dissolve tool to create individual polygons for each zoning classification.

The Dissolve tool is found in the **Data Management Tools** toolbox and the **Generalization** toolset. It works on points, lines, and polygons. Here is an illustration of the example that was just described:

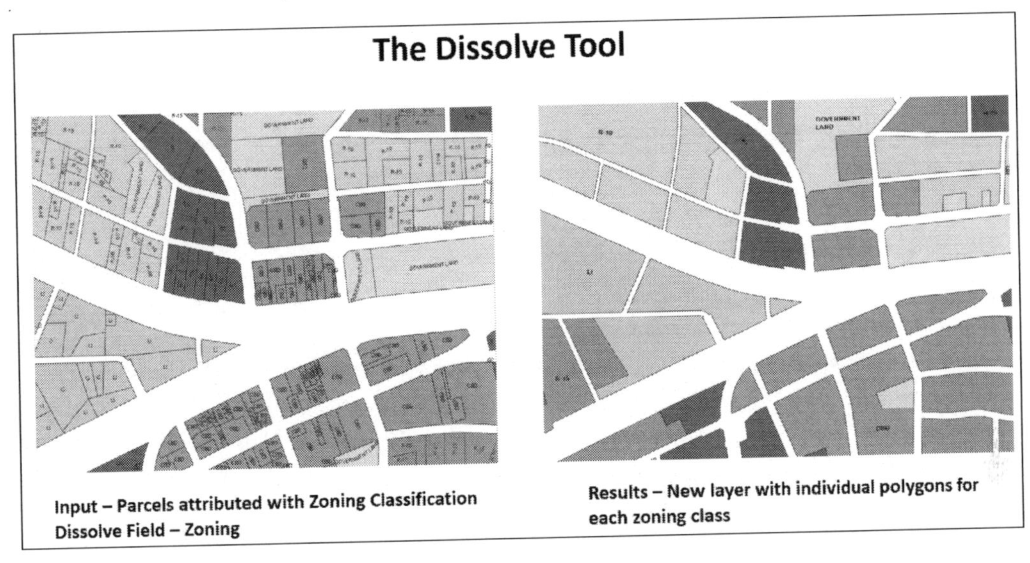

The Dissolve Tool

Input – Parcels attributed with Zoning Classification
Dissolve Field – Zoning

Results – New layer with individual polygons for each zoning class

As you can see, the initial parcel layer contained a large number of individual polygons that would have made determining the total area for each zoning classification difficult. Once you use the Dissolve tool, a new layer is created so that each zoning class has a single polygon. From there, it becomes much simpler to determine the total area for each zoning class.

The Project tool

The Project tool is used to transform or move a spatial layer from one coordinate system to another. It is important to know that when projecting data between coordinate systems, the actual coordinate values for your features must change. The Project tool will take your existing data and translate it into the new coordinate system using the proper math and translations to create a new layer that is in the designated coordinate system.

The Project tool is located in the **Data Management Tools** toolbox and the **Projections and Transformations** toolset. Certain tools and functions work best with different types of coordinate systems. There are two basic types of coordinate systems: Geographic and Projected. If you are trying to measure distances or areas then a projected coordinate system works best.

Even though ArcGIS Pro will project data on the fly so that even data that is in different coordinate systems will display together, it is a recommended best practice that you place all data you are analyzing into the same common coordinate system. This avoids errors caused by issues with differing units and transformations.

This tool should not be confused with the **Define Projection** tool, which is located in the same toolset. The Define Projection tool will assign a coordinate system to a feature class that is undefined or wrongly assigned. It does not actually project data to a new coordinate system. This is a common mistake for new ArcGIS users.

The Merge tool

The Merge tool will take data from two or more layers or tables and combine them into a new single output. This is useful if you get the same type of data from multiple sources or locations. For example, you are working with a regional emergency response group and you are trying to develop a regional evacuation plan. You receive road data from multiple jurisdictions. You can use the Merge tool to combine them all into a single layer.

The Merge tool can be found in the **Data Management Tools** toolbox and the **Generalization** toolset. The Merge tool can be used to combine points, lines, polygons, and even standalone tables. You can only merge like features, meaning you can only merge points with points, lines with lines, and polygons with polygons.

Here is another example of when you might want to use the Merge tool. You are responsible for inventorying all the fire hydrants within the city. So you go out over the course of several days collecting the locations of the fire hydrants. This results in a layer that shows all the locations collected daily. You wish to combine all the collected locations into a single layer.

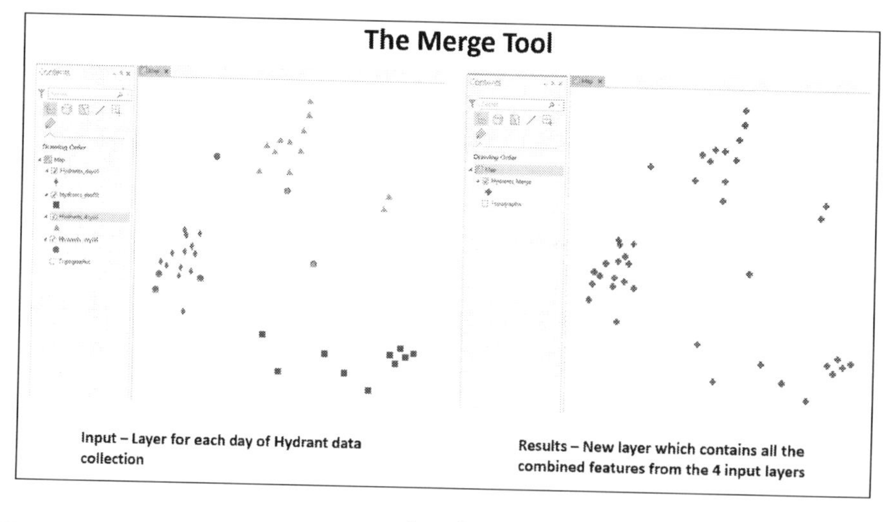

The Merge tool creates a new layer that includes all the features and attributes which were originally in the four separate layers. So now you have a single layer to manage, update, and analyze.

The Append tool

The Append tool is very similar to the Merge tool. It also combines data from multiple layers or tables into a single one. The big difference between these two tools is that Append is one of the few geoprocessing tools that actually changes the input data. It will add features or records to the target input.

You might use the Append tool if you have an existing layer of information and you just need to add newly acquired data to it. For example, continuing with the hydrant example, after merging the hydrants from the first four days you collected data, you go out and collect more hydrant locations. You wish to add those newly collected hydrants to the merged layer you created earlier.

The Append tool would work in this case. It would continue to add the newly collected hydrants to the existing data layer. It would not keep creating new layers that you would need to manage. This is illustrated in the following screenshot:

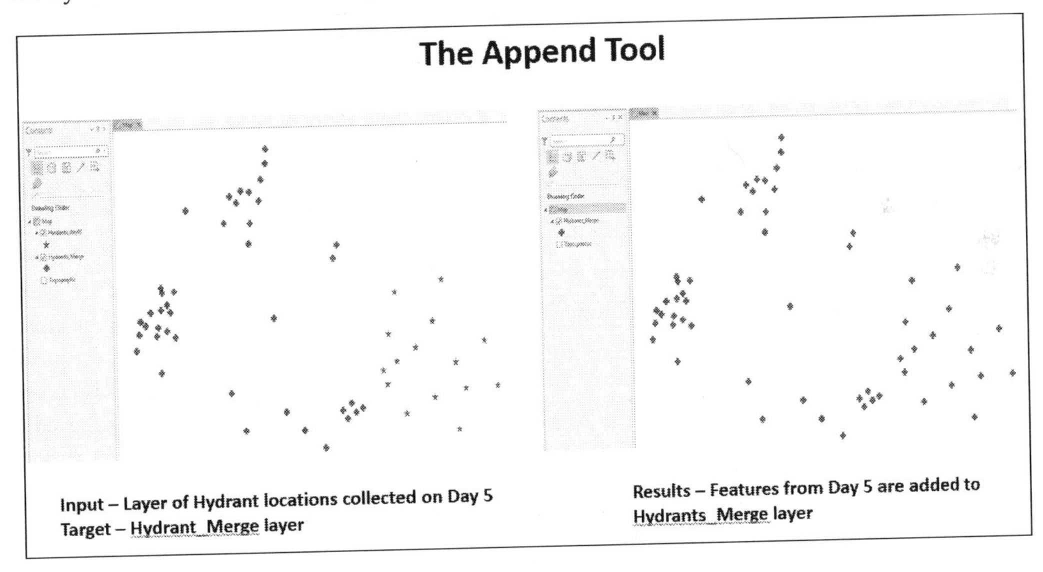

The Append tool is one of the geoprocessing tools that does edit or change your original input data.

Exercise 8B – using the Clip and Dissolve tools

The Public Works Director is working on a report that he must provide to the City Council to support his budget request. He needs to know the total length of each road within the City of Trippville. So he has asked you to provide him with those numbers.

Luckily, you already have the city limit and street data within your geodatabase. However, the street data extends outside the city limits and is broken down into individual road segments. So you will need to take some time to prepare the data before you can provide him with the numbers he needs.

Step 1 – evaluate the data

The Director has already provided you with the question and you have verified that you already have the data needed for the project. So now you just need to evaluate your data to verify the steps you need to complete the project.

In this step, you will open ArcGIS Pro and review the street and city limit data. You will make sure you have the information you need:

1. Start ArcGIS Pro and open the `Ex8B.aprx` file located in `C:\Student\IntroArcPro\Chapter8\Ex8B`.
2. Once the project opens, you should see a map with two layers: `City_Limit` and `Street_Centerlines`. Notice that the streets extend outside of the city limits.

Question: What geoprocessing tool, which you have read about in this chapter, do you think you should use to create a layer that only contains the streets inside the city limits?

____

____

3. Right-click on the `Street_Centerlines` layer and select **Attribute Table**.
4. The Director wants the total length for each road in the city. Review the table for the `Street_Centerlines` layer to see if there is a field that identifies what road each segment belongs to.

Question: What field identifies what road each segment belongs to?

____

____

5. Close the table and save your project.

You have now evaluated your data to determine its suitability to provide the information requested by the Director. You know you will have to extract portions of the street centerlines that are located inside the city limits and that you have an identification field you can use to dissolve the street segments by to easily calculate their length.

Step 2 – clipping the streets

In this step, you will clip the street centerlines, creating a new layer that only contains the portions of the streets which are located inside the city limits:

1. Select the **ANALYSIS** tab in the ribbon.
2. From the **Tools** group located in the center of the **ANALYSIS** tab, select **Clip**. This should open the **Geoprocessing** pane in the right side of the interface.
3. Click on the small drop-down arrow located to the right of the cell for the **Input Features** and select the `Street_Centerlines` layer.
4. For the **Clip Features**, select the `City_Limit` using the same process.
5. Make sure your **Output Feature Class** is set to `C:\Student\IntroArcPro\Chapter8\Ex8B\Ex8B.gdb\Street_Centerlines_Clip`.
6. Leave **XY Tolerance** blank.

Your **Geoprocessing** pane should look like this:

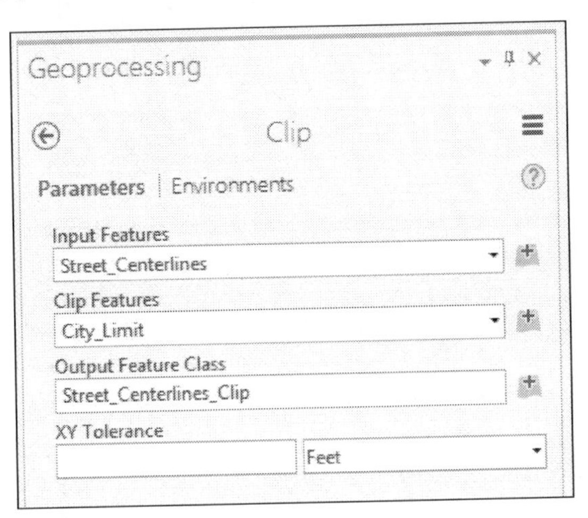

7. Once you have verified you have everything set correctly, click the **Run** button located at the bottom of the **Geoprocessing** pane.

When the **Clip** tool is complete, a new layer will be added to your map named `Street_Centerlines_Clip`.

8. Turn off or remove the `Streets_Centerlines` layer so you can better see the results of the **Clip** tool.
9. Right-click on the new layer you just created and select **Attribute Table**.
10. Right-click on the `ST_NAME` field and choose **Sort Ascending**. This will sort the records based on the name of each road.

Question: After sorting the records in the table, what do you notice about the number of segments for each road?

____

____

So now you can see how the Clip tool created a new layer that only contains those portions of the road which are located inside the city limits. Your original layer remains untouched. You are almost ready to provide the Director with the information he needs. You still need to simplify the data so you can more easily calculate the total length of each road.

Step 3 – simplifying the data and calculating the total length

In this step, you will use the Dissolve tool to simplify the clipped road centerline layer you created in the last step. This will create another new layer in your map.

1. Click on the **ANALYSIS** tab in the ribbon.
2. Click on the **Tools** button in the **Geoprocessing** group. This will reactivate or open the **Geoprocessing** pane once again.
3. In the top of the **Geoprocessing** pane, select **Toolboxes** located to the right of **Favorites**. This will display a list of all the toolboxes included in ArcGIS Pro plus any extensions you have access to.
4. Expand the **Data Management Tools** toolbox.
5. Expand the **Generalization** toolset located in the **Data Management Tools** toolbox.
6. Double-click on the **Dissolve** tool.
7. Set the **Input Features** to `Street_Centerlines_Clip` using the same process you used for the Clip tool.

8. Set the **Output Feature Class** to `C:\Student\IntroArcPro\Chapter8\Ex8B\Ex8B.gdb\Street_Centerlines_Dissolve_Name`.
9. Set the **Dissolve Field** to `ST_NAME`.

Your **Geoprocessing** pane should look like this:

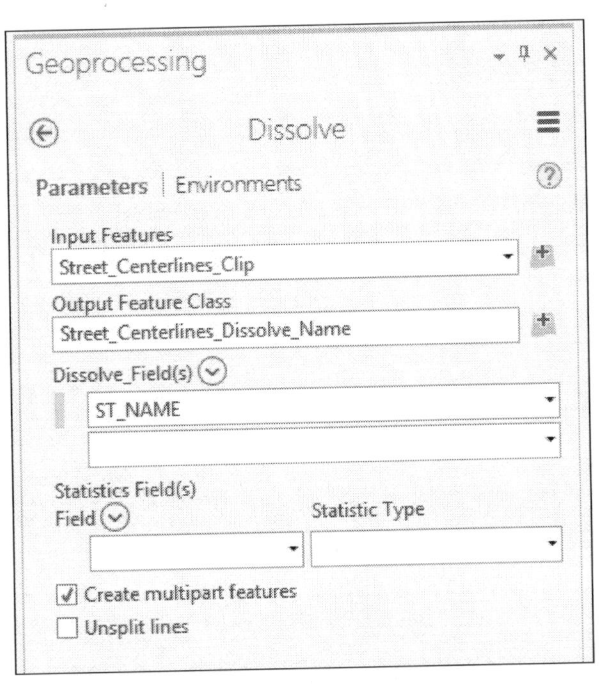

10. Once you have verified the Dissolve tool is properly configured, click the **Run** button.
11. A new layer is once again added to your map. Right-click on this layer in the **Contents** pane and select **Attribute Table**.
12. Right-click on the `ST_NAME` field and select sort ascending.
13. Scroll through the list of records. Pay attention to the number of records associated with each road name.

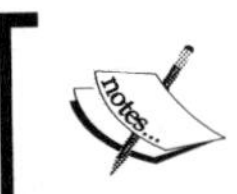

Question: How many records are there with the same road name?

____

____

You now have the information ready to give to the Director. With the dissolve completed, you have a list of each road and its associated total length using the `Shape_Length` field.

Step 4 – exporting a table to Excel

The Director appreciates your efforts. However, he does not have ArcGIS Pro. So he has asked if you can export your results to an Excel spreadsheet. This will allow him to more easily incorporate your results into his report.

In this step, you will export the results of your efforts to an Excel spreadsheet using tools found in the **Conversion** toolbox:

1. Return to the **Toolboxes** list in the **Geoprocessing** pane.
2. Expand the **Conversion Tools** toolbox and then expand the **Excel** toolset.
3. Select the **Table To Excel** script tool. This particular tool is actually a Python script. The scroll icon located next to the tool name identifies it as such.
4. Set the **Input Table** to `Street_Centerlines_Dissolve_Name`.
5. Set the **Output Excel File** to `C:\Student\IntroArcPro\Chapter8\Ex8B\Street_Lengths_by_Name.xls`.
6. Verify your **Geoprocessing** pane looks like the following image and click the **Run** button:

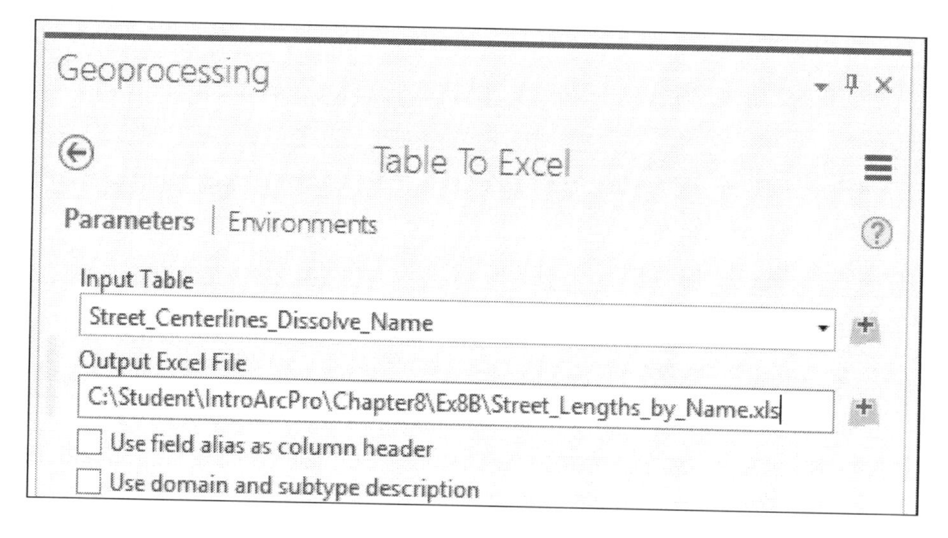

When the **Table To Excel** tool is complete, it does not add the resulting Excel spreadsheet to your map. If you wish to view your results, start Microsoft Excel and open the spreadsheet you just created. It should look very similar to the table you viewed in ArcGIS Pro.

7. Close the **Geoprocessing** pane and save your project.
8. Close ArcGIS Pro.

Congratulations! You have just completed your first analysis project using ArcGIS Pro.

Other commonly used geoprocessing analysis tools

With more than 300 geoprocessing tools, you have only begun to scratch the surface of the types of analysis that you can perform with ArcGIS Pro. ArcGIS Pro includes tools that allow you to perform spatial analysis of your data. This is broken down into several toolsets including Overlay, Proximity, and Statistics within the **Analysis** toolbox.

Overlay analysis

Overlay analysis compares two or more layers and locating areas where they overlap one another. Depending on which tool you use, you can determine only the areas where they overlap or erase the areas where they overlap or combine the total areas of all the inputs.

The **Overlay** toolset includes the following tools:

Tool name	Minimum licensing level	Short description
Erase	Advanced	Clips out areas of overlap from input features
Identify	Advanced	Calculates areas of overlap and no overlap
Intersect	Basic	Returns area of overlap only
Union	Basic	Combines total area of input polygons
Update	Advanced	Replaces area of overlap with new features
Spatial Join	Basic	Joins attributes from one feature to another based on spatial relationship
Symmetrical Difference	Advanced	Identifies areas where features do not overlap

ArcGIS Pro introduces a new **Pairwise** toolset that also performs overlay analysis. Tools in this toolset are designed to be used with extremely large datasets. They will provide simpler results than those created with the standard Overlay tools.

Now you will take a closer look at the two Overlay analysis tools that are available at all licensing levels: Union and Intersect.

Union

The **Union** tool takes the input of multiple polygon layers and combines all the information into a single feature class that contains all the data from the two or more input layers. It is important to remember this tool only works with polygons. It may not be used with points or lines. If you need to perform this type of analysis on points or lines, you would need to use the **Identity** tool.

You might want to use the **Union** tool if you wish to determine how much of each parcel was in a floodplain area and how much of each parcel was not in a floodplain area as illustrated in the following figure:

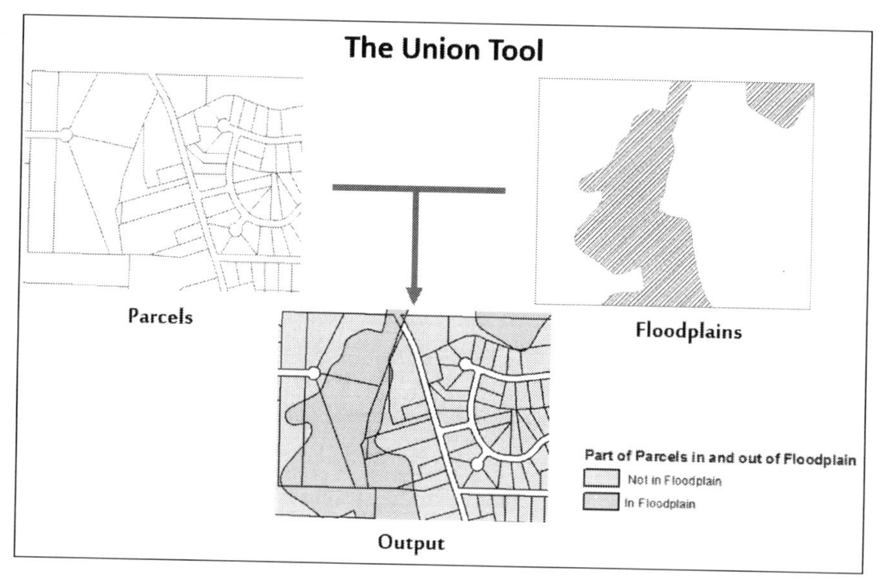

As you can see, the result is a new layer or feature class that attributes which part of each parcel is in the floodplain and which parts are not. Again, your original inputs are still intact.

Intersect

The **Intersect** tool takes multiple input layers and returns a new layer that shows where the inputs overlap. The resulting layer attribute table will contain the combined attributes of all the inputs. This tool works with all feature types, points, lines, and polygons. If you input multiple feature types, you get to choose what your resulting output type will be.

You might use the **Intersect** tool if you are working on an emergency evacuation plan for your community. You need to determine which roads might be blocked due to flooding so you need to know which segments are in the floodplain. You can use the **Intersect** tool to overlay the street centerlines with the floodplains to locate where and how much of each road is in the greatest danger of flooding as illustrated in the following figure:

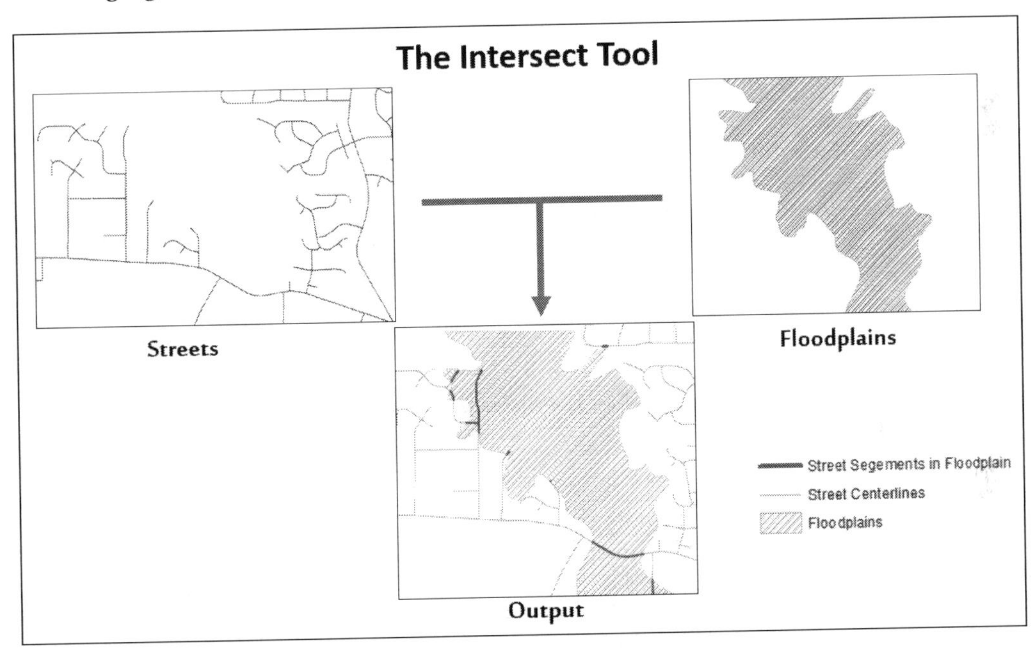

The resulting output of the Intersect tool in this scenario is a new layer that just contains portions of the streets which are located inside the floodplain area. As with other tools, your original input layers have not been changed in any way.

Proximity analysis

Proximity analysis compares, calculates, or shows distances between features in two or more layers. Proximity tools will generate distance buffers, locate nearest features, or calculate distances between features.

Tools included in the **Proximity** toolset include:

Tool name	Minimum license level	Short description
Buffer	Basic	Creates polygons around existing features at a set distance
Multiple Ring Buffer	Basic	Creates multiple buffer polygons at various distances
Create Thiessen Polygons	Advanced	Creates polygons around points showing areas of influence
Near	Advanced	Identifies how far nearest feature is between input and closest feature layers
Generate Near Table	Advanced	Creates a new standalone table that shows distances between features in two layers
Polygon Neighbors	Advanced	Identifies what polygons are next to source polygon along with calculating other associated information

We will now take a quick look at the Buffer and Multiple Ring Buffer tools.

Buffer

The **Buffer** tool is one of the most commonly used tools in ArcGIS. It creates a new polygon layer around the input layer based on a specified distance. The buffer distance can be a single value or be based on an attribute field in the attribute table of the features being buffered. You can choose to buffer any feature type. You can buffer points, lines, or polygons. However, the output will always be a polygon.

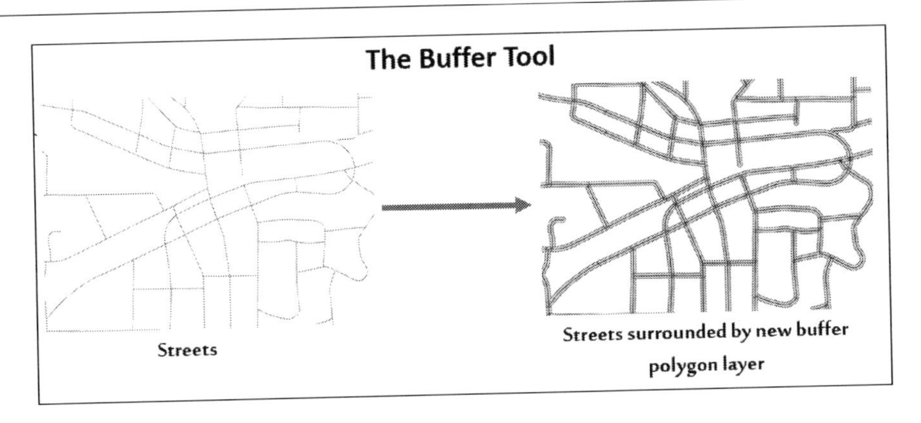

Buffers are very useful. They can be used to help determine if features in one layer are within a distance of another layer. They can also help create features for other purposes such as creating the rights of way for roads or railroads as just illustrated. In that illustration, you can see that a new polygon layer has been created around the existing street centerlines all at a uniform distance. This new layer represents the rights of way for those roads. Also, each new polygon inherits the attribute values of the street that was buffered. This means the new polygons are attributed with the road name and any other attributes that were linked to the street segments.

One of the options you have when using the **Buffer** tool is to dissolve the overlapping buffers. If you choose to dissolve the overlapping buffers, any buffers that overlap will be merged into a single polygon. This reduces the number of features that are in the resulting layer. Also, if you choose to dissolve overlapping buffers, the new polygons will not contain the attribute information that was associated with the features which you buffered.

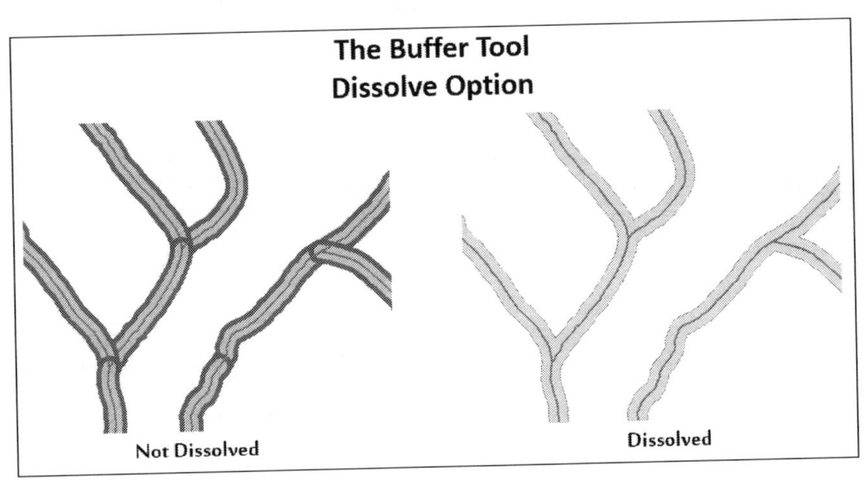

This illustration shows the difference between a buffer that is dissolved and one that is not. As you can see, the not dissolved example on the left contains many more polygons than the dissolved one on the right. The left example has many overlapping buffers so if they are dissolved, they become one.

Multiple Ring Buffer

The **Multiple Ring Buffer** tool is a Python script that runs the Buffer tool multiple times to create concentric buffer rings around the buffered features as illustrated in the following figure:

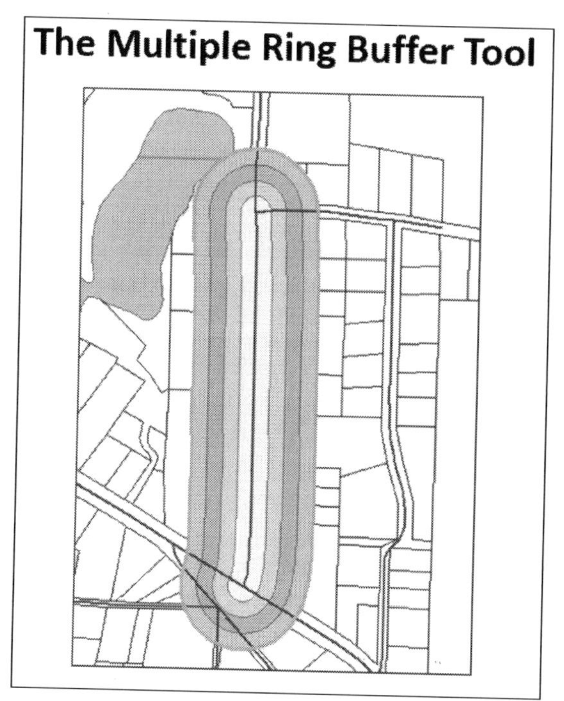

Like the standard Buffer tool, the **Multiple Ring Buffer** tool works with points, lines, and polygons but only outputs a new polygon layer. You also have the option to dissolve the overlapping buffers.

Now that you have had an opportunity to learn about some of the most commonly used analysis geoprocessing tools, let's give you a chance to put them into action.

Exercise 8C – performing analysis

Remember back in *Chapter 4, Creating 2D Maps*, when the Community and Economic Development Director asked you to prepare several maps that showed the location of commercial properties that were between 1 and 3 acres. After his meeting with the business owners, he needs some more assistance with this project.

He needs you to locate commercial properties that are within 150 feet of existing city sewer lines and have at least 1 acre that is not in a floodplain.

Step 1 – locating commercial properties near sewer lines

The first step of your analysis will be to locate all commercial properties that are between 1 and 3 acres in size and are within 150 feet of an existing sewer line. Luckily, you have already identified the commercial properties that meet the size requirements in *Chapter 4, Creating 2D Maps*, so that part is done. So now you just need to determine which of them are within 150 of the sewer lines.

In this step, you will create a 150 foot buffer around the sewer lines in the city. Then you will perform a spatial selection to select all commercial properties between 1 and 3 acres that are touched by or intersect the buffer you create:

1. Start ArcGIS Pro.
2. Open project `Ex8C.aprx` located in `C:\Student\IntroArcPro\Chapter8`.

When the project opens, you should see a map that looks very familiar. It should look like the one you created in *Chapter 4, Creating 2D Maps*. This map already contains all the basic layers you need to perform your analysis. You can see the commercial properties between 1 and 3 acres, the sewer lines, and the floodplains. Now you will create the 150 foot buffer around the sewer lines:

3. Select the **ANALYSIS** tab in the ribbon.
4. Select the **Buffer** tool to open the **Geoprocessing** pane and **Buffer** tool parameters.
5. Using the skills you learned in the previous exercise, set the **Input Features** class to `Sewer Lines`.
6. Set your **Output Feature Class** to `C:\Student\IntroArcPro\Chapter8\Ex8B\Ex8B.gdb\sewer_lines_Buffer`.
7. Set the **Distance** to `150` and the units to **Feet**.

8. Leave the **Side Type**, **End Type**, and **Method** with the default settings.
9. Set the **Dissolve Type** to **Dissolve all features into a single feature**. Since you do not need to know which sewer line is near which parcel, allowing ArcGIS Pro to dissolve the resulting buffer will make future analysis easier.
10. Verify your **Geoprocessing** pane looks like the following image and click **Run**:

Once the **Buffer** tool is complete, a new layer will be added to your map. This new layer will show the areas that are within 150 feet of the sewer lines. You will now use that new layer to select the commercial properties:

11. Click on the **MAP** tab in the ribbon.
12. Select the **Select By Location** button in the **Selection** group on the **MAP** tab.
13. If you left your **Geoprocessing** pane open, it will now display the **Select Layers By Location** parameters. Set the **Input Feature Layer** to `Commercial Properties 1 to 3 AC`.

14. Set the **Relationship** to **Intersect**. This will select all the commercial properties between 1 and 3 acres that are overlaid by the sewer lines buffer layer.
15. Set the **Selecting Features** to `sewer_line_Buffer`.
16. Leave all other parameters with default settings.
17. Verify your **Geoprocessing** pane looks like the following image and click **Run**.

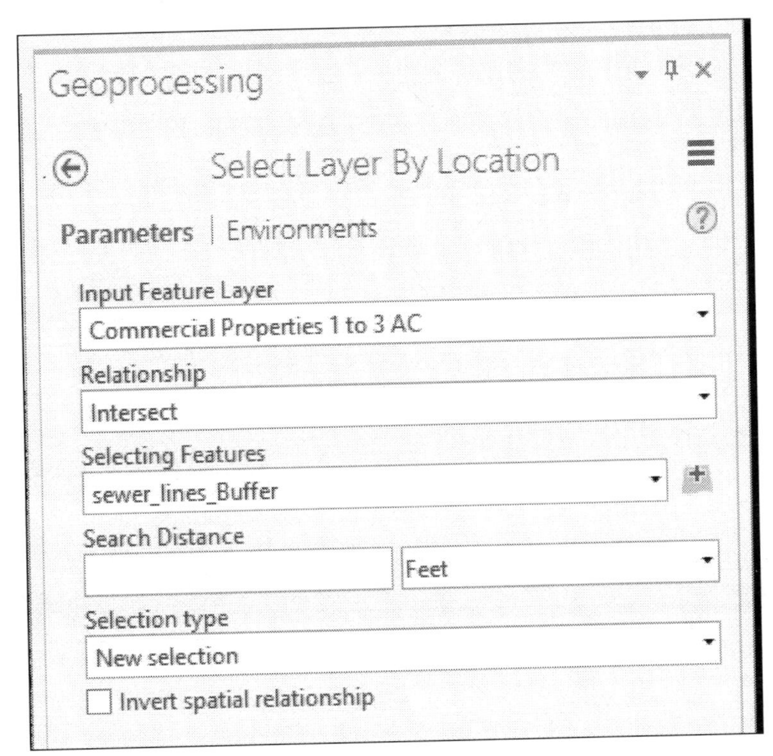

When **Select By Location** completes, you should have approximately 18 commercial properties selected. All of these are overlapped or touched by the sewer line buffer you created. This means they are all within 150 feet of an existing sewer line. You will now export those selected parcels to their own layer.

Step 2 – exporting selected parcels

Now that you have identified which commercial properties are within 150 feet of a sewer line, you will export those to a new feature class so you can use them for further analysis later. This will ensure you don't mistakenly change or corrupt the existing layer by accident:

1. Select the `Commercial Properties 1 to 3 AC` in the **Contents** pane.
2. Select the **Data** tab in the **Feature Layer** group.
3. Click on the **Export Features** button located in the **Export** group of the **Data** tab to display the parameters for the **Copy Features** tool in the **Geoprocessing** pane.
4. **Input Features** should automatically be set to `Commercial Properties 1 to 3 AC`. If not, set it to that layer.
5. Set the **Output Feature Class** to `C:\Student\IntroArcPro\Chapter8\Ex8B\Ex8B.gdb\CommercialProp_near_sewer`.
6. Verify your **Geoprocessing** pane looks like the following image and click **Run**:

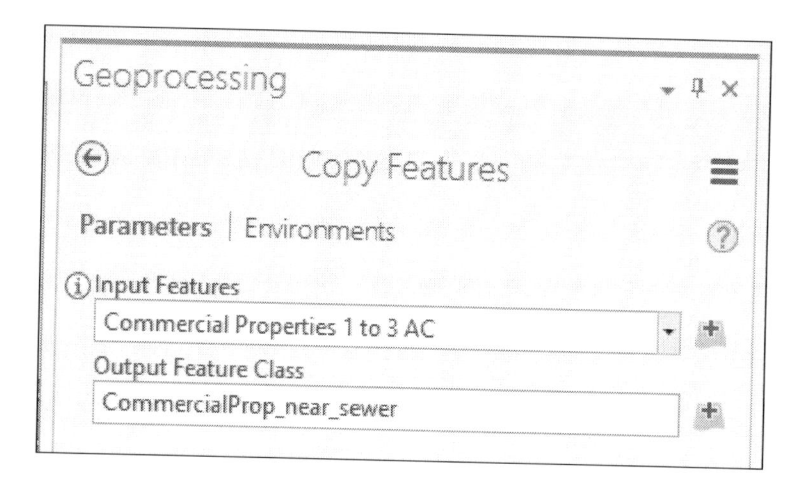

When the **Copy Features** tool completes, a new layer is added to your map that contains only those commercial parcels you had selected. If you have features selected in a map or table, most geoprocessing tools will automatically only use those selected records within that tool.

7. Open the **Attribute** table for the `CommercialProp_near_sewer` layer that was just added to your map.

8. Verify the table contains the same number of records as you had selected previously. It should be approximately 18.
9. Clear your selection by clicking on the **Clear** button in the **Selection** group on the **MAP** tab.
10. Close the table.
11. Turn off the sewer lines, sanitary sewer manholes, `sewer_line_Buffer`, and `Commercial Properties 1 to 3 AC` layers. You do not need to see those for the rest of your analysis. They might cause confusion.
12. Save your project.

Step 3 – determining how much of each commercial property is in the floodplain

Now that you have selected the commercial properties that are the required size and that are near the city's sewer system, it is time to calculate how much area of each of those parcels is within the floodplain. To do that, you will union the new layer you just created with the floodplains.

This will create a new layer that will split each commercial property into the part that is in the floodplain and that which is not:

1. Using skills you have already learned, open the attribute tables for both the `CommercialProp_near_sewer` and `Floodplains` layers. Take a moment to review what fields are located in each table and some of the values they contain. This will better help you understand the results produced by the **Union** tool.
2. Close the tables.
3. Click on the **ANALYSIS** tab in the ribbon.
4. Select the **Union** tool from the **Tools** group. The **Geoprocessing** pane will now display the parameters associated with the **Union** tool.
5. Set your **Input Feature** classes to `CommercialProp_near_sewer` and `Floodplains`.

6. Set your **Output Feature Class** to `C:\Student\IntroArcPro\Chapter8\Ex8B\Ex8B.gdb\Commercial_Floodplain_Union`.
7. Once you verify your **Geoprocessing** pane looks like the following image, click **Run**:

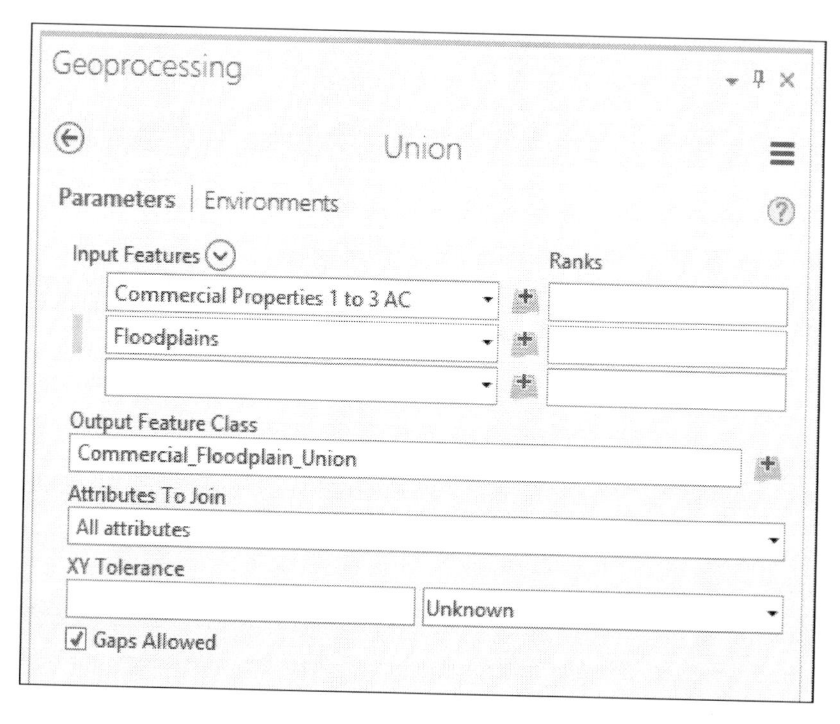

When the **Union** tool completes, it will add a new layer to your map that includes features which are the combination of the two input layers. Your map should look similar to the following illustration. Remember that your colors might be different.

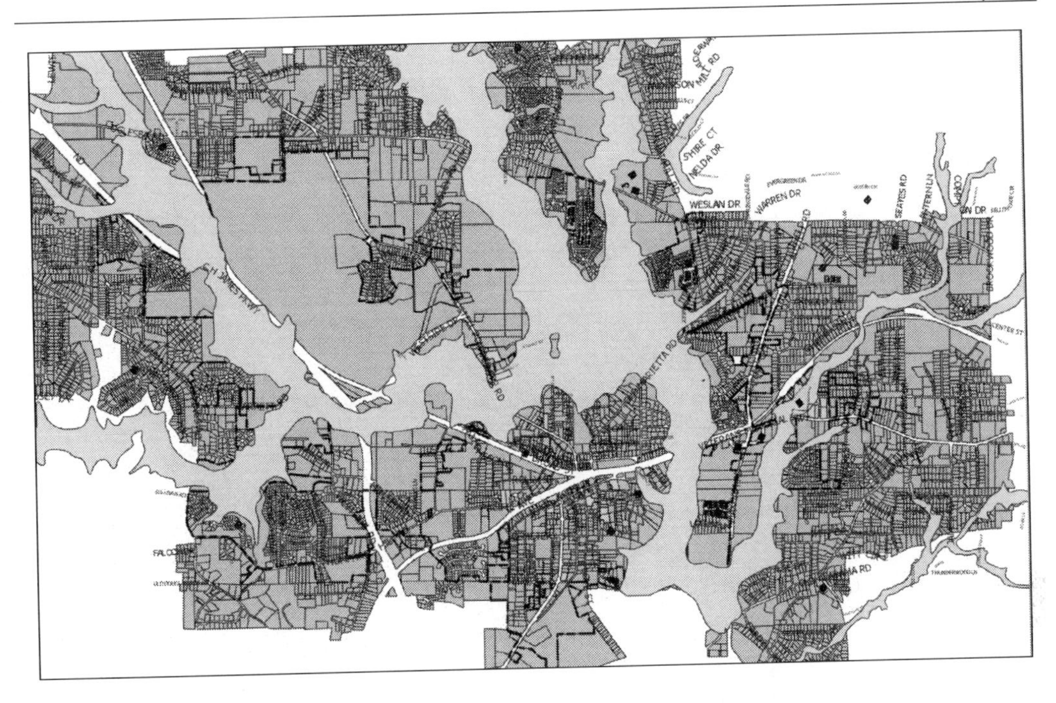

The green layer in the preceding map is the result of the **Union** tool. Now, perform the following steps:

8. Open the **Attribute** table for the layer you just created and that was added to your map.
9. **Sort Descending** on the **SFHA** field. All those polygons that are attributed with **IN** are inside the floodplain. All those that are **blank** or **NULL** are outside.
10. You need to update the `ACRE_CALC` field to reflect the new acreage for the `Commercial_Floodplain_Union`. Right-click on the `ACRE_CALC` field.
11. Select **Calculate Field**.

12. In the expression cell located below where it says `ACRE_CALC =`, type the following expression: `!Shape_Area! / 43560`. This will convert the `shape_area` field values that are in square feet to acres. It should look something like this when you are done:

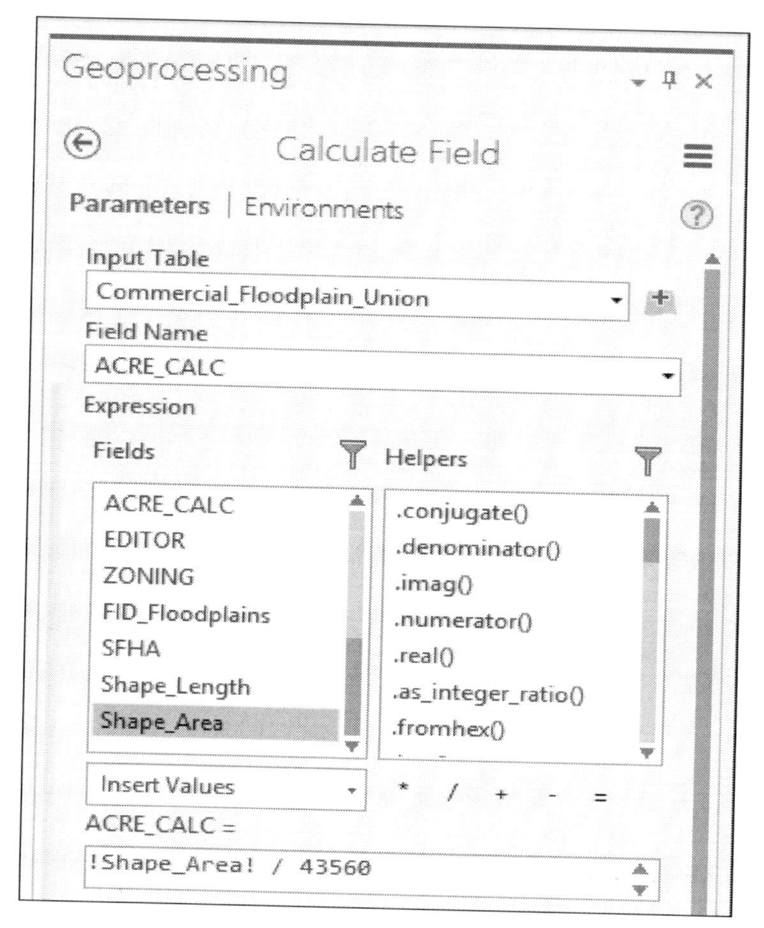

When converting units using the field calculator as just done, it is also possible to just specify the specific unit you are converting to within your expression. For example, you could have also used the following expression in the preceding task: `!Shape.Area@ACRES!`. The word acres replaces the conversion factor. This makes it easier to convert units if you do not know the proper conversion factor to apply.

13. Click **Run** once you have verified your expression.
14. Close the table and save your project.

Step 4 – selecting commercial parcels that are not in the floodplain

The one problem with using the Union tool in this process is that the resulting layer also includes parts of the floodplain polygons that did not overlap the commercial parcels. This means you need to either simplify the layer by removing those floodplain-only polygons or account for them in your query. If you had an Advanced license, you could have used the Identity tool instead; that would have avoided this next step.

In this step, you will select the commercial properties that have at least 1 acre or more not in a floodplain:

1. Click on the **MAP** tab in the ribbon.
2. Click on the **Select By Attributes** button. This will display the parameters for the **Select By Attributes** tool in the **Geoprocessing** pane.
3. Verify the **Layer Name** or **Table View** is set to `Commercial_Floodplain_Union`. If not, set it accordingly.
4. Selection type should be set to **New selection**.
5. Click on the **Add Clause** button.
6. Set the field to **Zoning** and the following operator to **Does Not Equal**. For the value, select the blank option at the top of the list and click **Add**. This will eliminate the polygons that just represent floodplain areas which do not overlap the commercial properties.
7. Click the **Add Clause** button again.
8. Set the **Field** to `ACRE_CALC` and the following operator to is **Greater Than** or **Equal to**. Type 1.00 for the value and click **Add**. This selects all commercial parcels that are the right size.
9. Click the **Add Clause** button one more time.
10. Set the **Field** to **SFHA** and the operator to **Does Not Equal**. Then set the value to **IN** and click **Add**. This removes any areas that are inside the floodplain from your final selection.

11. Click the green check mark to validate your query. Your **Geoprocessing** pane should look similar to this:

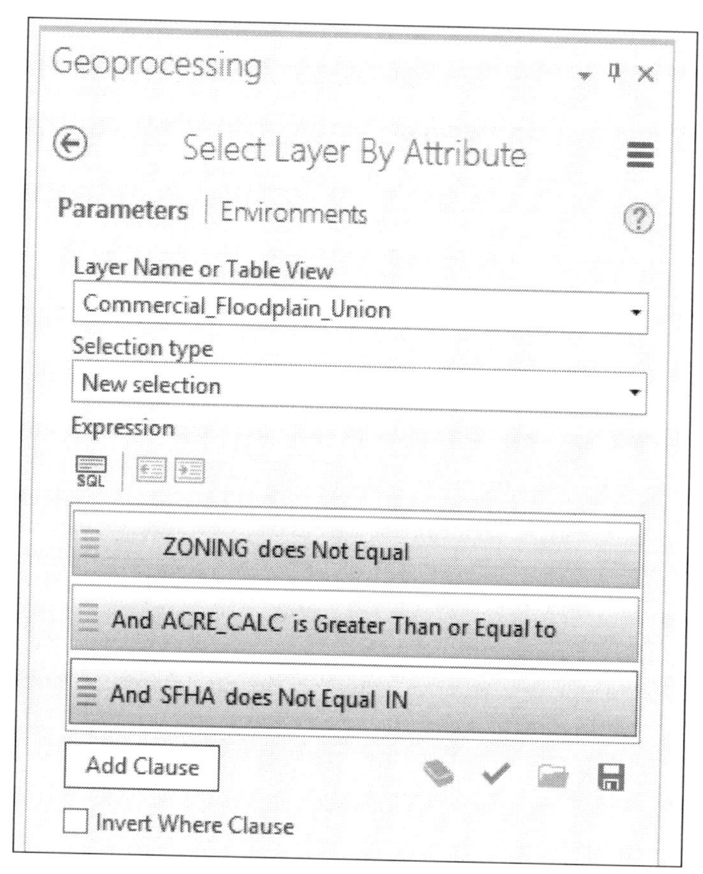

12. Click **Run** once you verify everything is set properly.

When it is complete, you should have 13 commercial properties selected. These should all be larger than 1 acre in size and outside the floodplain.

13. Using the skills you have learned, export your selection to a new layer. Set the symbology for the new layer to something that stands out.
14. Turn off the `Commercial_Floodplain_Union` layer so your new layer stands out even more.
15. Save your project and close ArcGIS Pro.

You have just identified the parcels within the city that meet the Director's requirements. You used various analysis and selection tools to answer his question. As you can see, it is not unusual to use multiple tools and methods to get the required answers to what may seem to be a simple question. Believe it or not, once you get familiar and comfortable with these tools, the process you just completed can be done in less than 10 minutes. It just takes practice.

Summary

ArcGIS Pro can be used to conduct spatial analysis to help answer a wealth of questions and concerns. It can help you see patterns and solutions. The powerful geoprocessing tools can be used with various types of data to get the answers you need to everyday problems.

In this chapter, you have learned what geoprocessing is and some of the tools that are available in ArcGIS Pro. You also learned how your licensing level and extensions can impact what tools are available to you when you need to perform analysis or manage GIS data.

This chapter also exposed you to some of the most commonly used analysis and data preparation tools. You conducted two separate analysis projects that provided hands-on experience with many of these tools and allowed you to see how they can be integrated with other tools you have already been exposed to in order to find answers.

9

Creating and Using Tasks

As you have now experienced firsthand, ArcGIS Pro contains a wealth of tools and methods for performing analysis, creating maps, and managing data. In many cases, there are two or three different ways to do the same thing. Also, many geoprocessing tools are very similar, such as Union, Intersect, and Identity. Although these are similar, each is designed to be used in specific circumstances.

All these different tools and methods make ArcGIS Pro a very powerful application. However, it can also make it confusing to new users and allows experienced users to perform functions very differently. This can result in inaccurate or wrong results. Things would be much similar if you could develop standardized workflows that everyone could use with step-by-step instructions required to complete specific tasks.

With ArcGIS Pro, you can do just this. They are called **tasks**. Tasks provide step-by-step workflows, which can be saved with your project. There is no limit to the number of tasks you can save with your project. You are not limited to saving them in the project. You can save and share them to a network folder, ArcGIS Online, Portal for ArcGIS, and more.

In this chapter, you will learn the following topics:

- What a task is and considerations to take into account before creating one
- How to create a task
- How to use a task

What is a task?

Simply stated, a task is a series of preconfigured steps required to complete a specific process. Tasks can be very simple containing only three or four steps, or they can be very complex containing group tasks within a task, and each group can contain multiple steps. It is entirely up to you. Tasks allow you to be as detailed as you believe you need to be for your user audience.

Tasks are stored as a Task Item within an ArcGIS Pro project. You will access your tasks from the **Project** pane. When you open a specific task, it will appear in a new pane named the **Task** pane.

Components of a task

Each task you create in ArcGIS Pro will consist of several components. The first is the Task Item, which is stored in the project. It is basically a folder to store related tasks within your project. You will access Task Items from the **Project** pane, as shown in the following image:

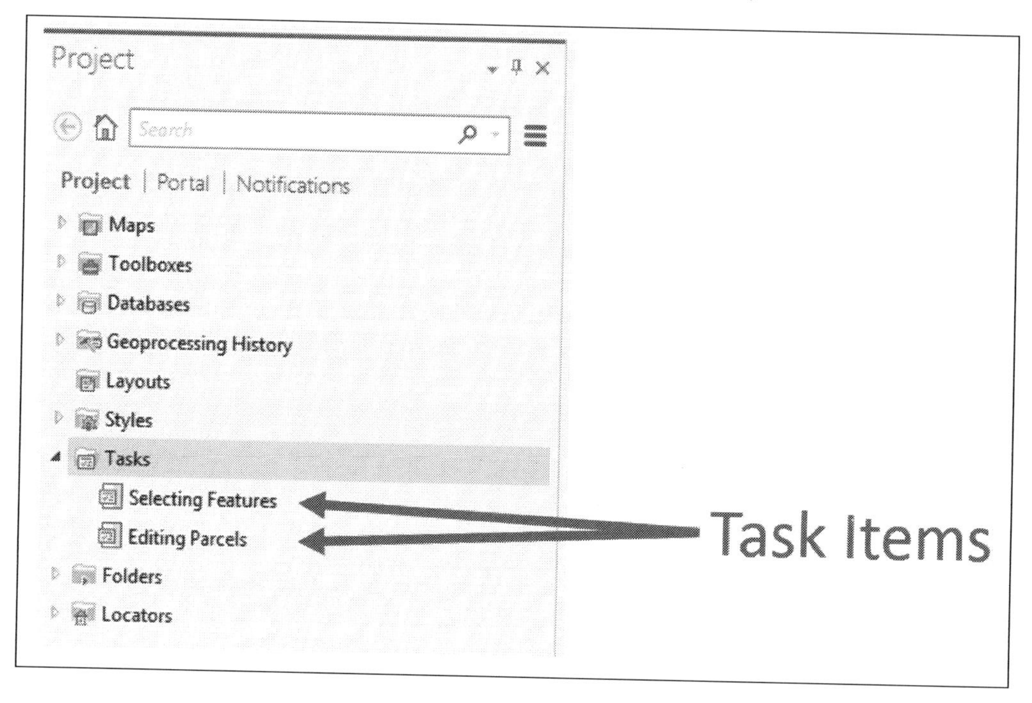

The second component is the Task Group. The Task Group is a subfolder within the Task Item in order to group related tasks by function or purpose. You can create Task Groups inside of other Task Groups just as you can create folders inside other folders in Windows. This allows you to create an organization structure to store your tasks, so they are easier to find and manage.

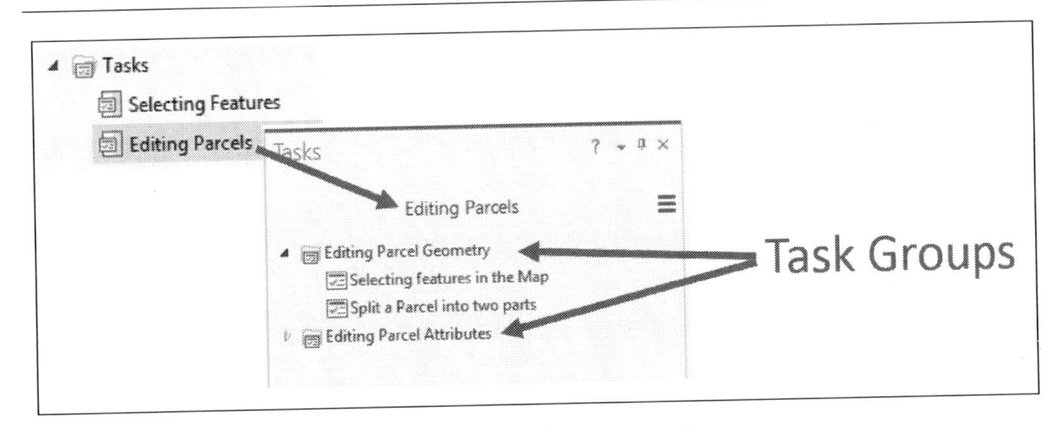

The third component is the task itself. This is a collection of steps need to complete a given process, such as splitting a parcel or adding a new water line or geocoding a new address. Tasks can be stored inside a Task Group or Stand Alone.

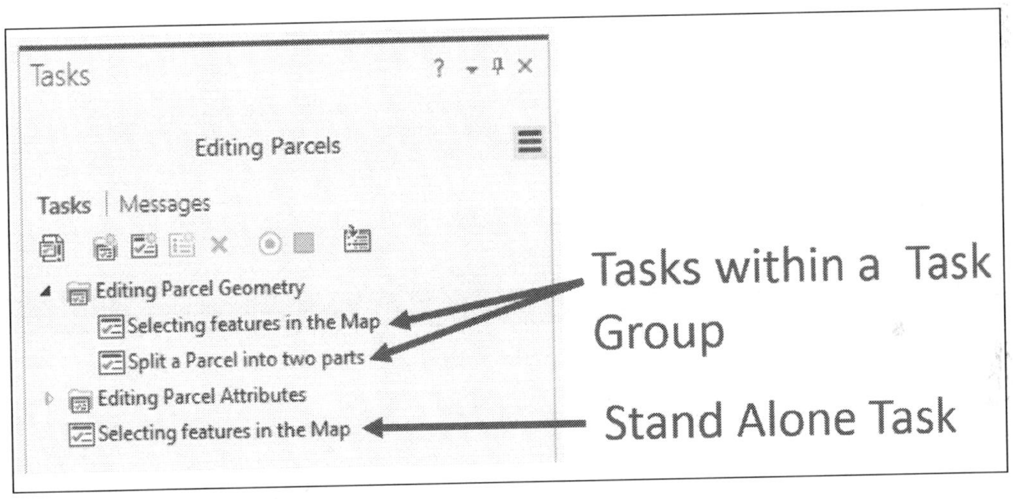

The last component is a step. A task will generally contain multiple steps. Steps refer to actions, buttons, geoprocessing tools, models, or scripts accessible in ArcGIS Pro. A common step would be to use the Explore tool to zoom to the location of a feature. Another example of a step would be once zoomed to the right area, selecting a given feature.

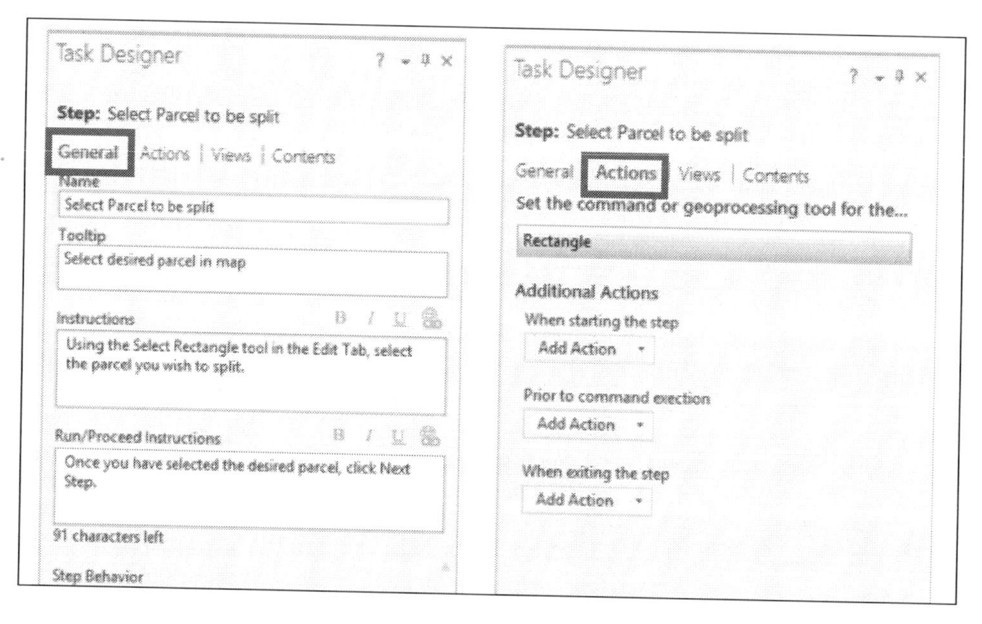

Each step is going to have several parameters you can set for that step, as illustrated later. You will provide general information regarding each step, such as a name, instructions for users to follow, and how the step will be run.

You can reference specific geoprocessing tools or commands within a step under actions. In the illustration, this step references the **Select by Rectangle** command. You have options to control which map and view the step should use. Finally, you can control the contents within that view via the step. You can decide which layers are visible, selectable, or editable.

You can have steps that simply provide instructions for the user and that run automatically. You can hide steps from users if desired. You will learn more about steps and how to create them later in this chapter.

Running a task

Tasks are run in the **Task** pane. The steps are presented to the user as they complete each one and more on to the next step. To get a task to open in the **Task** pane, you must first open the Task Item that contains the desired task you wish to run.

Once you open the Task Item from the **Project** pane, the **Task** pane will open, and you will be able to select which task stored in the selected task item you wish to run. You may need to expand a Task Group in order to find the specific task you are looking for. When you find the specific task you wish to run, simply double-click on it within the **Task** pane to run it.

Now, let's give you a chance to experience what it is like to run a task.

Exercise 9A – running a simple task

In this exercise, you will run a simple task that takes you through the process of selecting a feature in the map.

Step 1 – start ArcGIS Pro and open a project

You first need to open a project that contains stored tasks. In this step, you will start ArcGIS Pro and open a project that has several tasks:

1. Start ArcGIS Pro.
2. Open the `EX 9.asprx` located in `C:\Student\IntroArcPro\Chapter9`.

   When your project opens, you should see a single 2D map that contains layers representing the City Limits, Streets, and Parcels for the City of Trippville.

3. In the **Project** pane, expand the **Tasks** folder, so you can see the Task Items saved in this project.

> **Question**: What Task Items do you see included in this project?
>
> ____
>
> ____

Step 2 – opening and running a task

In this step, you will open a Task Item and then run a task that steps you through the process of selecting a feature in the map:

1. Double-click on the **Selecting Features Task Item** in the **Project** pane.
2. The **Project** pane should open on the left-hand side of the ArcGIS Pro interface. Note the tasks included in this Task Item.

Question: How many tasks are included in the Task Item you have opened and what are they?

____

____

3. Double-click on the **Selecting Features in the Map** task to open it.
4. Follow the instructions provided within the task's steps. When asked to zoom to an area in the map, you may zoom to any location for this exercise. Make sure to read and follow all the instructions.

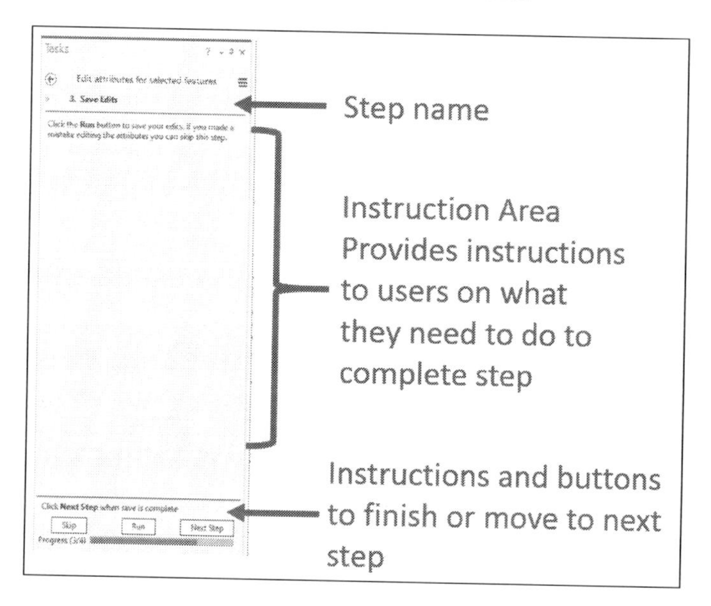

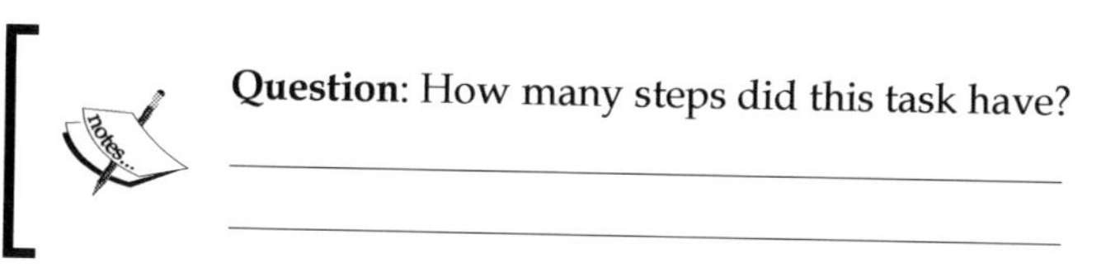

Question: How many steps did this task have?

____

____

5. When you are done running the task, clear your selection.
6. Close the **Task** pane.
7. Close ArcGIS Pro without saving the project.

You have just experienced how a task works from the user's perspective. This was a very simple task. Tasks can be as complex or as simple as you want them to be. Feel free to try some of the other tasks included in this project.

Creating tasks

Creating a task is not unlike creating a map or a layout. It requires some thought and planning to create a successful task. The time spent on the frontend of creating a task will alleviate a lot of frustration and headaches later for yourself and for those that use the tasks you create.

Just as you did before when creating a layout, you need to answer several questions:

- What is the purpose or goal of the task?
- What is the workflow that supports the purpose?
- How will it be used?
- Who is the audience for your task?

The answers to these questions will impact the design of your tasks. They will help determine the number of steps to include, which steps may need to be run automatically, the level of instruction you need to provide, what tools need to be included as a step, and more.

After answering those questions, you will then be ready to start creating your task. This may require you to start by creating a new Task Item and then creating tasks or Task Groups within that item or you can add new tasks to an existing Task Item.

As you create new tasks, you will use the **Task Designer** pane. The Task Designer allows you to add or modify steps. Steps are the heart of any task. They provide users with the instructions needed to complete the task. Steps themselves have several parameters, which you will need to configure as you create your task. Again, the answers to the questions mentioned earlier will help guide the creation of each step.

Before you begin creating a task, you will explore an existing task.

Exercise 9B – exploring a task

In this exercise, you will take a closer look at the task you ran in the *Exercise 9A – running a simple task* section. You will check whether it contains the number of steps you thought it did or whether there was more than met the eye.

Step 1 – open a project

In this step, you will open the same project you used in the last exercise. This will allow you to access the task you ran previously:

1. Open ArcGIS Pro and select `Ex 9` from the list of previously opened projects.

2. Expand the **Tasks** folder in the **Project** pane.
3. Verify that you see the **Selecting Features Task Item** in the **Tasks** folder.

Now that you have opened the correct project and verified that you see the Task Item you used in the last exercise, you will now learn how to open it for editing in the Task Designer.

Step 2 – opening a task in the Task Designer

You will now open the Task Item in the Task Designer. This will allow you to see all parameters, tasks, and steps included within the Task Item:

1. Right-click on the **Selecting Features Task Item** and select **Edit in Designer**. The **Tasks** pane opens in **Designer** mode on the left-hand side of the interface, and the **Task Designer** pane opens on the right-hand side of the interface.
2. Review the parameters for the Task Item in the **Task Designer** pane. Note that in addition to standard parameters, such as name, author, and description, you can also see what version of ArcGIS Pro is running and track the version of the task as you update and modify it.
3. Now you will make a couple of changes to this Task Item. Rename the Task Item as `Selecting Features in ArcGIS Pro` by typing that in to the cell located under **Name**.
4. Change the **Authors Name** to your own.
5. Change the description to the following: `This task item contains various tasks that demonstrate different methods to select features within ArcGIS Pro.`
6. Finally, enable **Auto Increment** of the Task Item version.

Note that there is no save button in the Task Designer. Changes are automatically applied when you click on another parameter, task or pane. This allows you to test the changes quickly. The actual changes are saved permanently when you save the project. If you close a project without saving, any changes you made to the Task Item and included tasks and steps will
be lost.

7. Save your project.

Step 3 – Reviewing a task's steps

Now that you have the Task Item opened in the Task Designer, you will now explore the steps of a specific task. In this case, it will be the same task you ran in the last exercise:

1. In the **Tasks** pane, select the **Selecting Features in the Map** task.
2. Click on the blue arrow that appears to the right of the task name to access the task's steps.

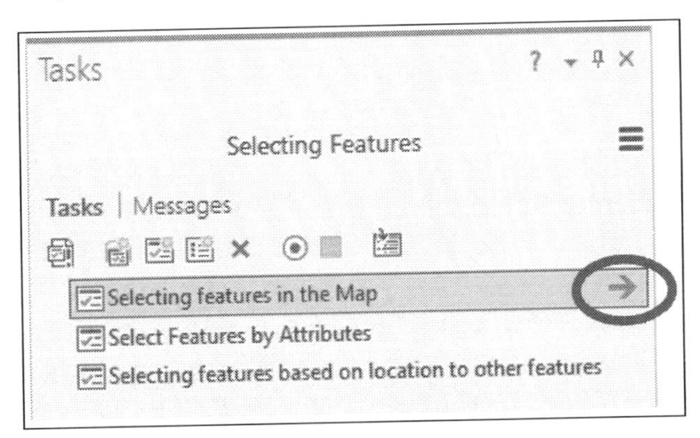

Question: How many steps do you see in this task?

____

____

Question: How does this compare to the number of steps you counted when you ran the task in the last exercise?

____

____

3. Select each step and review the **General**, **Actions**, **Views**, and **Contents** parameters associated with each step. They will be displayed in the **Task Designer** pane located on the right-hand side of the interface.

Question: Why do you think the number of steps you counted in the *Exercise 9A – running a simple task* section differs from the actual number of steps contained in this task?

____

____

4. Click on the back arrow located in the upper-left corner of the **Tasks** pane once you have reviewed each step.
5. Review the other tasks located within this Task Item and their associated steps.
6. Close the **Tasks** pane when you are done reviewing tasks.
7. Save your project and close ArcGIS Pro.

Now you have a much better idea of how a task is configured and the way steps work. Let's now start looking closer at the process to create a task.

Things to consider

As mentioned previously, you need to answer those four questions before you begin creating any task. The answers to those questions will guide much of the task design. So, how do those four questions impact or guide the design of your task?

What is the purpose or goal of the task?

Like a map, you are creating a task for a reason. It could be to show everyone in your organization the proper workflow to split a parcel so that everyone does it in the same way. It might be to develop a workflow to identify all parcels located within 300 feet of another parcel that is to be rezoned, so you can create a notification mailing list. Each of these represents a purpose.

The purpose will help you determine which tools will need to be referenced in your task. It will also help you answer the next question concerning the overall workflow that the task will address.

What is the workflow which supports the purpose?

The task workflow is the steps that will be included in the tasks. It is the tools and actions that will need to be performed in order to achieve the purpose and goal of the task. The workflow will determine the order of your steps within the tasks and when you use the tools your purpose has helped identify will be needed.

Think through your workflow carefully. Write down each step you believe will be required in the task and any tools associated with each step. Once you have the steps written down, you should verify that you have taken into account all the needed steps. It is not unusual for experienced users to forget a step because it is something they do automatically without even realizing that they do it. This means that a task may be incomplete or confusing depending on the audience using the task.

One of the best ways to validate or even develop the initial workflow for a task is to work through the entire process manually within ArcGIS Pro. If you are validating a workflow, make sure to follow it exactly as you have written it down. This will help you to identify any steps you may have missed. Following your own instructions will be harder than you think. You will want to just do it. Do not fall into that trap. Take your time and do each step as you have it outlined. Remember that those that may use the task you create may not have the same level of experience with ArcGIS Pro as you do. So missing a tool or step that is intuitive to you may not be to others.

If you work through the process manually to actually develop the steps, then record each step as you do it. Make sure to note the tools associated with that step. Creating a document with screenshots can also be helpful as you do this. This helps ensure you remember all the requirements associated with a given step when you go to create the task.

How will the task be used?

There are several reasons you create a task. You may want to standardize a common workflow within your organization to ensure that everyone is doing it in the same way. You might want to use it as a training tool for new users. You may want to establish the best practice for your organization to ensure accuracy and efficiency.

The *how* will often be tied into the purpose and will impact the complexity and level of documentation needed for the task. For example, if you are creating a task that will be used to train new users, you will need to make sure to include exact step-by-step instructions with very thorough documentation on exactly what the user needs to do to complete the step. You will require many of the steps to be manual interactive ones, so the user learns the process completely and understands the reasons for each step.

However, if you are just trying to establish a common workflow within an organization of experienced users, you can often reduce the number of interactive steps and use automatic steps since the users are already familiar with ArcGIS Pro functionality. Also, due to that experience level, you are able to reduce the level of instructional documentation required for each step. Tasks created for training or to establish the best practices need more time to create as they typically require more efforts due to the overlevel of the complexity of their design and documentation.

Another consideration, which falls under how the task will be used, is the license level required to perform the steps within a task. Steps often refer to the use of specific geoprocessing tools or ArcGIS Pro commands. As you have learned, some geoprocessing tools are only available with certain licensing levels or extensions. So, as you are developing a task, you need to always consider what license level or extensions are required to complete a task. If your task makes use of tools that require a specific license level or extension, you might want to include a step that has the user verify that they have the correct license level or extension needed to complete the task.

Who is the audience for the task?

Finally, you need to determine the audience for the task you are creating. Are they experienced ArcGIS Pro users or are they a new user? Are they someone who has used other GIS software? Are they computer savvy or not?

Obliviously, the level of experience and skill will impact your task design. The lower the skill and experience level, the more instructions you will need to provide. You may also need to include steps that would be intuitive to experienced users.

If your audience is a group that is familiar with another GIS software application such as ArcMap, you might need to include references to tools or processes in that application, so they can more easily relate ArcGIS Pro functionality to something they are familiar with. This may increase the amount of instructions you create for each step.

Now that you know what you need to consider before creating a task, it is time to learn how to actually create a task.

Creating a task

As you have learned, a task has several components. When creating a new task, you will need to create these components. The first step to create any task is to have a Task Item to contain it. This can be an existing or a new Task Item.

Once you create or identify an existing Task Item that will contain your task, you then need to decide if you want the task to be standalone or contained within a Task Group. When do you need to make a task standalone and when should you store it in a Task Group? If the task you are creating is not related to other processes or workflows, it is good to leave it as standalone. However, if the task is part of a larger workflow or process, then it is a good idea to store those related tasks within a Task Group. This makes them easier to find. If you want to store your task within a Task Group, you would need to create it after you create the Task Item.

At this point, you are now ready to create your task. You will use the Task Designer and the answers to those four questions to create the task along with all the included steps.

Creating a Task Item

There are several ways to create Task Items. If this is the first Task Item you are adding to your project, you will go to the **INSERT** tab on the ribbon and select the **Task** button in the project group. This will create a new blank Task Item in your project, as illustrated here:

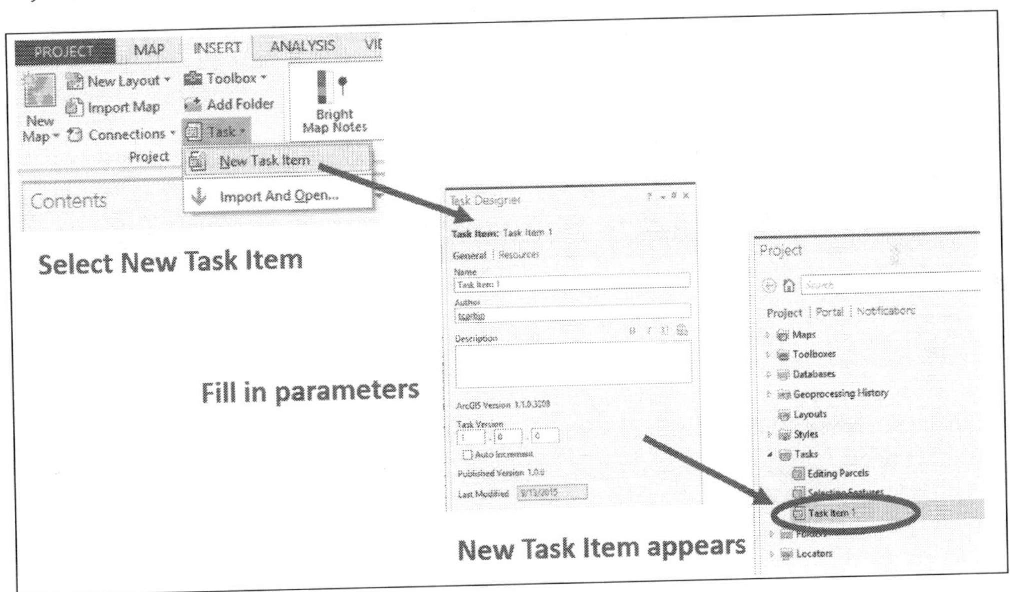

If you already have Task Items in your project, then you can add new ones from the **Project** pane. You simply right-click on the **Task** folder and select **New Task Item**.

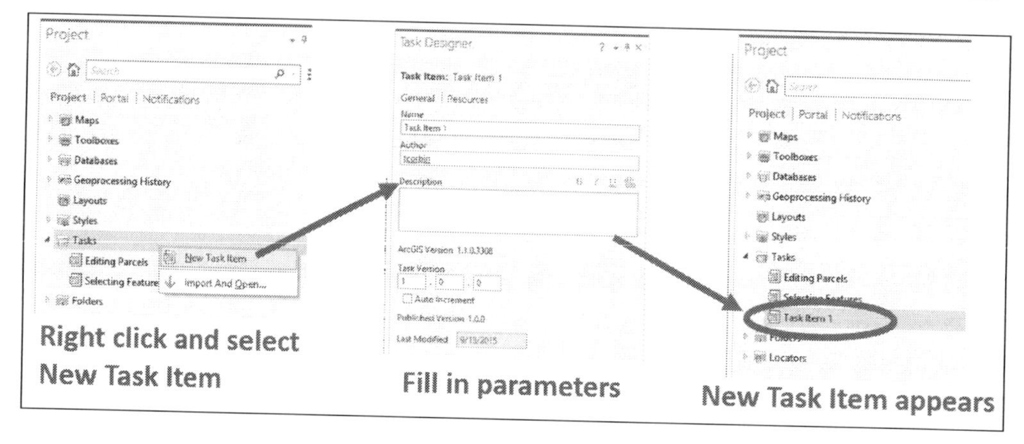

Creating a Task Group

If you decided that you want to store your task within a Task Group, you will need to create it within the Task Item. You will do this from the **Tasks** pane while in **Designer** mode.

Simply, click on the **New Group** button and fill in the parameters, as illustrated in the following screenshot:

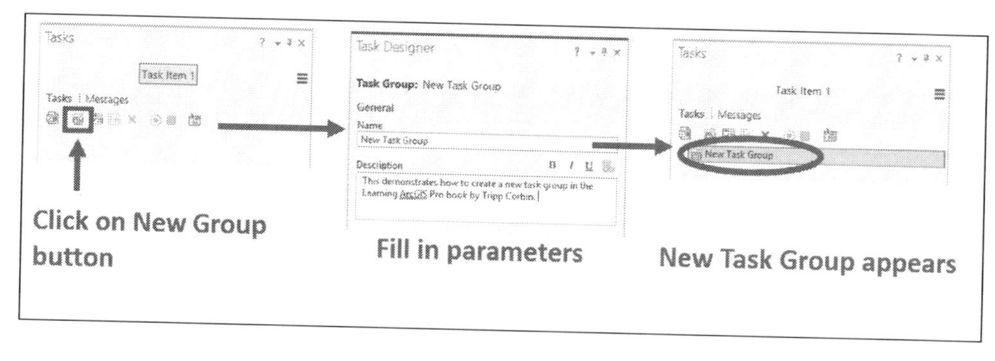

Creating a task

Now you are ready to create the actual task. Remember that you can add tasks to existing Task Items and Task Groups. Creating a task is very similar to creating a Task Group. You will click on the **New Task** button in the **Tasks** pane while in **Designer** mode. You will then fill in the associated parameters, and the new task is created, as illustrated in the following screenshot:

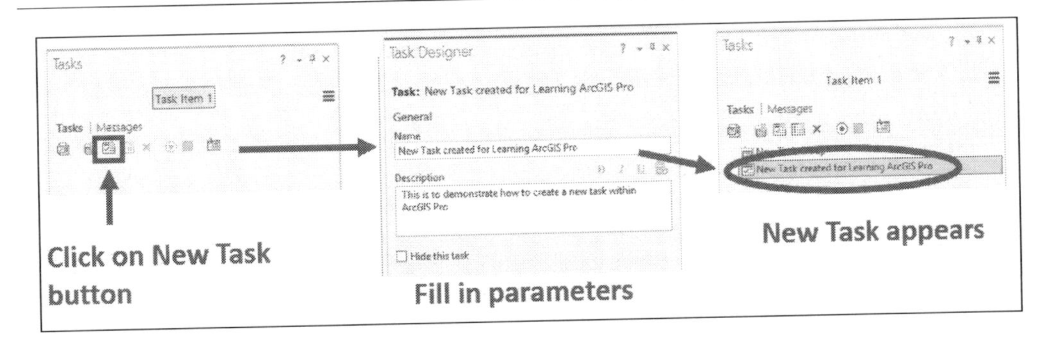

Exercise 9C – creating a task

In this exercise, you will create a new task in a project. In a later exercise, you will add steps to this task.

The purpose of the new task is to create a list of parcels located along a road. This will help the City notify those who live and work along the road when it is being repaired. This task will be used as a common workflow for existing and new GIS staff.

Step 1 – open is project and create a Task Item

In this step, you will open a project and then create the Task Item, which will contain the task created in a later step:

1. Start ArcGIS Pro and Open the `Ex 9` project from your list of recently opened projects.
2. Expand the **Tasks** folder in the **Project** pane.
3. Right-click on the **Tasks** folder and select **New Task Item**. The **Tasks** and **Task Designer** panes will open automatically.
4. In the **Task Designer** pane, fill in the parameters of your new Task Item as follows:
    - **Name**: **Road Repair Tasks**
    - **Author**: your name
    - **Description**: `This task item contains tasks associated with road repair projects, such as generating notification lists, locating nearby parcels, calculating total lengths, and more`
    - **Task version**: 1.0.0
    - Enable **Auto Increment**

5. Save your project.

You have just created your first **Task** Item. Now you need to add a task to that item.

Step 2 – creating a new task

You are now ready to create the task that will serve as the container for the steps needed to create the list of parcels located along a road. You will create a standalone task for this exercise since it is the only task you will create. If this was part of a large set of tasks you were creating, you might create Task Groups to help organize them:

1. In the **Tasks** pane on the left-hand side of the interface, click on the **New Task** button.
2. In the **Task Designer** pane, fill in the parameters for the new task as indicated here:
   - **Name** = **Create list of nearby parcels**
   - **Description** = `This task will step you through the process needed to create a list of parcels located along a road segment that will be repaired. The list will allow those who live and work along the road to be notified of the repair and how long it is expected to take.`
3. Close the **Task Designer** pane.

   Your **Tasks** pane should now look similar to this:

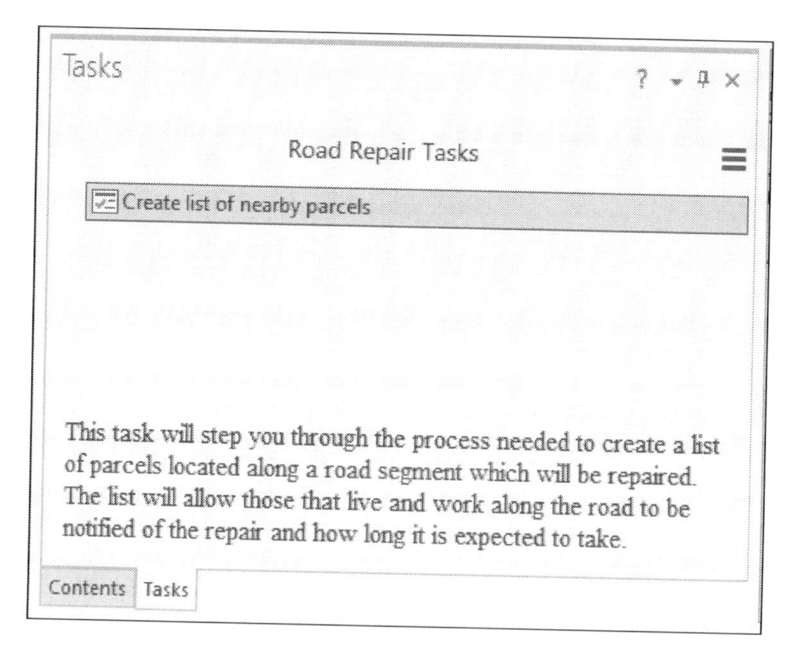

4. Save your project and close the **Tasks** pane.

You now have created a task, but it is empty. You still need to add the steps to the task that tell users the process needed to produce the list of parcels located near the road segments that will be repaired.

Task steps

Steps are really the meat of your tasks. They provide the instructions to complete a process or workflow to the users. Steps can be simple text instructions that tell how to perform an action such as how to use the Explore tool to zoom to the area of interest. Steps can also be very complex. You can include specific tools, layer behavior, and selection controls within a step.

It is entirely up to you. The complexity of the steps and how many you need will be driven by the answers to those four questions. If you are trying to standardize a specific workflow for experienced users, you may increase the complexity of the step structure to automate it as much as possible. This will increase the efficiency of your team while also ensuring that everyone performs the process in the same way. If you are designing the task to be a training tool, you may want the steps to be more manual and instructional to allow the user to gain a better understanding of what tools are used and how they work.

So, let's take a closer look at the components that make up a step and what purpose they serve. This will provide you with a good understanding, so you can build effective steps within your tasks.

Components of a step

A single step can include several components depending on its purpose. You can include instructions for the user, determine how the step will run, link it to geoprocessing tools or commands, control view and layer behavior, and manage selections within a single step. Let's look at these components or parameters more closely.

General

The first component of a step is the general information and behavior of the step. This includes the step **Name**, **Tooltip**, **Instructions**, and step run behavior as follows:

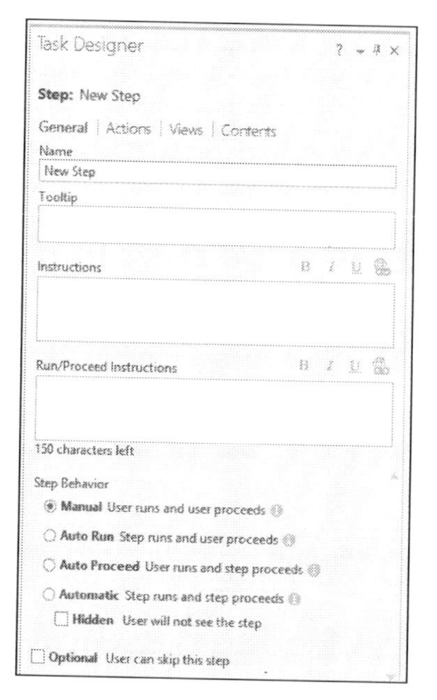

The **Name** of the step is what the user first sees listed in the task. The name should give the user a general idea what the step is supposed to accomplish such as Zoom to Area or Select parcel to edit.

Tooltips are pop-up windows that appear when you move your mouse pointer over the object in the question. This should provide more information for the user but still be limited to a short paragraph at the most. It will often be very similar to the name.

Instructions are one of the most important parameters of a step. They are where you provide the users with the instructions on what they need to do during the given step to complete it. These can be as long and detailed as you feel they need to be based on the purpose of the step and the intended audience. They should always be clear and easy to follow. Try to avoid using abbreviations or acronyms if possible since these can cause confusion.

Run/proceed instructions tell the user what they need to do once they have completed the instructions to proceed to the next step. For example, click on the **Run** button to start the step. Then, click on the **Next Step** button to continue.

Finally, you will need to determine how the step will run. You have four basic options to choose from **Manual**, **Auto Run**, **Auto Proceed**, and **Automatic**:

Step Behavior	Description	Example of use
Manual	The user must manually click on the **Run** button to start the step and the **Next Step** button to proceed.	You want the user to activate the **Explore** tool and then zoom to a specific area. You are using this as a training tool, so you need them to see the **Explore** tool being activated for use.
Auto Run	Runs the command or geoprocessing tool automatically. User performs action with an associated tool and manually clicks on **Next Step** to proceed.	You want the user to select features from the map. So, you have linked the **Select by Rectangle** tool to the step, and it runs automatically; the user does not need activate it. They just select features from the map.
Auto Proceed	User clicks on the **Run** button to run the step, and it automatically advances to the next step when completed.	You want the user to buffer a feature, but the distance will be different depending on circumstances. The user would enter the appropriate buffer distance and then click on the **Run** button. When the buffer is complete, it automatically advances to the next step.
Automatic	Enter step runs without any user interaction. Automatic steps have the option to be hidden.	Your user is working in one map, and you need them to change to another. You can include an automatic step, which switches the current active map to a different one before you proceed to the next step.

Actions

The next component of a step is **Actions**. Actions allow you to link an ArcGIS Pro command or geoprocessing tool to the step. An ArcGIS Pro command is any button or tool you see in a ribbon, such as **Add Data** or **Measure**. Geoprocessing tools are any tools available in a toolbox including custom Python scripts or models you or others may have created. A step is not required to have an action. If the step is strictly instructional, it may not have an action.

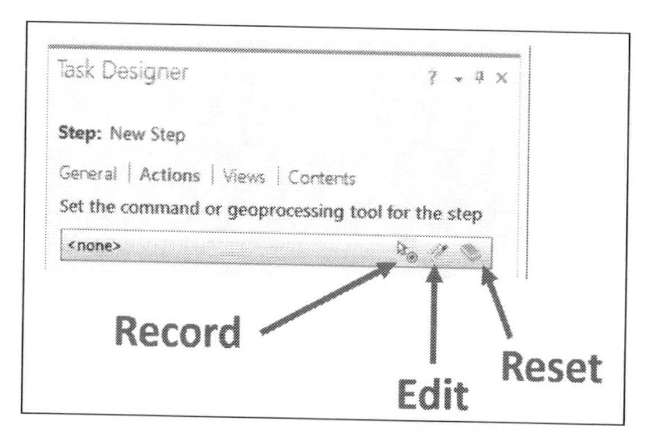

When adding a command or geoprocessing tool to a step, you can accomplish that in two ways. First is to edit the action. This allows you to pick a command or geoprocessing tool from a list. You have the option to search using keywords.

Another method is to record. When you record, you simply find the command or geoprocessing tool you wish to use from the ArcGIS Pro interface and click on it. This is a great option if you have performed the step manually in the past and know exactly which tool you need.

You can also manage selections associated with a step as additional actions when the step starts, before running a command or when exiting the step. Within the step you can save a selection, modify a selection, create a new selection, or clear a selection.

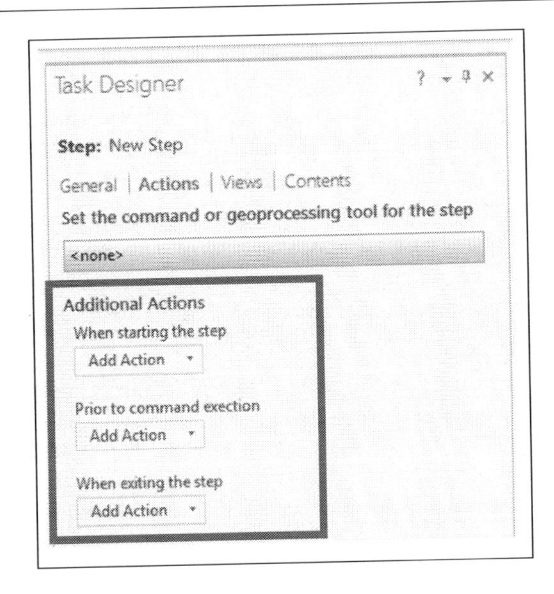

You can save currently selected features to a selection set. This selection set can then be used by later steps in the task. You can also save features created or modified by this or previous steps.

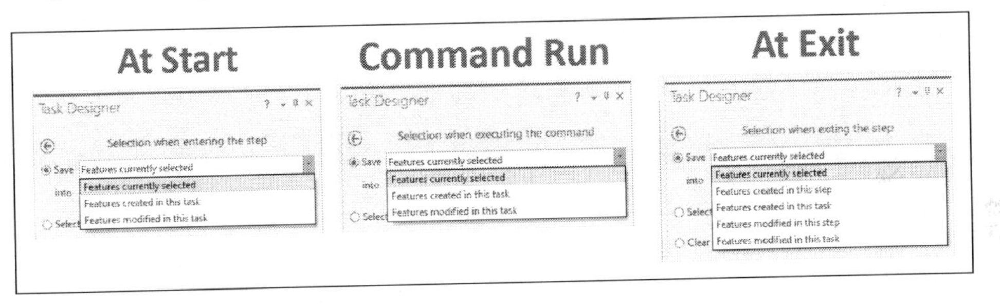

For example, if in a previous step you split a parcel into multiple parcels, which remained selected after being split and you wanted in a later step to edit the attributes for those new parcels, you could save the selection of those split parcels to a selection named `Split Parcels`, which could then be easily recalled in a later step.

You can also clear the selection or create a custom query to select features. The custom query is very similar to the **Select by Attribute** tool and a **Definition Query**, which you have used in previous chapters.

The use of additional actions is also optional. A step may have no actions or additional actions defined. The step may just have a command or geoprocessing tool indicated. It may have just additional actions defined or a combination of a command/geoprocessing tool and additional actions. This means that a step can be as simple or complicated as you wish it to be.

Views

Views allow you to control what maps, scenes, or layouts are open and active within your project as you run the step. You will see all views that are contained in the project you have open. While you can have many views open at any one time, only a single one can be active.

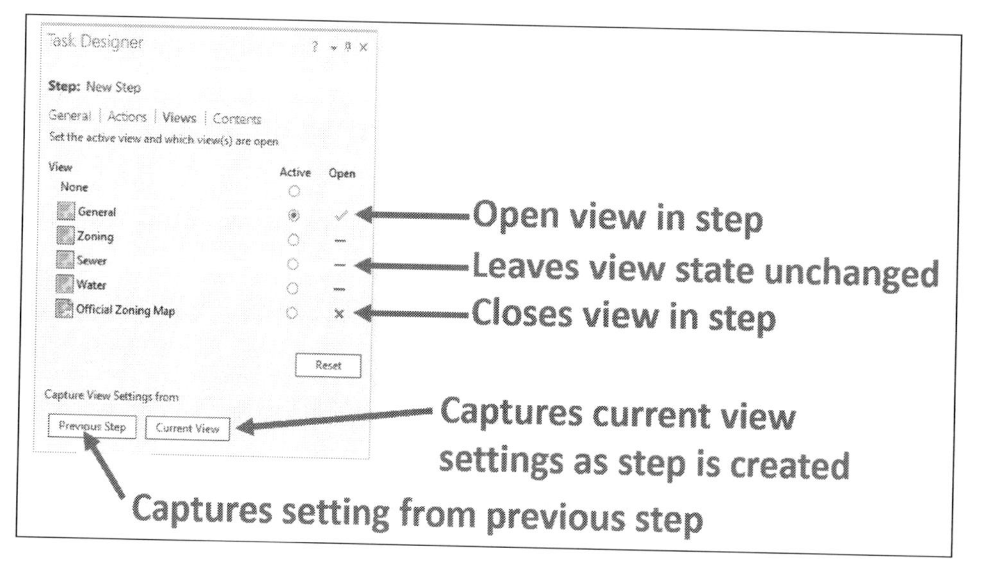

The active view will be the one the step and defined actions are applied to. So, if you are creating a step to select a zoning polygon, you might want to make sure that the **Zoning** view is active. If your step has someone adding a north arrow to a layout, you would want the proper layout to be open and active. You may want to close some views to remove possible confusion by the user and to reduce the amount of computer resources, which might be used.

As you can see in the illustration, you can set these parameters manually for the step, capture them from the previous step, or based on your project's current settings as you create the step.

Contents

Just as you can control which views are open and active, you can also control the layers within your active view. Within the step, you can control a layer's visibility, whether it is selectable or not, editable or not, snappable or not, is a selected layer and label visibility.

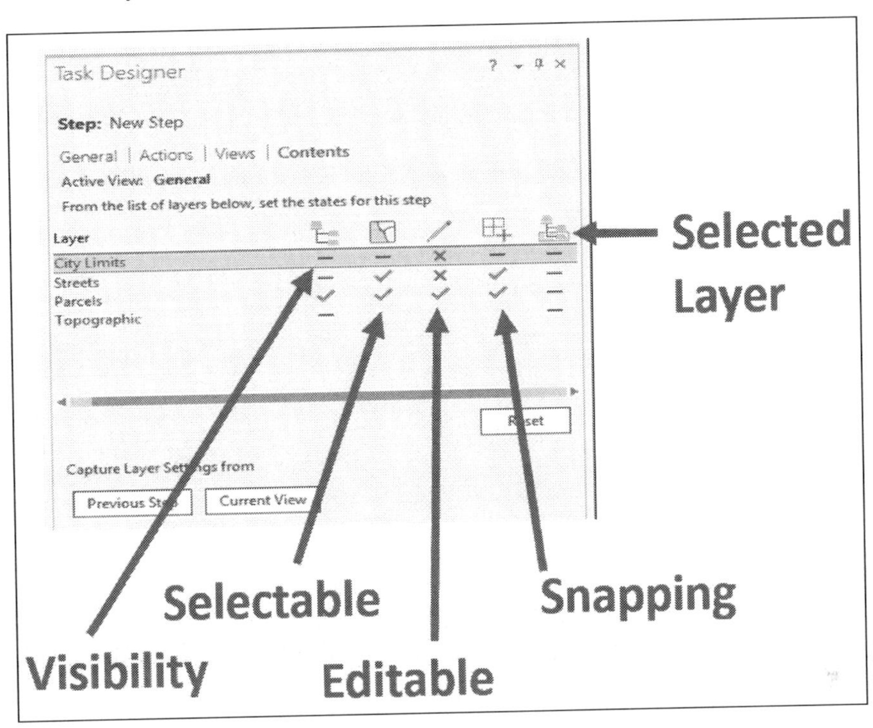

The level of layer control is very powerful. It ensures that you protect data and only work with the layers you need to in any given step. For example, if you are editing a parcel within a step, you would want to make sure that the parcels layer is set as the only editable layer. This will ensure that the user doesn't accidently edit the city limits or a street by accident.

Now that you have a much greater understanding of how steps are configured and work within a task, you are now ready to begin adding steps to the task you created in the previous exercise.

Exercise 9D – adding steps to a task

In this exercise, you will add the steps to the previous task you created to walk users through the process of generating a list of parcels near a selected road segment, which is to be repaired so that the owners can be notified.

Step 1 – open task in the Task Designer

In this step, you will open the task you created in the **Task Designer**, so you can begin creating your steps:

1. Start ArcGIS Pro and open `Ex 9` using the skills you have already learned.
2. Expand the **Task** folder in the **Project** pane.
3. Double-click on the **Road Repair Tasks** item you created in the last exercise. This will open the **Tasks** pane.
4. Right-click on **Create list** of the nearby parcels task and select **Edit in Designer**. The **Task Designer** pane should open on the right-hand side of the interface, and the **Tasks** pane should enter **Designer** mode.

You are now ready to begin adding steps. If this was not a classroom exercise, you would begin referencing your notes from the answers to those four questions you need to consider before creating a task. You would pay close attention to the workflow you outlined for question 2.

For the sake of the exercise, this has already been done, and you will just need to follow the exercise instructions.

Step 2 – adding a step to zoom

In this step, you will add a step that will instruct the user to zoom to the location of the road in question. You will have them use the **Explore** tool to zoom to the location.

1. Click on the **New Step** button in the **Tasks** pane. The new step is automatically listed in the **Tasks** pane, and the parameters for the step are displayed in the **Task Designer** pane.
2. Set the **General** parameters as follows:
    - **Name**: Zoom to street using **Explore** tool
    - **Tooltip**: Zoom to street to be repaired

  - **Instructions**: Using the **Explore** tool located on the **MAP** tab, zoom to the location of the street segment, which will be under repair in the map. If you are already zoomed to the correct area, you may skip this step.
  - **Run/Proceed Instructions**: Once you have successfully zoomed to the location of the street, click on **Next Step** to proceed.

I personally like to make the name of any tools or buttons I call out in the instructions in a bold font. This helps the user spot them as important. So, in the instructions and *Run/Proceed Instructions* mentioned previously, I would have made **Explore** and **Next Step** a bold font.

3. In the **Step Behavior**, set this as an **Auto Run** step. You will be connecting the **Explore** tool as an action to this step. By making it an **Auto Run**, the step will automatically enable the **Explore** tool without the user having to. All the user will need to do is begin zooming to the proper area within the map.
4. Finally, for the general settings, set this as an optional step. This will allow the user to skip the step if they already are zoomed into the proper location for the street that will be repaired when they start the task.

   You have configured the general parameters for this step. You will now link it to an action. The action will be the **Explore** tool.

5. Click on **Actions** located at the top of the **Task Designer** pane.
6. Move your mouse pointer to where it says **<none>**. Three icons should appear to the right. Click on the **Edit** icon.
7. Click on the drop-down arrow located to right of **<none>** in the **Command/Geoprocessing** pane.
8. Select **Command** from the drop-down list.
9. Click on the **Browser** button that appears to the right of **Selected Command**.

10. In the search area of the pop-up window, type `Explore` and select **Explore (Open current explore tool)** and click on **OK**.

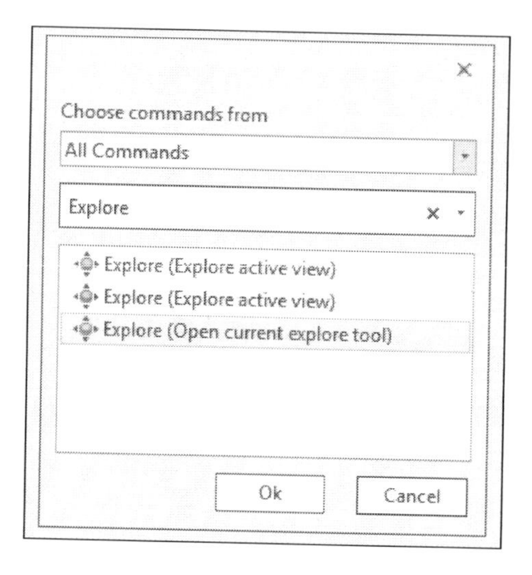

11. Click on **Done** located at the bottom of the **Task Designer** pane.

You do not need to configure additional actions or change the settings for views or contents in this step since you are just having the user zoom to a location. So, you are done creating this step. Remember that there is no save task or step button. They are saved when you save the project.

12. Save your project.

Congratulations! You have created your first step. You are not done yet. You have more steps that need to be created to complete this workflow.

Step 3 – selecting the street segment

Now you need to add a step that instructs the user to select the road segment, which will be repaired. You will use the **Select by Rectangle** tool and instruct the user to select the road from the map.

1. Click on the **New Step** button in the **Tasks** pane to create another new step.
2. Compete the general parameters as follows:
    - **Name**: Select **Road Segment**

  - **Tooltip**: Select road segment which will be repaired
  - **Instruction**: Click on the **Run** button to start this step. Then using the **Select by Rectangle** tool, select the road segment from the map which is scheduled to be repaired. If you have already selected the road segment, you may skip this step.
  - **Run/Proceed Instructions**: Once you have selected the road segment that is scheduled for repair, click on the **Next Step** button.
  - **Step Behavior**: Manual
  - Enable **Optional** to allow the user to skip if they have already selected a road segment.

3. Click on the **Actions** option at the top of the **Task Designer** pane.
4. Move your mouse pointer to the area that says **<none>** and click on the **Record** icon.
5. Click on the **Select** tool located in the **MAP** tab. Make sure that it is the **Select by Rectangle**. Note that this tool is automatically added as the action.

Since you will have the user select a feature on a specific layer in this step, you want to limit the selectable layers available when they run this step. As you learned earlier, you can do this through the **Contents** settings in a step.

1. Click on **Contents** located on the top of the **Task Designer** pane.
2. When this task is run, you want to make sure that the Streets layer is visible and selectable. Click on the - locate next to **Streets under the Visibility** column until it becomes a green check mark. Do the same for the **Select** column.
3. To ensure that no other layers are accidentally selected, you want to set the remaining layers to none selectable by clicking on the - in the **Select** column until they all have a red **X**.

Your **Task Designer** pane should look like this when you are done:

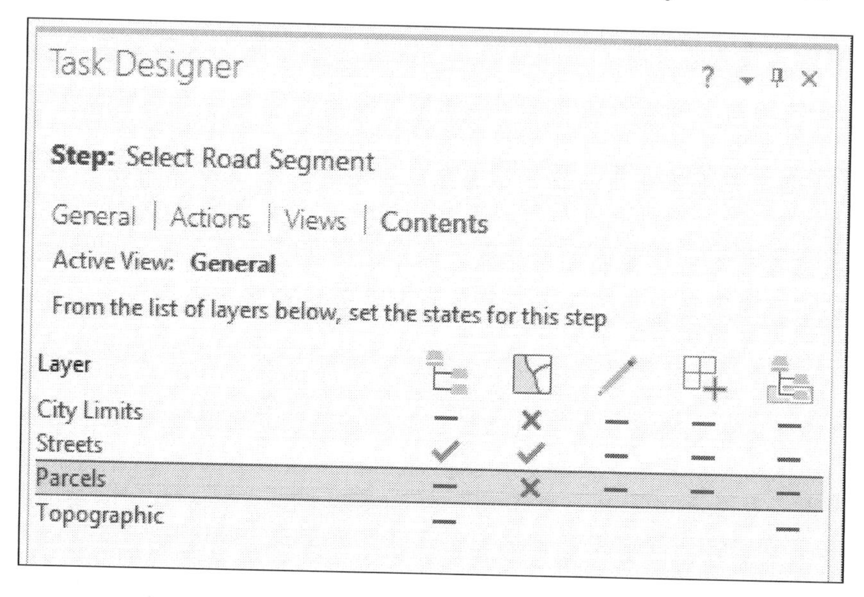

4. Save your project.

You have now provided users with the instructions they need to zoom to the location of the road scheduled for repair and how to select it. Now you need to tell them how to select the surrounding parcels.

Step 4 – selecting nearby parcels

In this step, you will create a step that tells users how to select the parcels that are near the road segment they selected in the last step. You will be using the **Select By Location** command to do this:

1. Click on the **New Step** button once again to create the new step.
2. Set the general parameters as follows:
   - **Name**: Select nearby parcels
   - **Tooltip**: Select parcels that are near the road segment you selected in the previous step
   - **Instructions**: Using the **Select by Location** tool, you will select parcels that are located within a distance of 100 feet from the selected road segment.

Ensure that this tool is configured as follows:

- Input Feature Layer = `Parcels`
- Relationship = `Within a Distance`
- Selecting Layer = `Streets`
- Selection Distance = `100 feet`
- Selection Type = `New selection`

Once you have verified that the settings are correct, click on the **Run** button

- **Run/Proceed Instructions**: Click **Next Step** to proceed.
- **Step Behavior**: **Manual**

3. Click on **Actions**, and using the **Record** option, set the command to **Select by Location** located in the **MAP** tab.
4. The **Geoprocessing** pane will open automatically. Since you are running this step manually, you can close it and return to the **Task Designer** pane.
5. Click on **Contents**. Set it so the Streets and Parcels will be visible and the Parcels will be the only selectable layer.
6. Save your project.

Now that the nearby parcels are selected, you have one last step to go. You will instruct users how to export the selected parcels to an Excel spreadsheet.

Step 5 – exporting selections to Excel spreadsheet

In this step, you will create a new step, which instructs users how to export the selected parcels to an Excel spreadsheet:

1. Click on the **New Step** button in the **Tasks** pane.
2. Set the general parameters as follows:
    - **Name**: Export to Excel
    - **Tooltip**: Exports selected parcels to an Excel spreadsheet
    - **Instructions**: Complete the parameters for the Table to Excel geoprocessing tool as indicated here:
        - **Input Table**: `Parcels`
        - **Output Table**: `C:\Student\IntroArcPro\Chapter9\Parcels_TableToExcel.xls`

- Click on **Run** once you have verified your settings.
- **Run/Proceed Instructions:** Click on **Finish** to complete the task.
- **Step Behavior**: Run manually

3. Click on **Actions** and click on the **Edit** icon.
4. Set the **Type of Command** to **Geoprocessing Tool**.
5. Click on the **Browse** button next to the **Selected Geoprocessing Tool**.
6. In the **Find** tools cell type `Excel`.
7. Select the **Table to Excel** Python script and click on **OK**.
8. Ensure **Embed** is enabled and set the parameters as described in your instructions. The **Task Designer** pane should look as follows:

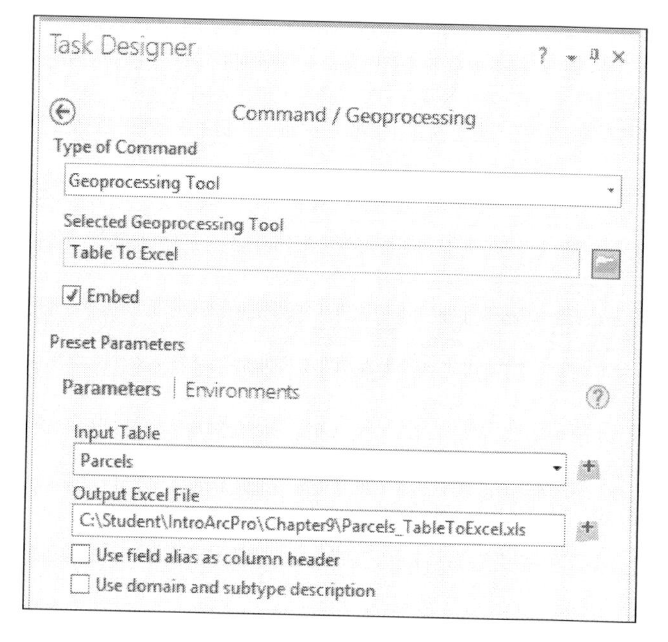

9. Once you have verified your settings, click on **Done**.
10. There is no need to change the views or contents settings, so close the **Task Designer** and save your project.

Your task is now complete. You have added all the steps required to select the parcels near a road segment that is scheduled for repair and then export them to an Excel spreadsheet. Now you need to test it.

Step 6 – run your task

Now that you task is complete, you will run it to see how well it works. This will verify that the task contains all the steps needed to complete the desired process and to verify your instructions are clear and the steps flow as expected:

1. Using the skills you have learned to locate and run the task you just created. Follow the instructions in each step closely. See if they make sense and you end up with the expected results.
2. Make any adjustments to the task and its steps you believe are needed.
3. Save your project and close ArcGIS Pro.

Assuming everything ran as expected, you have just created your first complete task. You developed a workflow, so others will be able follow to accomplish a specific process using your step-by-step instructions.

Challenge

Using the skills you have learned in this chapter, select a previous exercise from this book or a common process from your office and develop a task or tasks to replicate that workflow.

Summary

In this chapter, you learned how tasks can be used to standardize processes, train new users, and establish best practices within your organization. You also gained an understanding of all the components that make up a task and considerations you need to take into account when creating your own tasks.

Finally, you worked through the process of creating a new task from the beginning for example, creating a task that exports the selected data to an Excel spreadsheet.

10

Automating Processes with ModelBuilder and Python

As you have now learned, performing analysis or editing a feature requires many steps. The more you use ArcGIS Pro, the more you will find yourself doing the same process again and again. You may also realize that some of the processes you do repeatedly really require very little interaction on your part beyond selecting a feature and then telling ArcGIS where to save the outputs.

Would it not be extremely beneficial if you could automate processes you do repeatedly? Create the proverbial **Easy button** where you simply click on a single tool, fill in a few parameters, and off the tool goes providing you with the results when it was done. That would certainly make your job easier.

With ArcGIS Pro, you can create Easy buttons or tools using **ModelBuilder** and **Python** scripts. Both of these allow you to create automated processes that can run multiple tools together in sequence or at the same time to complete an operation. ModelBuilder uses a visual interface to create automation models without the need to be a programmer. Python is the primary scripting language for the ArcGIS platform. With it you can create very powerful scripts that cannot only be used within ArcGIS Pro but also integrate processes across all components of ArcGIS including Server, Online, Extensions, and more. Creating Python scripts does require writing code.

In this chapter, you will learn:

- Differences between tasks, geoprocessing models, and Python scripts
- How to create and run a model
- Components of a model
- How to export a model to a Python script
- Basics of the Python language

Tasks, geoprocessing models, and Python scripts – what is the difference?

Having just learned about tasks in the last chapter and reading the introduction for this chapter, you may be wondering what the difference is between a task, a geoprocessing model, and a Python script? That is a great question.

To answer it, you must first understand what each one is. You already know what a task is, so we will now focus on gaining a better understanding of what a geoprocessing model and Python script are. Once you understand that, you can then understand the differences between the three.

Geoprocessing model

A **geoprocessing model** is a custom tool created within the ModelBuilder window that contains multiple geoprocessing tools along with their various parameters (inputs, outputs, options, and other values) which work together as part of an integrated process that will run as if it is a single tool.

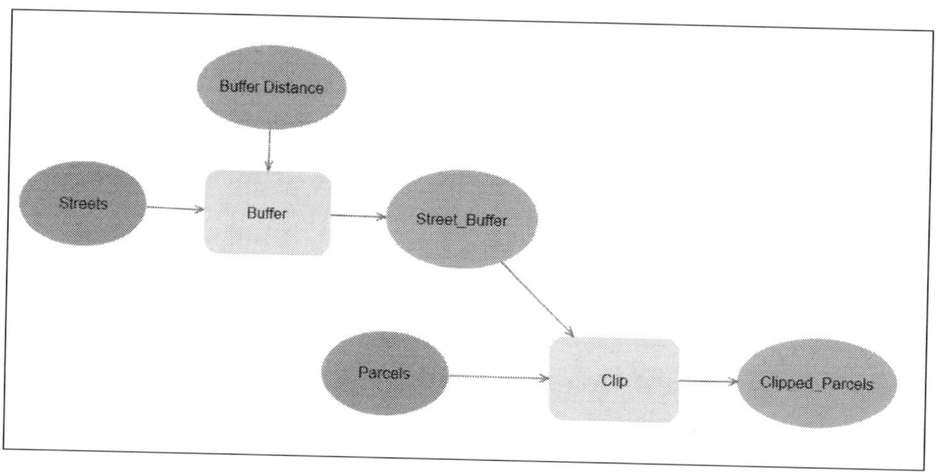

The preceding illustration is a sample of a very simple model. It contains two geoprocessing tools that you learned about in *Chapter 8, Geoprocessing*. In the model, the **Buffer** tool creates buffered polygons around the **Streets** input. The resulting buffer polygons are then used to clip the parcels that are located inside the street buffer polygons. Since both of these geoprocessing tools reside inside the model, the user only has to run the model instead of having to run each tool individually. The model automatically runs the tools based on the parameters specified within the model. You will learn more about the components of a geoprocessing model and how to create one later in this chapter.

Geoprocessing models can include geoprocessing tools, Python scripts, iterators, model-only tools, and other models. This allows them to be as simple or as complex as you need them to be to accomplish the process you have designed them to complete. The ModelBuilder window allows users to create geoprocessing models in a visual environment. No coding is required to build a model. Esri actually refers to ModelBuilder as a visual programming language.

Geoprocessing models can be created using ArcGIS for Desktop (**ArcMap** or **ArcCatalog**) or ArcGIS Pro. One downside to a geoprocessing model is that it can only be run from ArcGIS Pro or ArcGIS for Desktop. You cannot schedule them to run automatically at a specific date and time. At least not by themselves!

Python script

A Python script is also a custom tool that can run multiple geoprocessing tools along with their various parameters as part of an integrated process. However, unlike a model that does not require you to write programming code, Python scripts do. You must know the Python scripting language in order to create Python scripts. The following is a small snippet of a Python script created for ArcGIS:

```
#-------------------------------------------------------------------
# Name:         Union Tool Sample Script
# Purpose:      Runs the Union Geoprocessing tool from ArcGIS
# Author:       Esri & Tripp Corbin
#
# Created:      09/15/2015
# Updated:      09/15/2015
#
# Usage: Union two feature classes
#-------------------------------------------------------------------

# Import the system modules
import arcpy
from arcpy import env

# Sets the current workspace to avoid having to specify the full path
# to the feature classes each time

env.workspace = "C:\\student\\IntroArcPro\\Databases\\Trippville_GIS.
gdb"
#Runs Union Geoprocessing tool on 2 Feature classes
arcpy.Union_analysis (["Parcels", "Floodplains"], "Parcels_Floodplain_
Union", "NO_FID", 0.0003)
```

Python scripts have several advantages over a geoprocessing model. First, Python is not limited to ArcGIS. Python can actually be used to create scripts for many other applications such as Excel, SharePoint, AutoCAD, Photoshop, SQL Server, and more. This means you can use a Python script to run tools across multiple platforms to create a truly integrated process.

Second, Python scripts can be run from outside of ArcGIS. This means you can schedule them to run at specific times and days using your operating system's scheduler application. If your script does include ArcGIS geoprocessing tools, the script will require access to an ArcGIS license to run successfully but ArcGIS does not need to be open and active at the time the script is scheduled to run.

Third, Python can be used to create completely customized geoprocessing tools. It is not limited to just the geoprocessing tools you will find in the ArcGIS Pro toolboxes.

What is the difference?

Now that you have a much better understanding of what a task, geoprocessing model, and Python script are, you will be able to better understand the differences between them. Each can serve a purpose in standardizing and automating common workflows and processes.

The following grid will provide a clearer understanding of the differences between the three:

	Task	**Geoprocessing model**	**Python script**
Run a single geoprocessing tool automatically	Yes – can run a single tool automatically as part of a step	Yes	Yes
Allow users to provide input to tools before running	Yes	Yes	Yes
Run multiple geoprocessing tasks automatically and in sequence	No	Yes	Yes
Be included in a task	No	Yes	Yes
Be included in a geoprocessing model	No	Yes	Yes
Provided a documented workflow	Yes	Yes	No
Be run from outside of ArcGIS Pro	No	No	Yes
Integrate with other applications	No	No	Yes

	Task	Geoprocessing model	Python script
Be scheduled to run at specific times and days	No	No	Yes
Requires knowledge of programming language	No	No	Yes

Creating geoprocessing models

As mentioned earlier, geoprocessing models are custom tools you create from within ModelBuilder. ModelBuilder provides the graphical interface for building models as well as accessing additional model-only tools, iterators, environmental settings, and model properties.

Models are created for a number of reasons. The first and most common is to automate repetitive processes performed in ArcGIS Pro. If you have an analysis, conversion, or other process you perform on a regular basis, a model can be used as a way to automate it.

Secondly, you can use a model as a way to think through and flowchart a process within ArcGIS Pro. This can help you ensure you have considered all the tools and data you will need to complete a process. Once completed, the model then provides the tool for completing that process as well as the visual and textual documentation that explains how the process was performed.

You can share models with those in your organization so they can use it to perform the process. This can reduce your workload and allow you to concentrate on other tasks that require greater knowledge and skill levels. Because a model runs the geoprocessing tools contained within it automatically, you can create a model that is easy for other less GIS-savvy members of your organization to run themselves without needing a full understanding of ArcGIS Pro. This also helps standardize our methodologies, ensuring everything is done in a consistent and approved manner.

All of this helps to save us time and money through increased efficiency, which is ultimately the main power of ModelBuilder. Like tasks, models consist of multiple components and have their own terminology associated with them.

Model components and terminology

Before you can create a model, you need to understand the pieces and parts that make them up. Models include a series of connected processes. Each process includes a tool, which can be a geoprocessing tool, another model, or a Python script. Each tool has variables that serve as inputs or outputs.

The following figure is an example of a model with two processes:

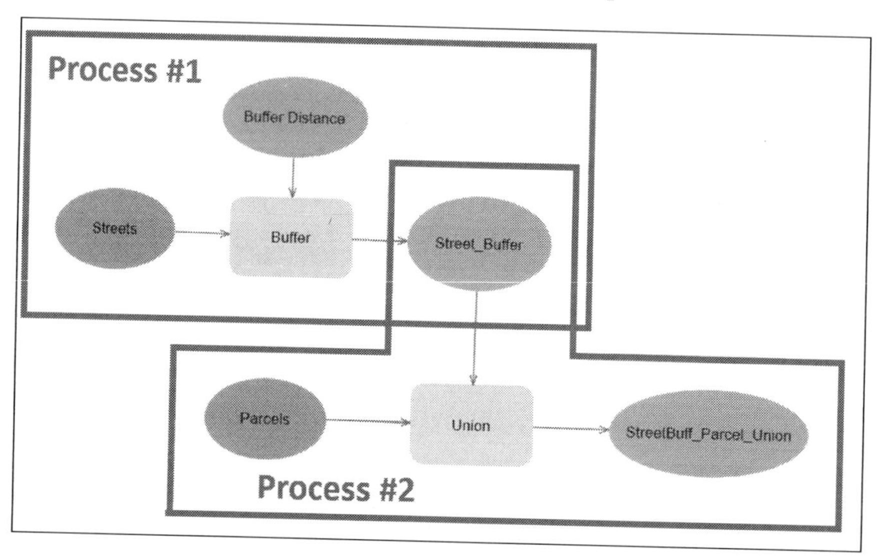

As you can see in the preceding figure, the model contains two processes built around the **Buffer** and **Union** tools. Each of these tools has a number of variables feeding into them. Variables are identified by the blue and green ovals. Notice that the two processes are sharing a variable, `Street_Buffer`. That variable is an output of the **Buffer** tool but an input for the **Union** tool.

There are three basic types of variables that are included in a model. They are data, value, and derived variables. Data variables are existing data that is used as an input to a tool, script, or model. These can be layers in a map, standalone tables, text files, feature classes, Shapefiles, and more.

Value variables are additional information that a tool may use to run. In the case of the **Buffer** tool, the distance used to create the buffer is considered a value variable as are the options to dissolve, end type, and the other parameters found that appear when you run the **Buffer** tool normally in the **Geoprocessing** pane.

Derived variables are the outputs of a process. Again, this can be a new layer, feature class, table, raster, or more, depending on the tool used in the process.

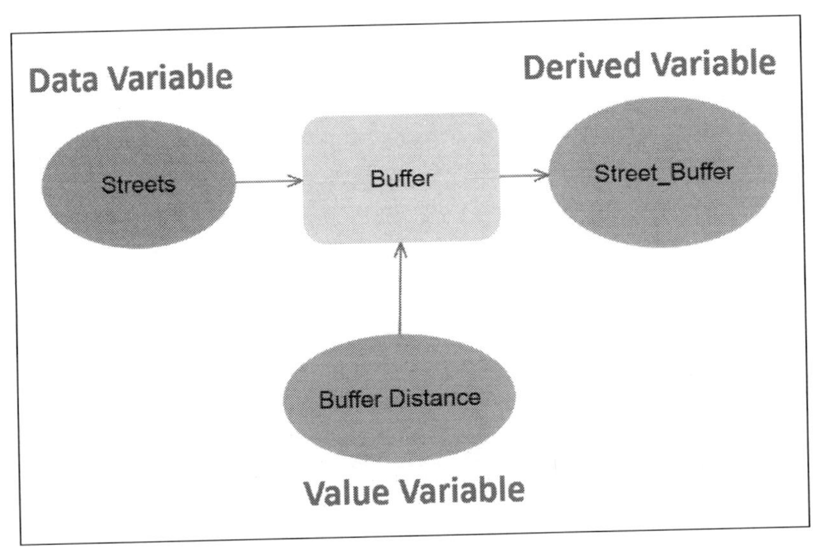

Since ModelBuilder is a visual programming language, you can distinguish variable types based on their colors. Although you can adjust these settings by default, data variables are a darker blue. Value variables are a lighter blue and derived variables are green.

Saving a model

If you wish to save a model you create so you can use it again or share it with others, you must save it in a custom toolbox that you create. Models cannot be saved in a system toolbox that is automatically included with ArcGIS Pro when it is installed.

When you create a new project, ArcGIS Pro automatically creates a custom toolbox for that project. It is stored in the project folder as a **TBX** file. This provides you with an easy to use place to store your models. This toolbox is also automatically linked to your project and accessible in the **Project** pane as shown in the following screenshot:

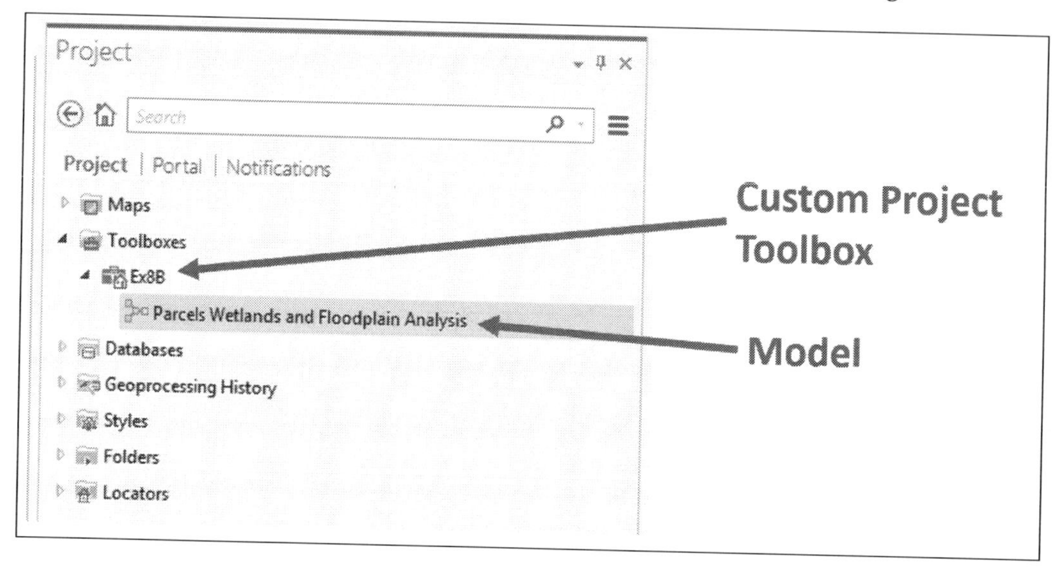

You can also create custom toolboxes within a geodatabase so that they, along with your models and Python scripts, are stored with your GIS data as illustrated next. This is a good option if the models or tools that you save to the toolbox will be used in multiple ArcGIS Pro projects.

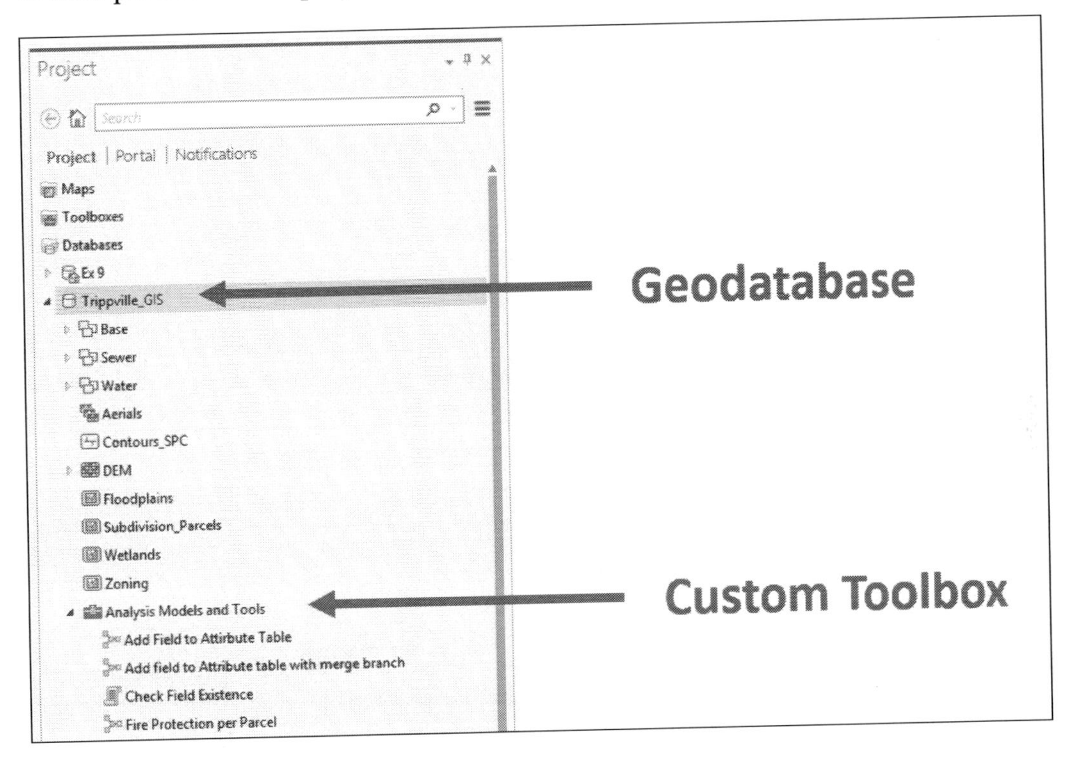

You can also create other custom TBX files besides the one that is automatically created with a new project. Using a custom TBX file is perfect if the tool and models you save to it will not only be used with multiple projects but also across multiple databases or in the case of consultants, multiple clients.

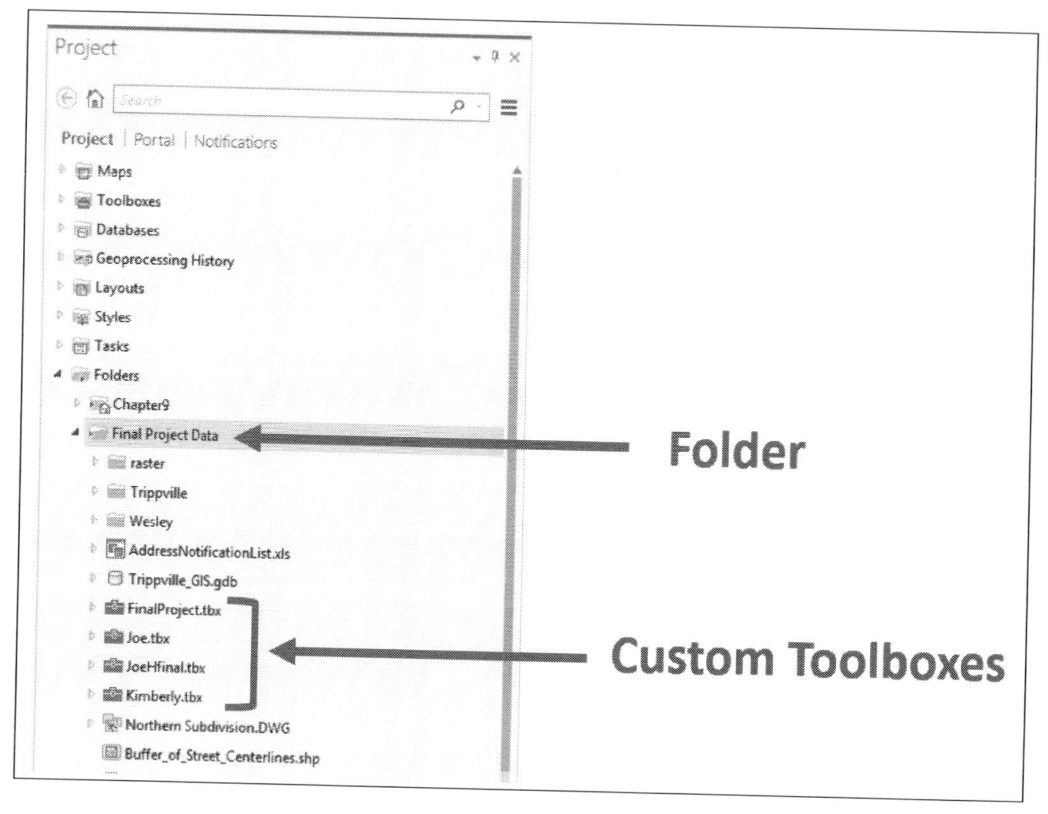

Using custom TBX files to store models also makes it easier to share them with others since they are smaller than a geodatabase, which also includes all the GIS data. TBX files can easily be e-mailed, uploaded to FTP sites, or placed in your ArcGIS Online account.

Now that you have a good general understanding of a model, its components, and how to save one, it is time to put that knowledge to use.

Exercise 10A – creating a model

A new ordinance was just passed to protect the streams in the Trippville area. It requires all new building or improvement projects to take place at least 150 feet from the centerline of all creeks or streams. This will hopefully protect the banks from erosion and reduce polluted runoff from reaching them.

The Community and Economic Development Director has asked you to calculate the total area of each parcel that is inside the non-disturb buffer and how much of each parcel is outside. Since you will need to update this analysis anytime a new subdivision or commercial development is added, you have decided to create a model that you can run every time you need to perform these calculations.

In this exercise, you will create a simple model that will calculate how much of each parcel is inside and outside a non-disturb buffer area around the streams. This model will include a couple of geoprocessing tools and their associated variables.

Step 1 – open the project and the ModelBuilder window

The first step is to open the project and then the ModelBuilder window so you can begin creating the model performing the following steps:

1. Start ArcGIS Pro and open the `Ex10.aprx` project found in `C:\Student\IntroArcPro\Chapter10`.
2. When the project opens, expand the **Toolboxes** folder in the **Project** pane.
3. Right-click on the `Ex10` toolbox you see.
4. Select **New** and then **Model**.

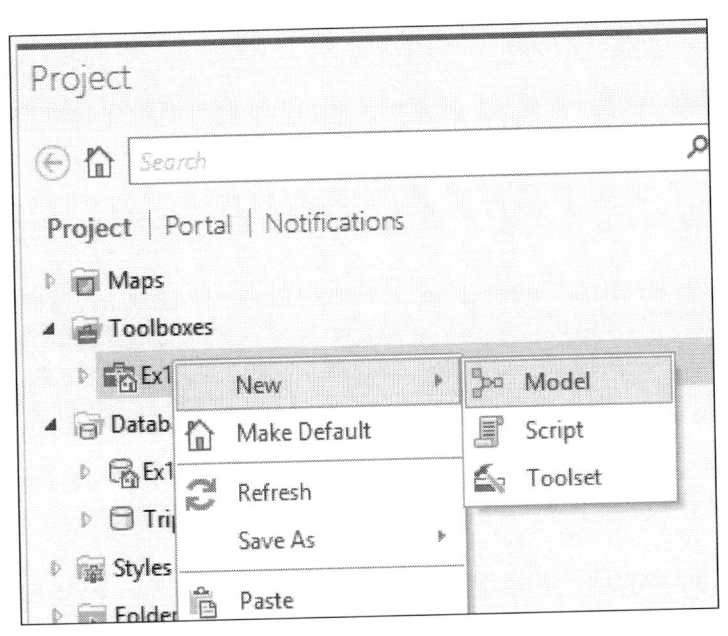

ModelBuilder should now be open and the **MODELBUILDER** tab has appeared in the ribbon. The ModelBuilder window and tab are used together to create or edit models. The tab contains tools for saving the model, navigating in the ModelBuilder window, and adding content to the model.

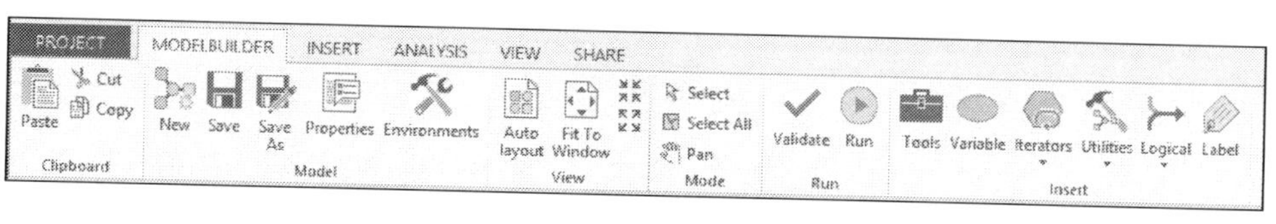

You will now begin using these tools to create your model.

Step 2 – adding model components

In this step, you will begin adding tools and variables to your model. You will explore some of the different methods that can be used. You will start with adding the process that will generate the non-disturb buffers around the streams.

1. Click on the **Tools** button on the **Insert** group on the **MODELBUILDER** tab in the ribbon. This opens the **Geoprocessing** pane on the right side of the interface.
2. Click on **Toolboxes** at the top of the pane to expose the various toolboxes within ArcGIS Pro. These are your system toolboxes.
3. Expand the **Analysis** toolbox and then the **Proximity** toolset.
4. Drag and drop the **Buffer** tool from the **Toolbox** into the ModelBuilder window so that it looks like this:

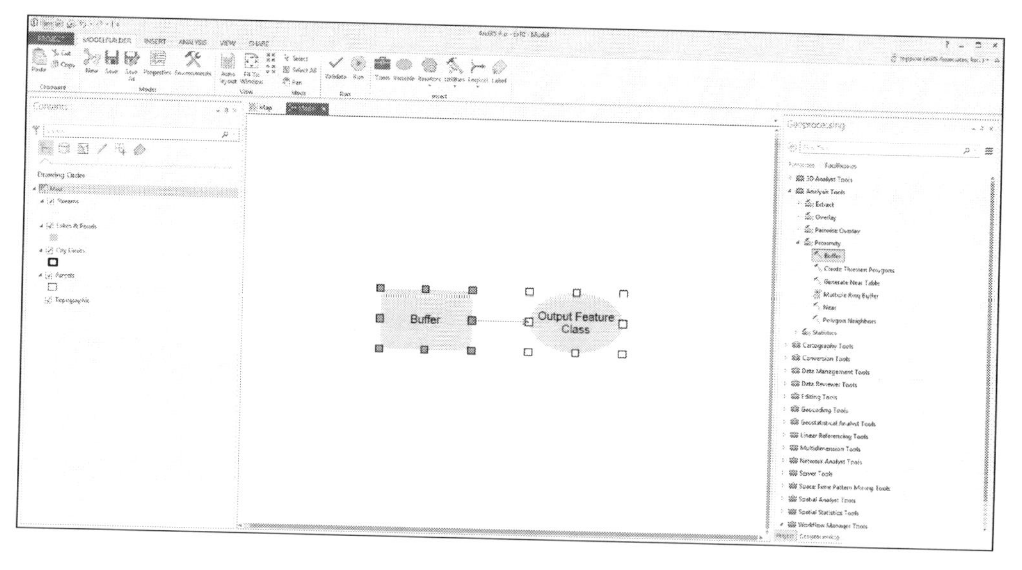

You have just added your first process to a model. Model processes will exist in one of three states: not ready to run, ready to run, and have been run. The process you just added is in the not ready to run state. ArcGIS Pro indicates that visually by displaying the tools and variables in gray. A process will be in the not ready to run state until all required variables have been defined. In the case of the **Buffer** tool, you have not yet defined the three variables required: input feature class, buffer distance, and output feature class. You will now do that:

5. The input feature class for the **Buffer** tool used in the model will be the **Streams** layer in your map. So you will now add that layer as a variable to the model. Select the **Streams** layer and drag it into the ModelBuilder window. It will be added as a blue oval.
6. Click somewhere within the blank space in ModelBuilder to deselect the **Streams** variable.
7. You now need to connect the **Streams** variable you just added to the **Buffer** tool. Click on the **Streams** variable and with your left mouse button still depressed move your mouse pointer until it is over the **Buffer** tool. Then release your mouse button.
8. A small popup menu should appear; select **Input Features**. You have just defined the input feature class for the **Buffer** tool.
9. Now, double-click on the **Buffer** tool in the ModelBuilder window. This will open the tools dialog window in **MODELBUILDER** so you can define additional variables.
10. In the distance field, type `150` and verify the units are **feet**.
11. The output should automatically be set to `Natwtr_Stream_Buffer`, which is being saved to `C:\Student\IntroArcPro\Chapter10\Ex10.gdb\`. Verify this is true by hovering your mouse over the output feature class name. This will be fine for this exercise so you will leave it as is without changing it.

12. Since the Director has not indicated that any of the stream attributes are important for the calculations, you will have the resulting buffers dissolve. Click on the drop-down arrow under **Dissolve Type** and select **Dissolve all output features into a single feature**.

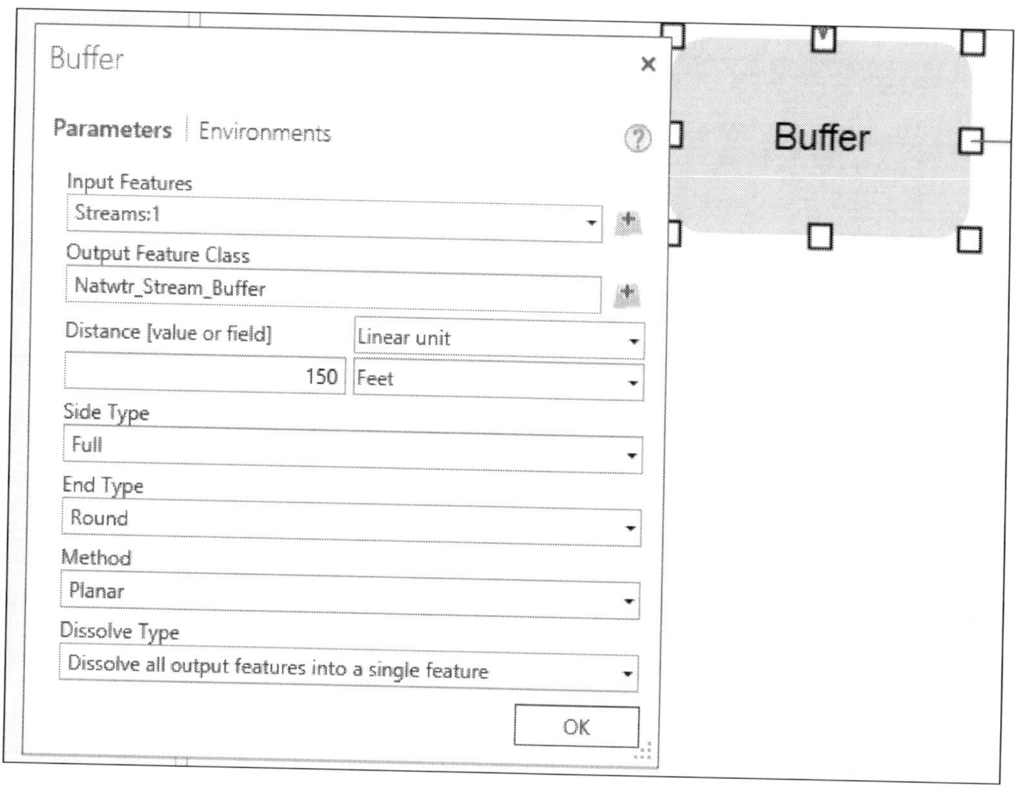

The **Buffer** tool window should now look similar to this. Depending on what you may have done previously, the name of the input feature may be slightly different:

13. Click **OK** once you have verified your settings.
14. Click on the **Auto Layout** button on the **MODELBUILDER** tab.

Your model should now include a single completed process that is in the ready to run state. You can tell it is ready to run because the tool and all lined variables are now displayed with a colored fill that is not gray.

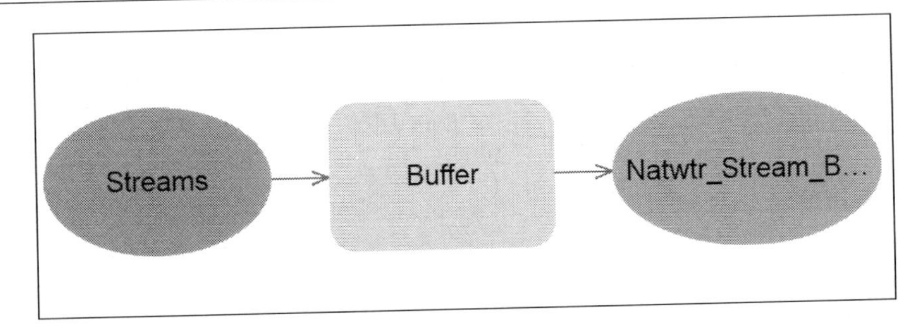

Now, let's save your model to ensure your hard work is not lost in case something happens.

15. Click on the **Properties** button in the **Model** group on the **MODELBUILDER** tab.
16. Complete the properties as indicated here:
    - **Name** = `ParcelsStreamProtectionBuffer`
    - **Label** = `Parcels Stream Protection Buffer Analysis`
    - Leave all other properties with default settings
17. Click **OK**.
18. Click the **Save** button located in the **Model** group on the **MODELBUILDER** tab to save the model. If you still have the **Project** pane open, you should see the name of the model change from **Model** to the **Label** you just entered.

The process you just created in the model will generate the buffer areas around the streams. Now you need to add another process that will calculate how much of each parcel is in and outside that buffer area. You will use the **Union** tool to union the parcels with the newly created stream buffer. This will create a new feature class that will split each parcel where it is overlapped by the stream buffer, thus allowing you to determine how much is in and outside the buffer.

Step 3 – adding another process

In this step, you will add another process to your model. This process will include the **Union** tool. You will then link this new process to the one you created in the last step:

1. Click on the **Tools** button in the **MODELBUILDER** tab again.
2. Expand the **Overlay** toolset in the **Analysis** toolbox.
3. Add the **Union** tool to your model using the same method you used to add the **Buffer** tool.

4. Using your scroll wheel, zoom out in the ModelBuilder window until you have room to move the **Union** tool so it and the **Buffer** tool will be visible.
5. With the **Union** tool and the **Output Feature Class** variable selected, use your mouse to move them so that they are located below the **Buffer** tool as shown next:

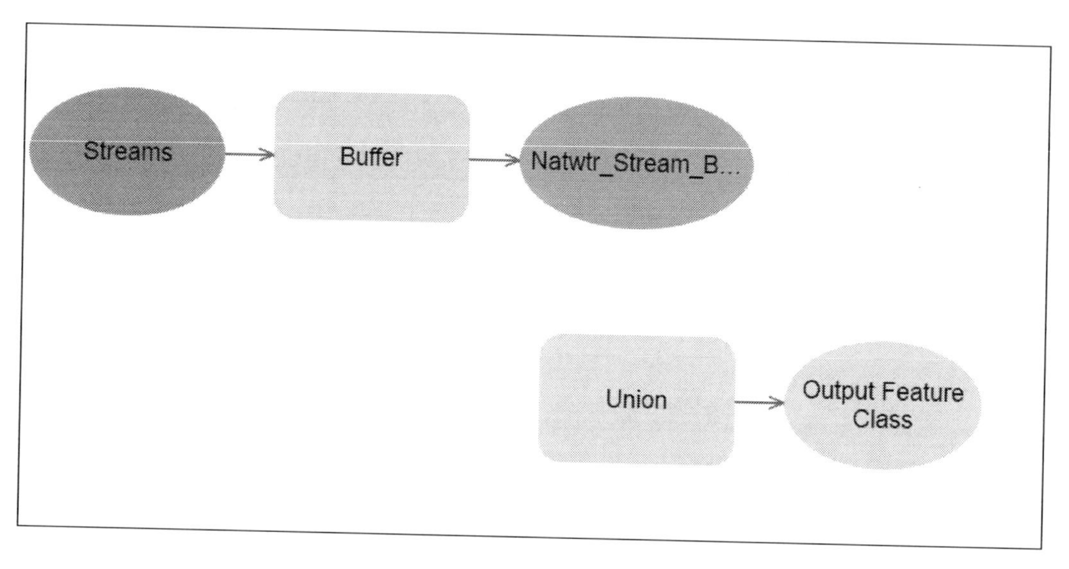

You have now added the **Union** tool to the model. Now you need to link it to the output from the **Buffer** tool and define the rest of its required variables:

6. Double-click on the **Union** tool to open the tool window.
7. In the **Union** tool window, click on the small arrow next to **Input Features**.
8. Check the **Parcels** layer and the `Natwtr_Stream_Buffer` model variable. Click **Add** as shown in the following screenshot:

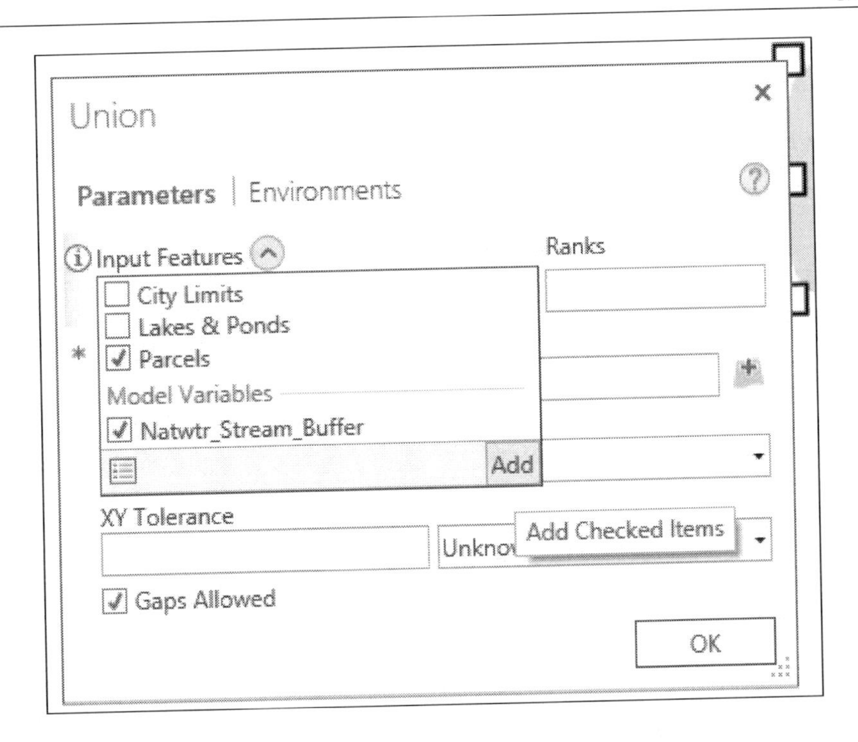

9. Make the output `C:\Student\IntroArcPro\Chapter10\Ex10.gdb\Parcels_StreamBuff_Union`.
10. Click **OK**.

Your entire model should now be in the ready to run state and look similar to this:

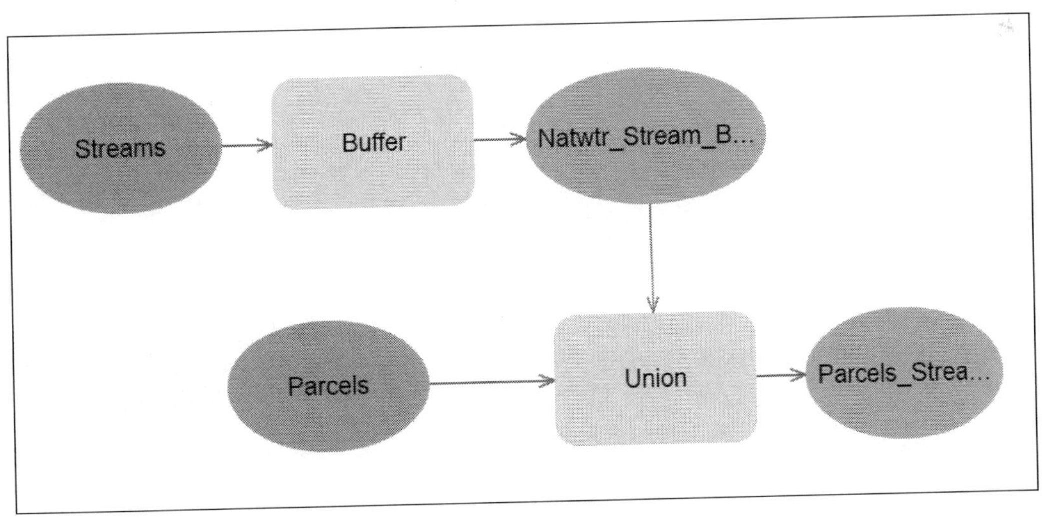

Your layout may be different. That is acceptable as long as the proper connections are made and the processes are in the ready to run state.

11. Save your model and your project.
12. You may close ArcGIS Pro or leave it open if you plan to continue.

Running a model

After creating a model, of course you will want to run it. There are many ways to run a model. You can run the entire model, those processes in a ready to run state, or just a single process in the model.

If you wish to run the entire model, the easiest way to do that is to simply double-click on it from the toolbox it is stored in. This will run all processes in the model that are in the ready to run or have been run states. If you have allowed the users to provide values for some of the variables within the model, they will be prompted to enter those before the model is run. Otherwise, if you did not allow for user input, the model will just indicate that there are no parameters in the geoprocessing window and all you need to do is click the **Run** button. You will learn how to make a model interactive a little later in this chapter.

You can also choose to run the model or processes in the model from the ModelBuilder window. Clicking on the **Run** button in the **MODELBUILDER** tab will run all ready to run processes within the model. It will not run processes that are in the have been run or not ready to run states. This allows you to build and test a model as you go without being required to run the entire model.

Exercise 10B – running a model

Now that you know a little more about how to run a model, you will get to put that knowledge into practice. In this exercise, you will first run your model from within ModelBuilder. Then you will get to run it directly from the toolbox so you can see what your users will experience when they run the model.

Step 1 – running the model from ModelBuilder

In this step, you will run the model you created in the *Exercise 10A – creating a model* section from within ModelBuilder. You will also explore how to run individual processes so that you can test your model as you create it:

1. If you closed ArcGIS Pro after the last exercise, start ArcGIS Pro and open the `Ex10.aprx` project.
2. Expand the **Toolboxes** folder in the **Project** pane and then the `Ex10` toolbox.

3. Right-click on the model you created in *Exercise 10A – creating a model* and select **Edit** from the displayed context menu. This will open the ModelBuilder window.

[If you created your model successfully in the last exercise and saved it, all processes should be in the ready to run state. This is indicated by all tools and variables having a solid color fill applied. If any are filled with gray or empty then you need to go back to the *Exercise 10A – creating a model* section and work back through the exercise.]

4. Right-click on the **Buffer** tool in ModelBuilder. Select **Run** to run the **Buffer** tool with the connected variables you defined in the model. A small window will pop up inside ModelBuilder that displays the progress of the **Buffer** tool and will let you know when it is complete. When the tool is finished, notice what happens to the graphics for the **Buffer** tool and its associated variables.

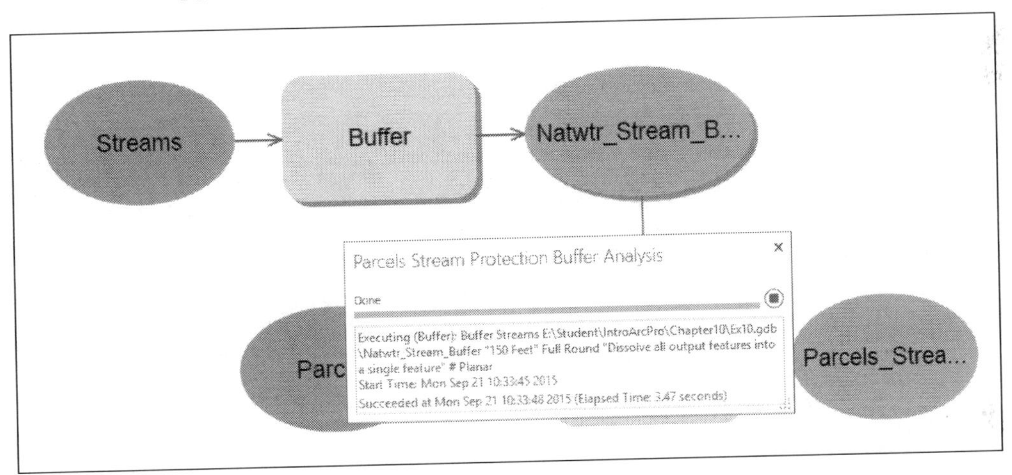

[**Question**: How have the graphics for the **Buffer** tool and its associated variables changed?

____

____]

The **Buffer** tool process is now in the *has been run state*. This means you have successfully run that process in the model. As you are beginning to learn, the state of a process will impact how it runs. Now that this process is in the has been run state, it will not run again if you click the **Run** button in the ribbon. The **Run** button will only run those processes that are in the ready to run state. Let's verify that though:

5. Close the popup window that appeared when you ran the **Buffer** tool.

6. Click the **Run** button in the **Run** group on the **MODELBUILDER** tab. Watch what happens to the model as it is run.

> **Question**: Did the model try to re-run the **Buffer** tool?
>
> ____
>
> ____
>
> **Question**: What tool or tools did the model run when you clicked on the **Run** button and why?
>
> ____
>
> ____

Step 2 – resetting the run state

In this step, you will learn how to reset the run state of all the have been run processes back to ready to run:

1. Click on the **Validate** button in the **Run** group on the **MODELBUILDER** tab in the ribbon.

> **Question**: What happens to all the processes in the model that were in the have been run state?
>
> ____
>
> ____

2. Click the **Run** button on the ribbon and watch how your model runs this time. All the processes will run this time because all of them are in the ready to run state.

Now you will actually verify that your model ran and created the feature classes it was supposed to in the project database.

3. Expand the **Databases** folder in the **Project** pane and then expand the `Ex10` geodatabase.

> **Question**: What do you see in this database?
>
> ____
>
> ____

4. Right-click on each feature class you see in the `Ex10` geodatabase and select **Delete** until the database is empty. This will allow you to verify the model runs properly when you run it directly from the toolbox in the next step.
5. Close the ModelBuilder window. If asked to save the model, do so.

Step 3 – running the model from a toolbox

In this step, you will now run the model directly from the toolbox. This will be how most users will access and run models you create. Running the model using this method will allow you to have the same experience your users will when they run the model:

1. Make sure the **Map** view is active.
2. In the **Project** pane, expand the **Toolboxes** folder and the `Ex10` toolbox.
3. Double-click on the `Parcels Streams Protection Buffer Analysis` model you created.

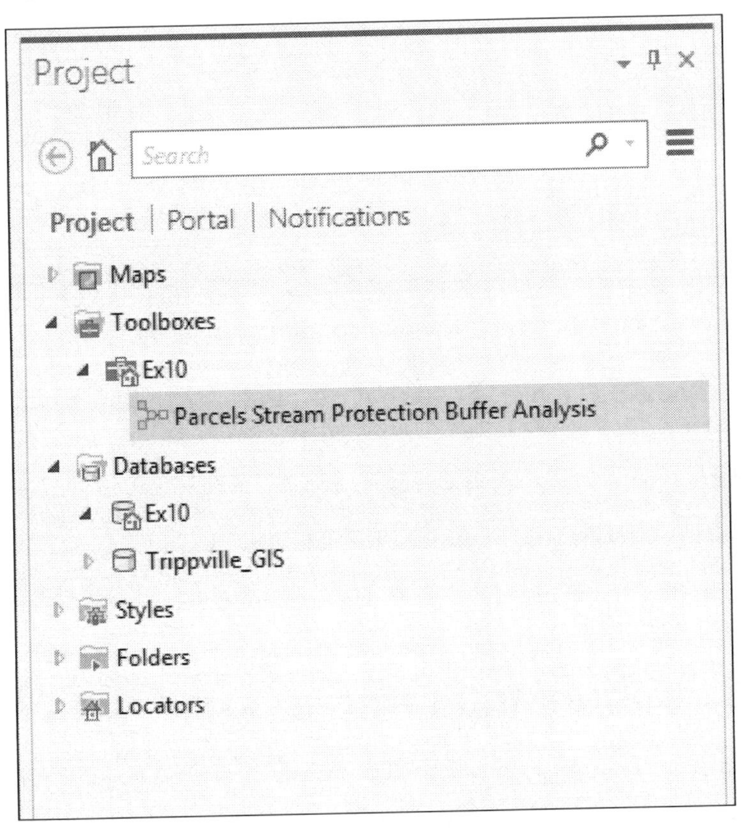

4. When you double-click on your model, it should open in the **Geoprocessing** pane. It will state that there are **No Parameters**. This is expected because you have not defined any variables as parameters that will accept user input. Click the **Run** button at the bottom of the **Geoprocessing** pane.
5. When the model is finished, return to the **Project** pane.
6. Expand the **Databases** folder and the `Ex10` geodatabase. You may need to refresh the geodatabase view to see the results of your model. To do that, simply right-click on the geodatabase and select **Refresh**.

Question: What feature class is now in the `Ex10` geodatabase and how does that compare to when you ran the model from inside ModelBuilder?

____

____

When you ran the model from inside ModelBuilder, it produced two different feature classes within the `Ex10` geodatabase. However, when you ran it from the toolbox, it only produced one. Why is that?

The answer is the feature class that was created by the **Buffer** tool in the model is considered as intermediate data. Intermediate data is any feature class or table that is created within a model that is then used by other tools and is not a final result of a series of linked processes.

When you run a model from a toolbox, it will automatically clean up after itself. This means it automatically deletes intermediate data that is created as the model runs. The only data it leaves is the final results of any processes in the model that are not intermediate data. The end result is you have the data that you need without also being left with a lot of partial datasets or layers that can clutter your database.

7. Save your project and close ArcGIS Pro.

Now you have created and run your first model. You can now run this model anytime you need to update the calculations for the areas of each parcel in and outside a floodplain.

Making a model interactive

So you have created your first model. It is a very efficient tool that will help you quickly update information. However, what happens if the buffer distance changes or the Director wants to look at different layers such as land use or just the commercial properties?

Right now your model is hardcoded to a specific set of variables. If something changes, you will be forced to edit the model before it can be used. Wouldn't it be more effective to allow others to specify different values for the variables in the model when they run it? You can allow that. It simply requires you to designate a variable as a parameter within the model. This allows the user to provide a value before they run the model.

To designate a variable as a parameter so a user can specify a value when it is run, you simply right-click on the variable in ModelBuilder and select **Parameter**. When you do, a small capital **P** will appear next to the variable indicating it is now a model parameter as illustrated in the following screenshot:

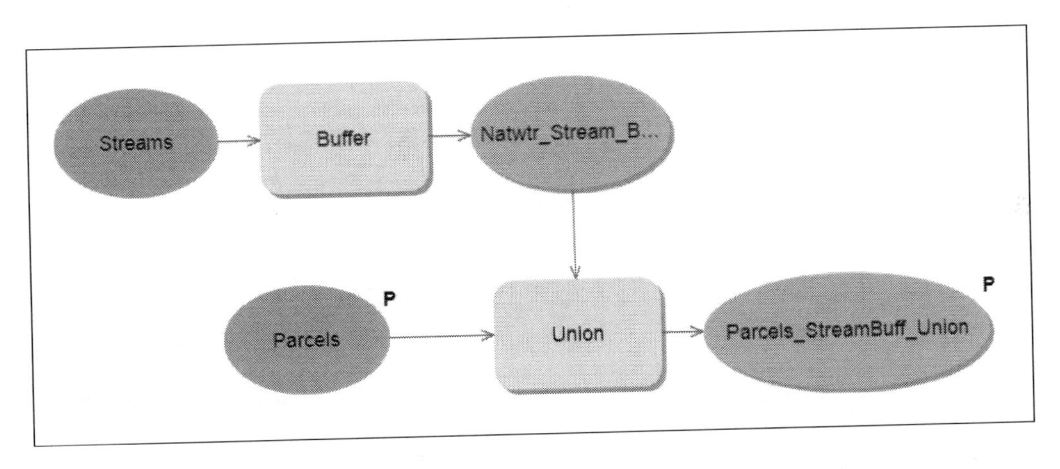

In this example, you can see that the **Parcels** and **Parcels_SteamBuff_Union** variables are both marked as parameters. This will allow the user to select the values they wish to use for those variables. This means they will be able to union the stream buffers with another layer besides just the parcels layer and control where the results are saved to and the name it is given.

Making a model interactive can greatly increase its functionality. It will allow the model to be used in different scenarios and datasets. The downside is the more interactive you make a model, the greater the chance to introduce operator errors. Users may select the wrong input layer for a tool or forget where they set the final results to be saved. This can result in more problems than the model was designed to solve. So it is always a balancing act between flexibility and hardcoding to eliminate error sources.

Now, let's give you an opportunity to make your model interactive.

Exercise 10C – making a model interactive

The Director was impressed with the model you created. It allowed him to easily calculate the area of each parcel that was in and out of the stream protection area. The council is considering changing the buffer distance for the non-disturb area and the Director wants to look at the impact of several different distances. So he will need to be able to run the model in a way that allows him to specify different buffer distances and be able to save the overall results with different names so he can review the results of the different distances.

In this exercise, you will make the previous model you created interactive so that users can provide their own values to variables within the model. You will allow users to specify the buffer distance they want to use and the final output of the model.

Step 1 – marking variables as parameters

In this step, you will learn how to designate variables as parameters within a model. You will make the buffer distance and the output of the **Union** tool parameters within your model:

1. Open ArcGIS Pro and the `Ex10.aprx` project.
2. Expand the **Toolboxes** folder in the **Project** pane.
3. Expand the `Ex10` toolbox and right-click on the **Parcels Stream Protection Buffer Analysis** model you created in a past exercise. Select **Edit** to open it in ModelBuilder.
4. Right-click on the output variable for the **Union** tool and select **Parameter**. A small **P** should appear beside the variable as shown in the following screenshot:

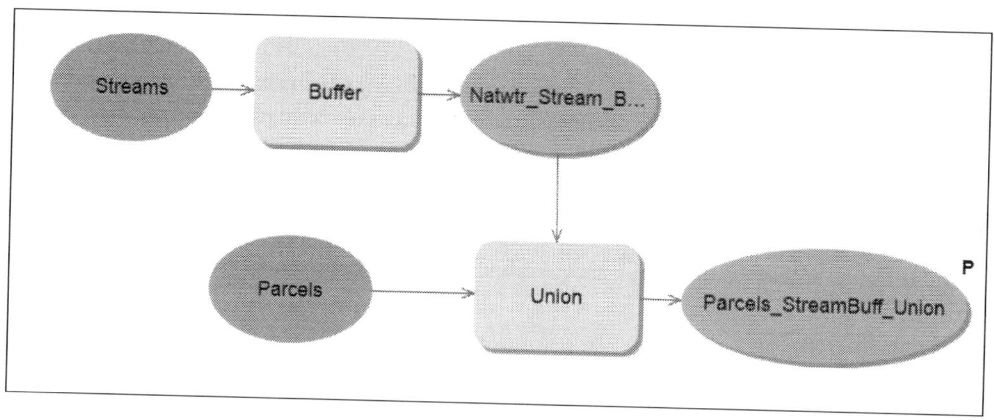

5. Save your model.

By making the output of the **Union** tool a model parameter, users will now be able to choose where they will save the final output of the model and what it will be named. This is one of the two requirements the Director asked for. Now you need to allow users to specify a buffer distance.

The buffer distance is currently hardcoded into the model. You need to make it a parameter as you did the output of the **Union** tool. However, the buffer variable is hidden. So first you will need to make it visible within the model and then designate it as a parameter.

Step 2 – exposing hidden variables

In this step, you will expose the distance variable for the **Buffer** tool so you can make it a parameter:

1. Right-click on the **Buffer** tool in ModelBuilder.
2. Select **Create Variable** and then **From Parameter**. This will display a list of all hidden variables associated with the **Buffer** tool.
3. Select **Distance [value or field]** as shown in the following image:

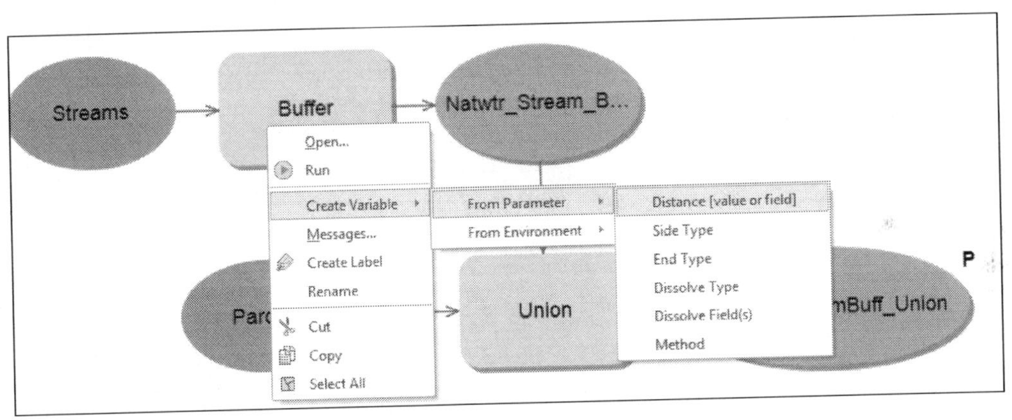

The **Distance** variable is now visible in your model. Now that it is visible, you will be able to designate it as a parameter.

4. Move your mouse pointer so it is over the **Distance** variable you just added to our model. When your pointer changes to two crossed arrows indicating it is now in move mode, drag the **Distance** variable so it is above the **Buffer** tool as shown in the following screenshot:

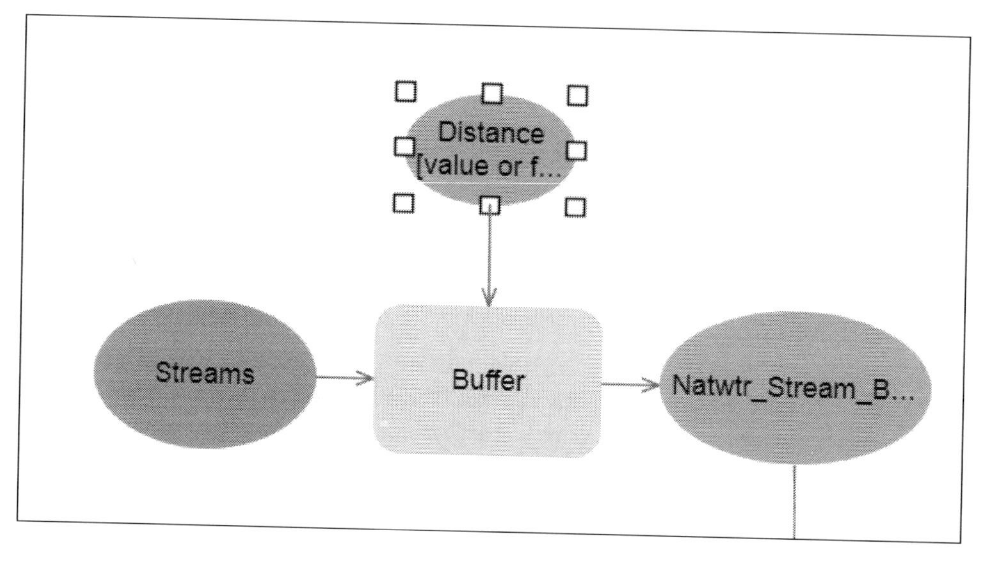

5. Right-click on the **Distance** variable and select **Parameter**. The small capital **P** should now appear next to the distance variable indicating it is now a model parameter.
6. Save your model.

Your model should now look very similar to this:

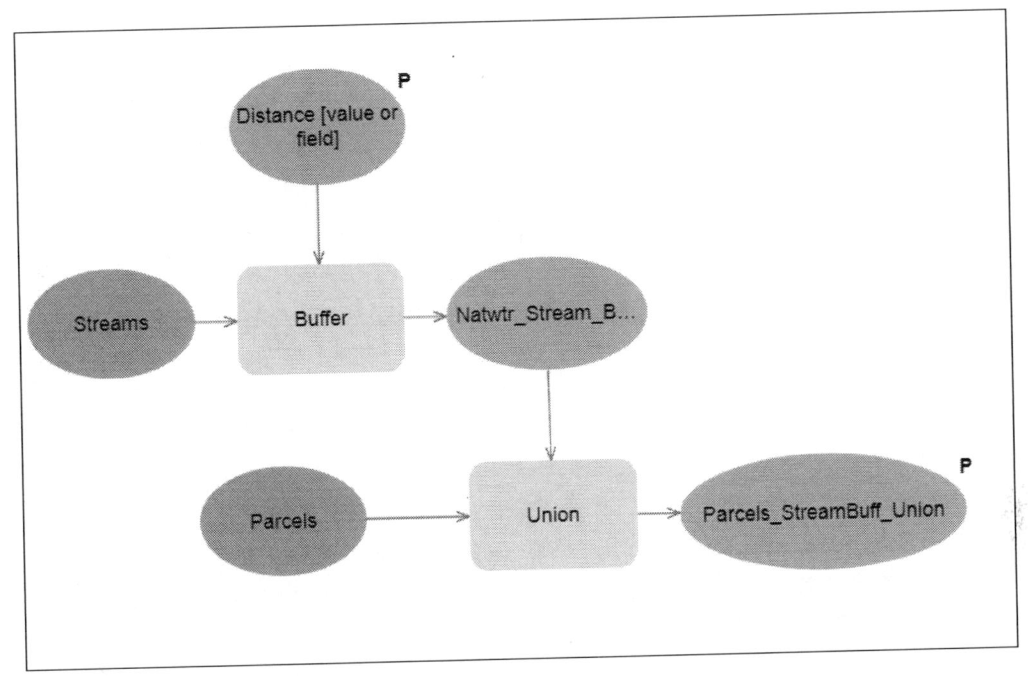

This model should now meet the requirements the Director asked for. He will now be able to use different distances from the streams and see the impact it will have on the parcels. He can save the result to a different name and location each time he runs the model.

The last step is to verify your work. You need to test run the model to see if it allows users to specify a distance and the output values.

Step 3 – running the model

In this step, you will run the model from the toolbox to make sure it will allow the Director to input a distance and specify where the output will be saved. Since you have not changed the overall logic or functionality of the model, there is no need to test the processes inside the model again:

1. Close the ModelBuilder view.
2. If needed, expand the **Toolboxes** folder and the `Ex10` toolbox in the **Project** pane.
3. Double-click on the model you created to open it in the **Geoprocessing** pane.

> Notice this time that when you open the model in the **Geoprocessing** pane, it looks different. Instead of saying **No Parameters**, it is asking for user input. The user can provide values for the two variables you designated as **Parameters**.

4. Change the value for the `Parcels_Stream_Union` variable to `C:\Student\IntroArcPro\Chapter10\Ex10.gdb\%Your Name%_Results` (i.e. `Tripp_Results`).
5. Change the **Distance** value to any value you wish that is not `150` feet. You can even change the units if you desire.
6. Click the **Run** button at the bottom of the **Geoprocessing** pane when you are done changing the values of the variables.
7. Once the model has finished running, close the **Geoprocessing** pane.
8. In the **Project** pane, verify the resulting output feature class is located in the `Ex10.gdb`.
9. Save your project and close ArcGIS Pro.

You have now created your first interactive model. This model provides more flexibility for the user, allowing them to investigate different scenarios.

Python

Python is the primary scripting language for the ArcGIS platform. It has replaced others, such as VB Script. ArcGIS Pro is currently compatible with Python 3.4, which is automatically installed when you install ArcGIS Pro.

Python has been fully integrated with the ArcGIS Geoprocessing **Application Programming Interface** (**API**) via the ArcPy module. This means you can use the geoprocessing tools from within ArcGIS Pro within your scripts, allowing you to automate and schedule tasks.

Unlike ModelBuilder, Python is not limited to just the ArcGIS platform. It can be used to create scripts that access functions in other applications, the operating system, and the computer. This gives you the ability to create scripts that extend and integrate ArcGIS Pro's functionality across platforms and applications. As a result, Python is a very versatile tool in the GIS developer's arsenal.

Python scripts can be stored within ArcGIS toolboxes or in standalone folders as `.py` files. Unlike other programming languages such as C++ or Visual Basic, creating Python scripts doesn't require special application development software. You can use simple text editors such as Notepad or WordPad. There are several free **integrated development environment** (**IDE**) applications for Python such as PythonWin or IDLE. IDE applications provide a better development environment over text editors because they include automatic coding hints and debugging tools. When you install ArcGIS, it automatically installs Python and IDLE.

ArcGIS Pro also includes a Python window that can be used to write Python scripts, run tools using Python, and load a Python script to view the code. New Python developers often find the Python window helpful because of its integrated interface and autosuggest functionality that helps guide proper syntax.

Some Python basics

Since this is your first introduction to Python, it is a good time to introduce some fundamentals and best practices. These will serve you well as you begin to write your own scripts.

Commenting and documenting your scripts

When you begin creating a Python script, it is considered a best practice to include documentation within the code that will help other developers understand what is happening within the code and the purpose of specific parts of the script. This can also prove helpful to yourself if you have to come back to a script you wrote some time ago and need to make changes.

This code documentation is traditionally accomplished using commenting. Think of commenting code as a form of metadata stored within the code itself. It provides users and other programmers with the who, what, where, when, and whys they may need to successfully use, integrate, or edit a script you create. Different programming languages use different methods to comment code. Python uses the pound sign (#) to identify comment lines within its code as illustrated next:

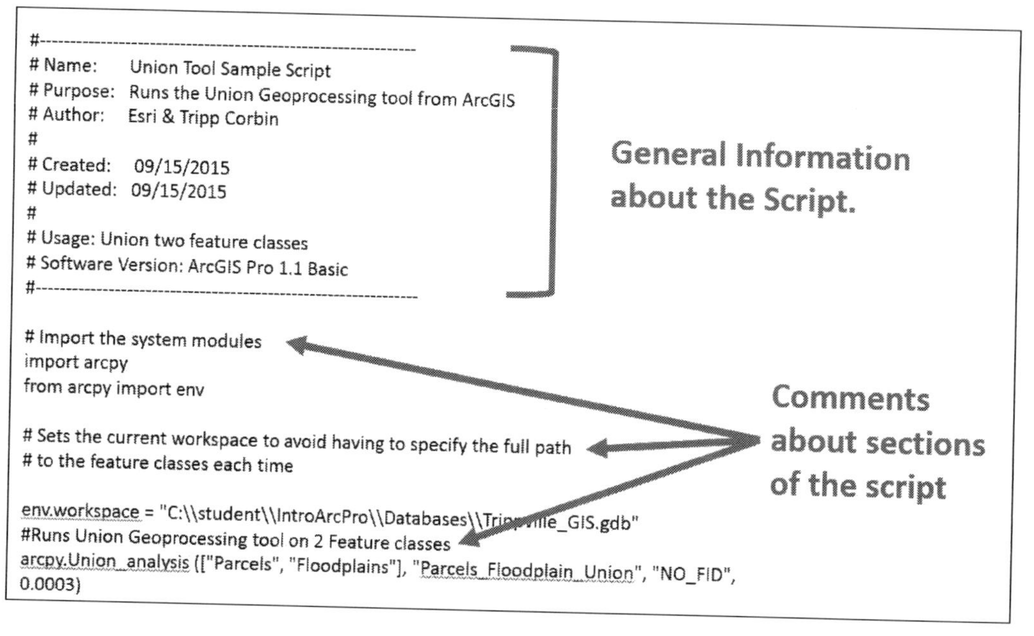

Anytime Python encounters a line that starts with a #, it ignores that line and moves to the next. It will keep ignoring lines until it encounters one that does not have a # at the beginning.

It is also possible to use triple double quotations (""") to indicate a multiple line comment. Any text between the opening and closing triple quotes will be ignored.

Traditionally, the first group of lines in a Python script is used to provide basic information about the script such as its purpose, who created it, when was it created, what ArcGIS version it was created for, and so on. Providing this basic information is considered an industry best practice.

Variables

Like a model, a Python script can contain variables. When you define a variable in Python, you give it a name and a value. Also similar to a model, the value assigned to a variable can be hardcoded, reference the result of another process, or be a function of the **Arcpy** or other module.

For example, you could define a variable that would be used as input for the **Buffer** tool as:

```
In_buf_fc = "C:\\GIS\\Trippville.gdb\\Base\\Natwtr_Stream"
```

This variable would then be used by the **Buffer** tool in a Python script as follows:

```
arcpy.Buffer_analysis (In_buf_fc,
  "C:\\GIS\\Trippville.gdb\\Streams_Buffer", "125 Feet", "FULL",
  "ROUND")
```

You can see the use of the defined variable has been highlighted in the sample Python code. In an actual script, you would not embolden the variable. That was just done in this example to help you see the use of the variable more easily.

Another very important thing to keep in mind when writing scripts is that Python is case sensitive. This means a variable named `Mapsize` is not the same as one named `mapsize`. To Python, those are two different and distinct objects. This is one of the most common causes of problems when writing and running Python scripts.

Python also has other restrictions when defining a variable within a script:

- Variable names must start with a letter. They cannot start with a number.
- Variable names cannot include spaces or other special characters. The one exception is an underscore (_).
- Variable names cannot include reserved keywords such as:
    - `Class`
    - `If`
    - `For`
    - `While`
    - `Return`

Data paths

Often when you define a variable, access data, or save the results of a tool, you need to reference a specific file or data path. In a traditional Windows environment, this typically requires you to define the path using backslashes. For example, you have been accessing the data and exercises for this book by going to `C:\Student\IntroArcPro`. This is an example of a path.

Unfortunately, you cannot use this common method of defining a path within a Python script. Backslashes are reserved characters within Python that are used to indicate an escape or line continuation. So when specifying a data path, you must use a different method. Python supports three methods for defining a path:

- Double backslashes - `C:\\Student\\IntroArcPro`
- Single forward slash - `C:/Student/IntroArcPro`
- Single backslash with an `r` in front of it - `rC:\Student\IntroArcPro`

You can use either of these methods when creating your own scripts. Although it is acceptable to use any of these within a single script, it is recommended you try to use the same method throughout the entire script. This will help you locate possible errors and fix them more quickly.

The ArcPy module

The ArcPy module is a Python site package that allows Python access to ArcGIS functionality. The level of functionality is limited to the ArcGIS Pro license level and extensions available to the user running the script.

Through the ArcPy module, Python cannot only be used to perform geoprocessing tasks using tools in the ArcGIS Pro system toolboxes or other custom tools but it can also execute other functions such as listing available datasets within a given location or describing an existing dataset. It can also create objects such as points, lines, polygons, extents, and more.

The ArcPy module contains several sub modules. These sub modules are specific-purpose libraries containing functions and classes. These sub modules include data access (`arcpy.da`), mapping (`arcpy.mp`), a spatial analyst module (`arcpy.sa`), and Network Analyst (`arcpy.na`) modules. The spatial analyst and network analyst modules require access to ArcGIS extensions of the same name.

The ArcPy module must be loaded into a script in order for Python to access ArcGIS Pro functionality. This is typically done at the very beginning of a new script using the following syntax:

```
import arcpy
```

This one line allows Python to access ArcGIS Pro tools and functions. Additional modules can also be loaded using that same line such as the operating system (`os`) or system (`sys`).

Locating Python syntax for a geoprocessing tool

To find the Python code needed to execute a specific geoprocessing tool, it is as easy as opening help for the tool. Esri has included sample Python code for all the geoprocessing tools in ArcGIS Pro and its extensions. This includes the proper syntax to use within a script along with a description of the variables that can be used with the tool. Help for a specific tool can be accessed in the **Geoprocessing** pane when the tool is opened by clicking on the small blue question mark located in the upper-right side of the pane.

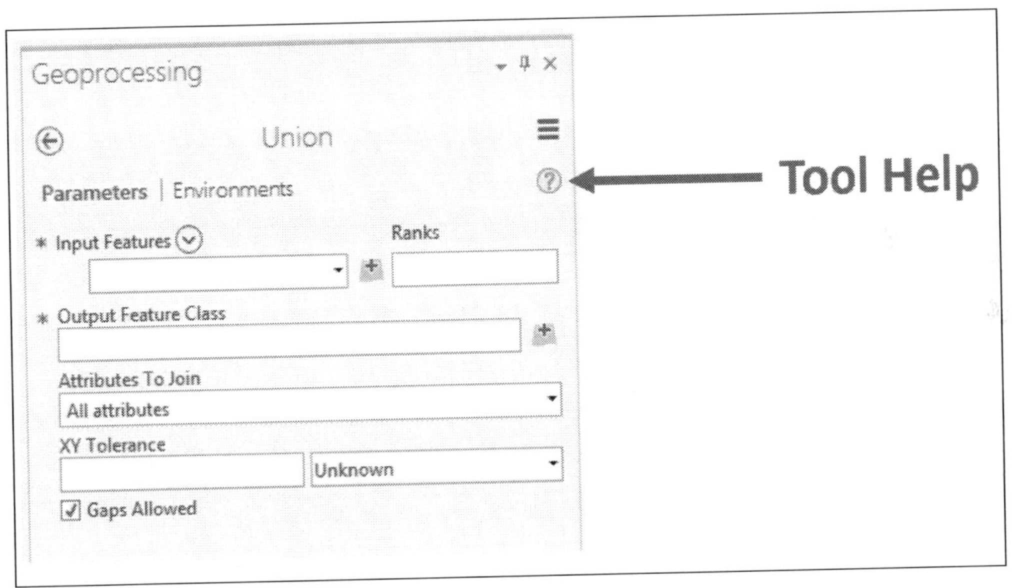

The syntax in the Help tool will show you the proper format for the code along with a description of the possible variables that may be included. The following is an example of the syntax for the **Union** tool from the Esri help:

Syntax

```
Union_analysis (in_features, out_feature_class, {join_attributes}, {cluster_tolerance}, {gaps})
```

Parameter	Explanation	Data Type
in_features [[in_features, {Rank}],...]	A list of the input feature classes or layers. When the distance between features is less than the cluster tolerance, the features with the lower rank will snap to the feature with the higher rank. The highest rank is one. All of the input features must be polygons.	Value Table
out_feature_class	The feature class that will contain the results.	Feature Class
join_attributes (Optional)	Determines which attributes from the input features will be transferred to the output feature class. • ALL —All the attributes from the input features will be transferred to the output feature class. This is the default. • NO_FID —All the attributes except the FID from the input features will be transferred to the output feature class. • ONLY_FID —Only the FID field from the input features will be transferred to the output feature class.	String
cluster_tolerance (Optional)	The minimum distance separating all feature coordinates (nodes and vertices) as well as the distance a coordinate can move in X or Y (or both).	Linear unit
gaps (Optional)	Gaps are areas in the output feature class that are completely enclosed by other polygons. This is not invalid, but it may be desirable to identify these for analysis. To find the gaps in the output, set this option to NO_GAPS, and a feature will be created in these areas. To select these features, query the output feature class based on all the input feature's FID values being equal to -1. • GAPS —No feature will be created for areas in the output that are completely enclosed by polygons. This is the default. • NO_GAPS —A feature will be created for the areas in the output that are completely enclosed by polygons. This feature will have blank attributes and its FID values will be -1.	Boolean

Help for all tools in ArcGIS Pro can be accessed from the ArcGIS Pro help online via the *Tool Reference*. The address to access it is `http://pro.arcgis.com/en/pro-app/tool-reference/main/arcgis-pro-tool-reference.htm`.

The help will also include code sample snippets that help put the syntax into context with a larger process. It is often possible to copy the sample code from the help and then paste it into your script. Then you can easily adjust the copied code to meet your needs.

Union Example 2 (Stand-alone Script)

The following stand-alone script shows two ways to apply the Union function in scripting.

```
# unions.py
# Purpose: union 3 feature classes

# Import the system modules
import arcpy
from arcpy import env

# Set the current workspace
# (to avoid having to specify the full path to the feature classes each time)
env.workspace = "c:/data/data.gdb"


# Union 3 feature classes but only carry the FID attributes to the output
inFeatures = ["well_buff50", "stream_buff200", "waterbody_buff500"]
outFeatures = "water_buffers"
clusterTol = 0.0003
arcpy.Union_analysis (inFeatures, outFeatures, "ONLY_FID", clusterTol)

# Union 3 other feature classes, but specify some ranks for each
# since parcels has better spatial accuracy
inFeatures = [["counties", 2],["parcels", 1],["state", 2]]
outFeatures = "state_landinfo"
arcpy.Union_analysis (inFeatures, outFeatures)
```

The preceding screenshot is an example of a sample code snippet for the **Union** tool that is found in the help from Esri. As you can see, it provides an understandable example of the code in a real-world context. This provides a much better understanding of how the tool can be used within a custom script you might create. Notice the comments included within the code sample and how they help to provide a better understanding of the purpose of the various parts of the code.

Now it is time for you to try your hand at writing a simple Python script.

Exercise 10D – creating a Python script

The City of Trippville operates a GIS web application that allows citizens and elected officials to access parcel data. This GIS web application combines data from the city with other data layers from ArcGIS Online and Google Maps. As a result, the parcels must be projected from the local state plane coordinate system to the **WGS 84 Web Mercator (Auxiliary Sphere)** system. This is the common coordinate system used by Esri, Google, and Bing for GIS web applications and data.

You also update the **Acres** field as new parcels are added or combined before the new data is added to the web application. You use the **Calculate Field** tool to accomplish this with an expression that converts the `Shape_Length` field that is in feet to acres.

In the past, you have manually performed these operations. However, you will be going on vacation and the Director wants the parcel data to still be updated regularly while you are gone. He is able to copy the data to the web application but does not know how to perform the other operations. So he wants you to create an automated routine that can perform these operations automatically on a regular schedule.

Since he wants this routine to run on an automated schedule, you will need to write a Python script. A model will not work in this case. In this exercise, you will write a basic Python script that will calculate the acreage of each parcel and update the `Acres` field and then project the data from the state plane coordinate system it is currently in to the WGS 84 Web Mercator (Auxiliary Sphere).

Step 1 – open IDLE

In this step, you will open the IDLE application so that you can begin creating your script:

1. Click on your Windows *Start* button. This is normally located in the lower-left corner of your screen in your task bar. Depending on your operating system, it may appear as four colored squares with or without the word Start next to it. In Windows 8.1 or 10, it appears as just four white squares.
2. In Windows 7, click on all programs. In Windows 8.1 or 10, click the small downward pointing arrow to access all installed programs or apps.
3. Navigate to the ArcGIS program group in the list of all programs. In Windows 7, you may need to expand the group to see the programs inside.
4. Locate the IDLE (Python GUI) application and click on it to launch the program.

You have now opened the Python IDLE application. You will write your script within this application. It will open with the shell window.

The shell window displays messages and errors generated by a script when it is run from IDLE. You do not actually write scripts within this window. This window is used primarily for testing and verifying results of scripts. To begin writing a script, you will need to open a new code window:

5. Click **File** and **New Window**. This will open the code window you will use to write your script. You should now see the following:

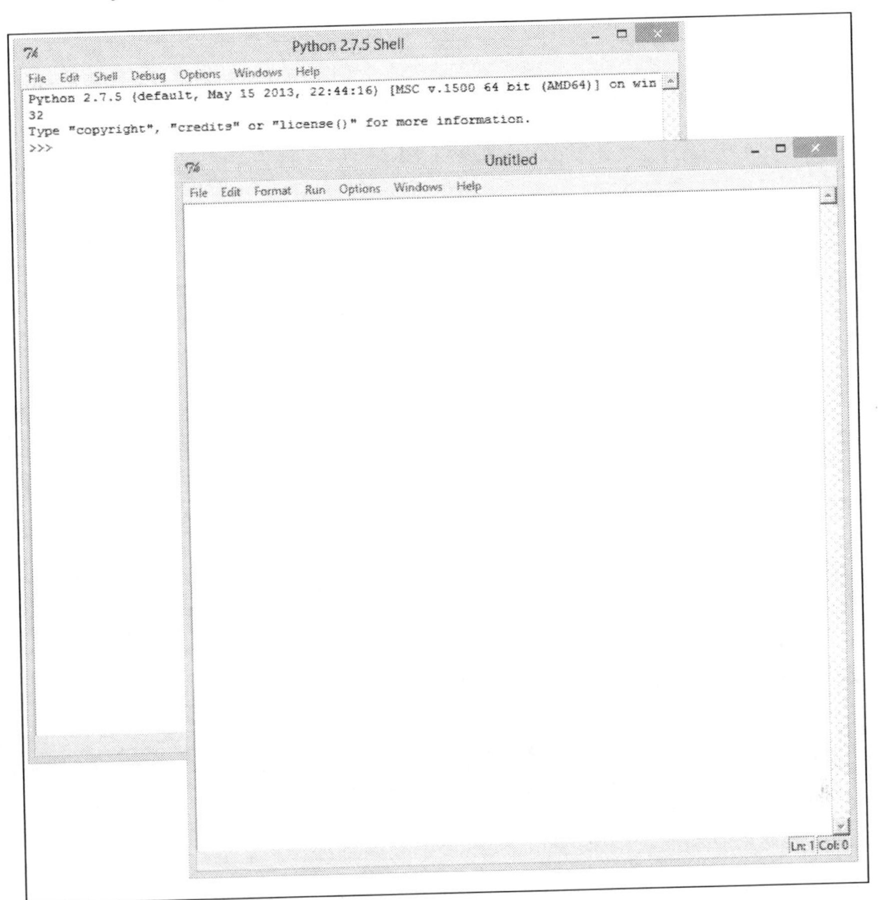

6. Click **Options** in the top menu.
7. Select **IDLE Preferences**.
8. Click on the **General** tab and set **Default Source Encoding** to **UTF-8**.
9. Click **Apply** and **OK**.

Step 2 – writing the script

Now you will begin writing the script you need to accomplish the tasks you performed manually before. To start, you will insert some basic information concerning your script in accordance with best practices. Then you will import the ArcPy module and lastly write the code for the script:

1. First, you will save your empty script so it has a name. Click on **File** and then **Save**.
2. In the **Save As** window, navigate to `C:\Student\IntroArcPro\Chapter10` and name your file `AcresWebProject.py`. Click **Save**.

You have just saved your empty script. You should see the new name and path shown in the top of the code window

3. Now you will add the general information at the beginning of the script as comments. Remember the # identifies a comment within Python code. Type the following example code into the IDLE code window. The purpose should all be typed on a single line. If you split it on to multiple lines, you will need to place a # at the start of each line:

```
#*********************************************
#Script Title: AcresWebProject.py
#Script Author: Your Name
#Script Created on: Today's date
#Last Updated on: Today's date
#Last Updated by: Your Name
#Purpose: This script calculates the parcels area in acres and
updates the acres field. It then projects the parcels to the WGS
84 Web Mercator coordinate system so it can be used within the
City's web application.
#Software: ArcGIS Pro 1.1
#**************************************************
```

4. Now you need to add the code line that imports the ArcPy module so the script can access the ArcGIS Pro tools. Add the following code to your script in the code window:

```
#Imports the ArcPy module for ArcGIS
import arcpy
```

5. Save your script by clicking **File** and **Save**. If you get a warning, just click on **OK**.

Now you will define some variables within your script that specify the location of the parcels data and where to save the results of the **Project** tool.

6. Type the following code after the import statement in the code window:

```
#Specifies the input variables for the script tools
#If the data is moved or in a different database then these paths
will need to be updated
Parcels = "C:\\Student\\IntroArcPro\\Databases\\Trippville_GIS.
gdb\\Base\\Parcels"
Parcels_Web =
  "C:\\Student\\IntroArcPro\\Chapter10\\Ex10.gdb
  \\Parcels_Web"
```

7. Save your script.
8. Now you need to begin adding the code for the tools you will need to run in the script. You will use the ArcGIS Pro help to get the proper syntax for the **Calculate Field** and **Project** tools. Then modify it so it works properly in your script.
9. Open ArcGIS Pro and `Ex10.aprx`.
10. Click on the **ANALYSIS** tab and **Tools** button.
11. In the **Geoprocessing** pane, click on **Toolboxes** located near the top of the pane.
12. Expand the **Data Management Tools** toolbox and the **Fields** toolset.
13. Select the **Calculate Field** tool.
14. Click on the **Help** button. It is the blue question mark in the upper-right corner.
15. You have opened the online tool reference for this tool. Click on **Syntax**.
16. Highlight and copy the syntax for the tool. It should read: `CalculateField_management (in_table, field, expression, {expression_type}, {code_block})`
17. Go to the IDLE code window and paste the copied syntax onto a line below the variables you defined earlier.
18. Add a comment above the code you just pasted into the script that says: `Calculates the area in acres for each parcel and updates the Acres field.`
19. Now edit the code sample syntax you just pasted into the script as follows:

```
arcpy.CalculateField_management (Parcels, "Acres",
  "!Shape_Area! / 43560", "PYTHON_9.3", "")
```

20. You have now defined the **Calculate Field** tool within a Python script so it includes all the variables it needs to run. Save your script.

21. Now you need to add the **Project** tool to the script and define its syntax properly. Using the same process you used for the **Calculate Field** tool, open the help for the **Project** tool and copy the syntax into your script. The **Project** tool is located in the same toolbox but is in the **Projections and Transformations** toolset.

22. Add an appropriate comment to the script above the code for the **Project** tool that will let others know its purpose similar to the comment you added for the **Calculate Field** tool.

23. Modify the **Project** tool code as follows (to make things easier, you can copy the syntax from the **Project** tool `Sample.txt` file in the `Chapter10` folder):

```
arcpy.Project_management(Parcels, Parcels_Web, "PROJCS['WGS_1984_
Web_Mercator_Auxiliary_Sphere',GEOGCS['GCS_WGS_1984',DATUM['D_
WGS_1984',SPHEROID['WGS_1984',6378137.0,298.257223563]],PR
IMEM['Greenwich',0.0],UNIT['Degree',0.0174532925199433]],
PROJECTION['Mercator_Auxiliary_Sphere'],PARAMETER['False_
Easting',0.0],PARAMETER['False_Northing',0.0],PARAMETER['Centr
al_Meridian',0.0],PARAMETER['Standard_Parallel_1',0.0],PARAMETE
R['Auxiliary_Sphere_Type',0.0],UNIT['Meter',1.0]]", "WGS_1984_
(ITRF00)_To_NAD_1983", "PROJCS['NAD_1983_StatePlane_Georgia_
West_FIPS_1002_Feet',GEOGCS['GCS_North_American_1983',DATUM['D_
North_American_1983',SPHEROID['GRS_1980',6378137.0,298.25722
2101]],PRIMEM['Greenwich',0.0],UNIT['Degree',0.01745329251994
33]],PROJECTION['Transverse_Mercator'],PARAMETER['False_East
ing',2296583.333333333],PARAMETER['False_Northing',0.0],PARAM
ETER['Central_Meridian',-84.16666666666667],PARAMETER['Scale_
Factor', 0.9999],PARAMETER['Latitude_Of_Origin',30.0],UNIT['Foot_
US',0.3048006096012192]]")
```

24. Save your script.

    Your script should look similar to this:

```
#*************************************************
#Script Title: AcresWebProject.py
#Script Author: Tripp Corbin, GISP
#Script Created on: 09/24/2015
#Last Updated on: 09/24/2015
#Last Updated by: Tripp Corbin. GISP
#Purpose: This script calculates the parcels area in acres and updates the acres field. It the projects the parcels to the WGS 84 Web Mercator coordinate system so it can be used withi
#Software: ArcGIS Pro 1.1
#*************************************************

#Imports the ArcPy module for ArcGIS
import arcpy

#Specifies the input variables for the script tools
#If the data is moved or in a different database then these paths will need to be updated
Parcels = "C:\\Student\\IntroArcPro\\Databases\\Trippville_GIS.gdb\\Base\\Parcels"
Parcels_Web = "C:\\Student\\IntroArcPro\\Chapter10\\Ex10.gdb\\Parcels_Web"

#Calculates the area in acres for each parcel and updates the Acres field.
arcpy.CalculateField_management (Parcels, "Acres", "!Shape_Area! / 43560", "PYTHON_9.3", "")

# Projects Parcels from State Plane to Web Mercator Coordinate System creating a new feature class
arcpy.Project_management(Parcels, Parcels_Web, "PROJCS['WGS_1984_Web_Mercator_Auxiliary_Sphere',GEOGCS['GCS_WGS_1984',DATUM['D_WGS_1984',SPHEROID['WGS_1984',6378137.0,298.257223563]],P
```

25. Once you have verified your script and saved it, close IDLE.

An example of the completed python script has been included in the Python Sample folder located in the Chapter10 folder. You can use this sample to verify your code if needed.

Step 3 – adding the script to ArcGIS Pro and running it

Now that you have created a Python script, you need to add it to ArcGIS Pro and test it. In this step, you will add the script you just created to a toolbox in your project and then run it.

1. If necessary, start ArcGIS Pro and open the `Ex10.aprx`.
2. In the **Project** pane, expand the **Toolboxes** folder.
3. Right-click on the `Ex10` toolbox and select **New | Script**.
4. Fill out the information for the new script as follows:
5. **Name**: `CalcAcresProject`
6. **Label**: `Calculate Parcel Acres and Project to Web Mercator`

7. Click on the **Browse** button located next to the **Script File** cell. Navigate to `C:\Student\IntroArcPro\Chapter10` and select the `AcresWebProject.py` script you just created. Your window should now look like this:

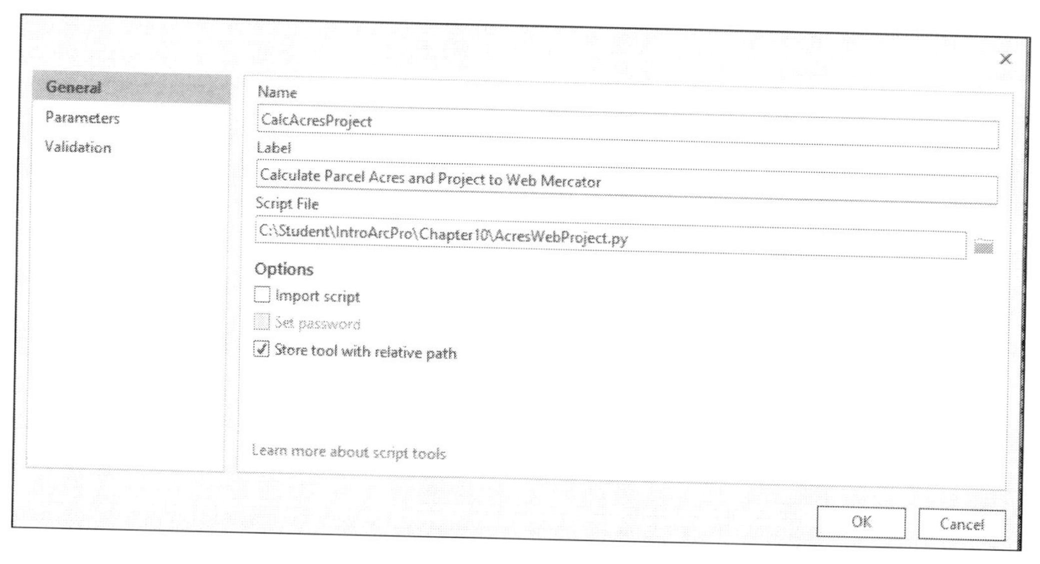

8. Click **OK** once you have verified everything is set correctly.

The script will appear in the `Ex10` toolbox. This means you can now run it in ArcGIS Pro. You must add all Python scripts you create to a toolbox before they are able to be used in ArcGIS Pro. For those with experience with ArcGIS for Desktop, it should be noted that ArcGIS Pro 1.1 does not support Python add-ins yet. That functionality may be added to a future version.

Now you will run the script to test and make sure it works as expected:

9. Double-click on the Python script you just added to the toolbox. This will open it in the **Geoprocessing** pane. Since you hardcoded all the variables into the script, it has no parameters that the user needs to define.
10. Click the **Run** button at the bottom of the **Geoprocessing** pane.
11. Once the script has completed running, return to the **Project** pane.
12. Expand the **Databases** folder and the `Ex10.gdb`. If the script ran successfully, you should see a new feature class named `Parcels_Web`. You may need to right-click on the `Ex10` geodatabase and choose **Refresh** in order to see the new feature class that was created.

If you do see the new feature class, congratulations! You have just created and run your first Python script. If your script did not run successfully, you may wish to compare it to the sample included in the `Python Sample` folder located in `C:\Student\IntroArcPro\Chapter10\`.

13. Save your project and close ArcGIS Pro along with IDLE if it is still open.

Summary

ArcGIS Pro contains two methods for automating and streamlining tasks: you create a model or a Python script. Which will work best will largely depend on your skills and how they will be used.

Models are created in ModelBuilder, which provides a graphical interface for creating tools that will automate a series of processes required to accomplish an analysis or other workflow. Each process within a model will include a tool that can be a geoprocessing tool, script, or another model, along with their associated variables. As you create a model, you can choose to make it interactive by designating variables as parameters. The biggest limitation of a model is that it can only be run from inside of ArcGIS Pro. This means it cannot be scheduled to run automatically.

Python scripts can be used to automate processes that can then be scheduled to run at specified times and dates. Unlike a model, creating Python scripts does require knowledge of the Python language and the ability to write code. In addition to the ability of Python scripts to be run on a schedule, they can also be used to access functionality from other applications other than ArcGIS Pro. This allows you to create scripts that can integrate the functionality of several different applications into a single automated script.

If you would like to learn more about creating Python scripts for ArcGIS, you might want to get *Programming ArcGIS with Python Cookbook, Eric Pimpler, Packt Publishing*.

11
Sharing Your Work

As you have learned throughout this book, ArcGIS Pro has very powerful tools to create amazing 2D and 3D maps and to perform a wealth of spatial analysis. This whole functionality means little if you cannot get the results of all this effort into the hands of those that need it.

This need to access the results of your GIS efforts is also being fueled by the growing use of geospatial and mobile technologies by the masses. Mostly everyone nowadays has a smart phone or tablet and has used some sort of mapping application, such as Google Maps, Waze, Bing maps, or the local county tax parcel application. This means that people are becoming much more geospatially and technology savvy. They expect to be able to access and use the data and analysis you provide.

ArcGIS Pro provides several methods and tools in order to share your GIS content with others. Which of the available tools or methods will work best depends on several factors. First is what is it you are trying to share. Is it data, an entire map, or a tool? The second is what software and skills your audience possesses and what you want to allow them to do with the shared content. Third consideration is whether they are able to connect to your GIS directly via your network.

In this chapter, you will learn the following topics:

- How to share GIS content with those on your network
- How to share GIS content with those not on your network
- How to export to other GIS data formats
- How to export to non-GIS data formats

Sharing content with those on your network

Sharing content with those on your network is easy if they also have ArcGIS Pro, ArcGIS for Desktop, or some other GIS application. They will also need permissions and access to the databases, files, and folders used to store your GIS content.

If they have ArcGIS Pro

If other users on your network have ArcGIS Pro, they will be able to open the projects you create as long as they have access and permissions to the project and data sources used in the project. You can share layer, map, tasks, and layout files as well to standardize your content.

A layer file

A layer file in ArcGIS Pro has a `.lyrx` file extension. It stores all the property settings associated with a layer in a map, such as source, symbology, field visibility, labeling, definition queries, and more. Layer files allow you to standardize these settings, so the layer will appear the same across multiple maps and projects.

Using a layer file to add a new layer uses the same process as adding a layer using a feature class. However, instead of going to the geodatabase, Shapefile, CAD file, or raster, you select the layer file instead. When you do this, your layer will be added to your map with all the properties preconfigured, so you don't need to go back and set all the layer properties manually. This can save you a lot of time.

Creating a layer file is relatively easy. You simply select the layer in the **Content** pane. Then, you activate the **SHARE** tab and choose **Layer File**. From there, you fill out the required information and click on **Save**. It is that easy.

Why don't you give it a try?

Exercise 11A – creating a layer file and using it

In this exercise, you will create a layer file and then use it to add a new layer to a map within a project.

Step 1 – creating the layer file

In this step, you will create the layer file for the Trippville Zoning layer. It appears that several people have been using the Zoning layer in several maps but have been doing so with their own personal settings. This is causing confusion. So, the Community Development Director wants you to develop a standard based on the official zoning map for the city.

1. Start ArcGIS Pro and open the `Ex11.aprx` project file located in `C:\Student\IntroArcPro\Chapter11`.
2. Select the **Zoning** layer in the **Contents** pane.
3. Select the **SHARE** tab in the ribbon.
4. Click on the **Layer File** button in the **Save As** group. This will open the **Save Layer(s) As LYRX File** window.
5. Navigate to `C:\Student\IntroArcPro\Chapter11` using the tree on the left-hand side of the window.
6. Name the new LYRX file `Trippville Official Zoning` and click on **Save**.

You have just created a layer file for the **Zoning** layer as it appears in the official zoning map in the `Ex11` project.

Step 2 – using the layer file to add a layer to a map

In this step, you will test the layer file you created to ensure that it works as expected. You will use it to add a new layer to a new blank map in your project.

1. In the **Project** pane, expand the `Folders` folder. Then, expand the `Chapter11` folder, so you can see its contents.
2. Right-click on the `Trippville Official Zoning.lyrx` file and select **Add To New Map**, as follows:

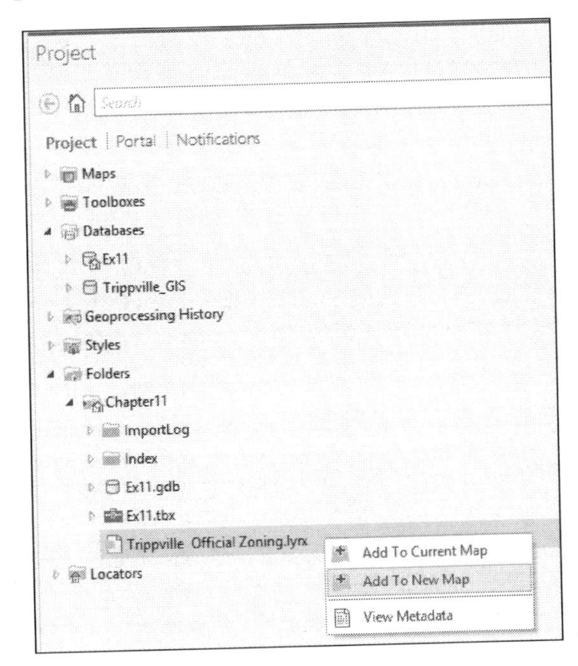

A new map has just been added to your project, and it should contain the **Zoning** layer automatically symbolized the same way it was in the Official Zoning Map. Because you added the layer to the new map using the layer file, the symbology was already configured.

3. Save your project and close ArcGIS Pro.

Map files

Map files are similar to a layer file except they save the settings for an entire map. This includes all the layers shown in the **Contents** pane when that map is active and the settings for those layers. They can be based on a 2D or 3D map from an ArcGIS Pro project. Map files have a `.mapx` file extension. Similar to a layer file, map files also do not store or save the source data displayed in the map. Users must still have a connection to the data.

Map files allow you to share maps you create in ArcGIS Pro, so others can include them in their projects. This can save a lot of time because users will not need to create a map from scratch. They can simply import the map you created into their project and then make whatever adjustments they deem necessary.

Here is a common example of where a map file may be useful. You are working on a project within a city. You are creating a layout for the project that will show the primary project area and also contain a general location map that shows the project area in relation to the entire city. One of your colleagues is working on another project in the city and has already created a general location map of the city, which contains all the layers you need in addition to having all the symbology and label settings defined. You ask your colleague to create a map file for the general location map he is using, which you then import into your project. The imported map file is added to your project with all the layers preconfigured. You do not have to then spend your time creating this map and configuring the layers.

Adding a new map to your project from a map file is not difficult. Simply click on the **INSERT** tab in the ArcGIS Pro ribbon and then, select **Import Map**. From there, you simply navigate to the location of the map file and import it.

Creating a map file follows the same basic process as creating a layer file. You go to the **SHARE** tab in the ribbon. Instead of selecting **Layer File**, you select **Map File**.

For those who have been using ArcGIS for Desktop, remember that a map in ArcGIS Pro is not exactly the same as a map document. In an ArcGIS Pro project, you may have several maps along with separate layouts. When you create a map file in ArcGIS Pro, it will only contain the active map that was selected when you create the map file. This will not include any layouts that may include the map.

The map file challenge

Using the skills you learned creating a layer file and the information mentioned earlier, create a map file based on the official zoning map in the `Ex11` project. Then, open a new project and try to import the map file to create a new map within that project.

Layout files

Layout files allow you to share layouts you create so that they can be used as templates by yourself and others. Layout files will include all the elements you see in the layout, including maps, legends, borders, north arrows, scale bars, titles, logos, disclaimers, and more.

If the data referenced in the maps included in the layout is not accessible or in a different location, you will need to repair your data sources. Repairing a data source is simply resetting the properties of a layer to point to an accessible data sources. Layout files, such as layer and map files, do not store any GIS data.

> To learn more about how to repair broken data links, you can go to `https://pro.arcgis.com/en/pro-app/help/mapping/map-authoring/repair-broken-data-links.htm`.

The process to create and import layout files is very similar to the process used for map files. Again, it starts on the **SHARE** tab in the ribbon. Then, the **Layout File** button in the **Save As** group. Of course, you can only create a layout file if your project contains a layout.

The layout file challenge

Try exporting the layout included in the `Ex11` project to a layout file. Then, open a new blank project or one of the previous projects you have worked with and import the layout file into the project. Here's a hint: try using the **New Layout** tool on the **INSERT** tab in the ribbon. Note what happens when you import the layout file.

If they don't have ArcGIS Pro

If you want to share your GIS content with those who don't have ArcGIS Pro, your options are much more limited. Depending on the software and abilities of your potential users, you can export your content to other GIS formats, non-GIS formats, or try publishing to ArcGIS Online, Portal for ArcGIS or ArcGIS Server as a web map or layer.

Exporting to other GIS formats

ArcGIS Pro will allow to you to export your data to many different GIS formats, such as **Shapefiles, Keyhole Mark-up Language (KML/KMZ)**, or **CAD files (DWG, DXF**, or **DGN**). Some of these allow you to export multiple layers into a single file, whereas others only support single layers.

Shapefiles are a very common GIS vector data format that originated with Esri. A Shapefile only stores a single layer or feature class. So, if you wish to use this format, you may need to export multiple layers each to its own Shapefile.

Most GIS-enabled applications, such as QGIS, Grass, MapWindow, and Map3D, are able to read and display Shapefiles. Some even have the ability to edit. In addition, most GPS/GNSS software applications and data collectors are also able to import and export Shapefiles. As a result, Shapefiles have become the *de facto* data sharing format for many.

The **KML** format is also popular. It is the format used by Google Earth, which is a free application. Exporting to KML would allow you to view your data in relationship to the data you see on Google's maps.

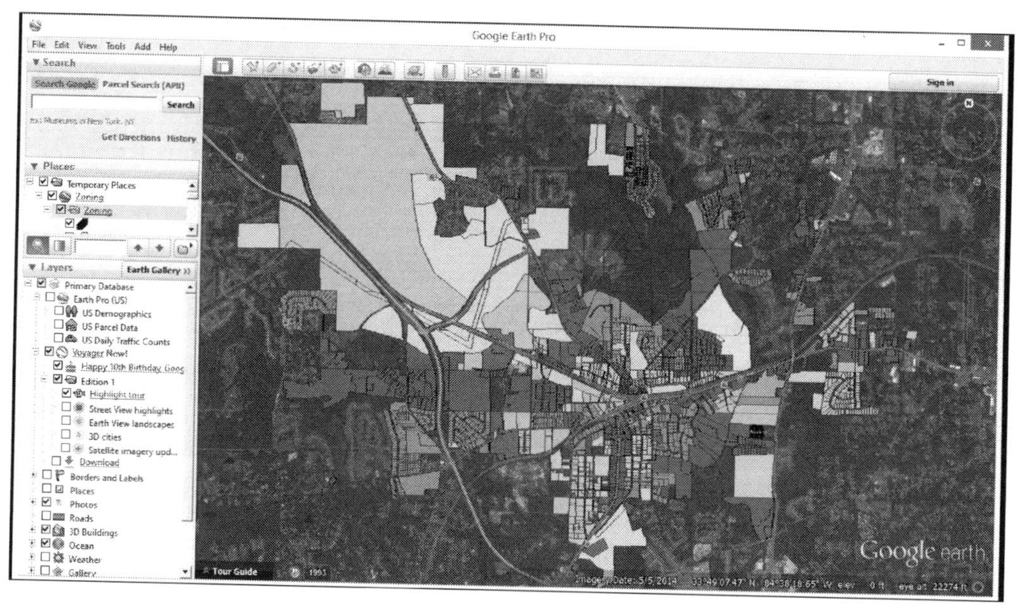

You will find the tools to export to these other GIS formats in the **Conversion Tools** toolbox and their individual toolsets. Each format has its own toolset.

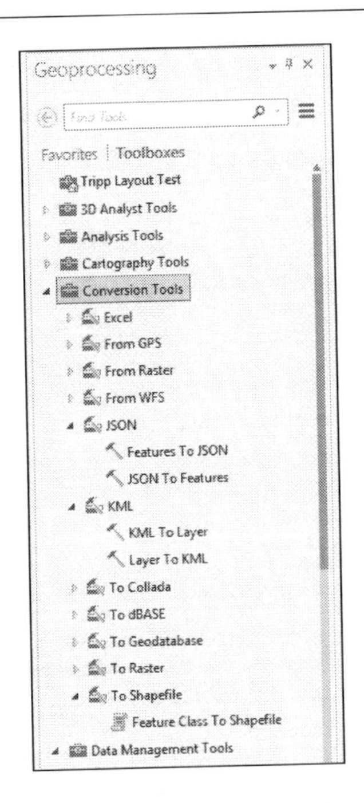

Exercise 11B – exporting to KML

The Director wants to be able to display the city's zoning layer over the imagery in Google Earth. So, he has asked you to export it to a KML file that he can use to accomplish this.

Step 1 – download and install Google Earth or Google Earth Pro

You will need to download and install Google Earth or Google Earth Pro. Both of these applications are free and available for download from the Internet. If you are installing the application on a work computer, you may need to get permission or help from your IT staff. If you already have it installed, you can skip this step.

1. Open your web browser (such as Google Chrome, Firefox, or Internet Explorer).
2. Go to `https://www.google.com/earth/`.
3. Click on the **Explore** option under either Google Earth or Google Earth Pro.
4. Click on the **Desktop** option located in the upper-right corner of the web page.
5. Read the short descriptions for both applications and choose which one you wish to download and install. For this exercise, either will work. However, I recommend going for Pro since it is now free and has greater capabilities.

6. Accept the license agreement and download.
7. Once the install file downloads, you need to run it. Double-click on the downloaded install file located at the bottom of your browser if you use Chrome or Firefox as follows:

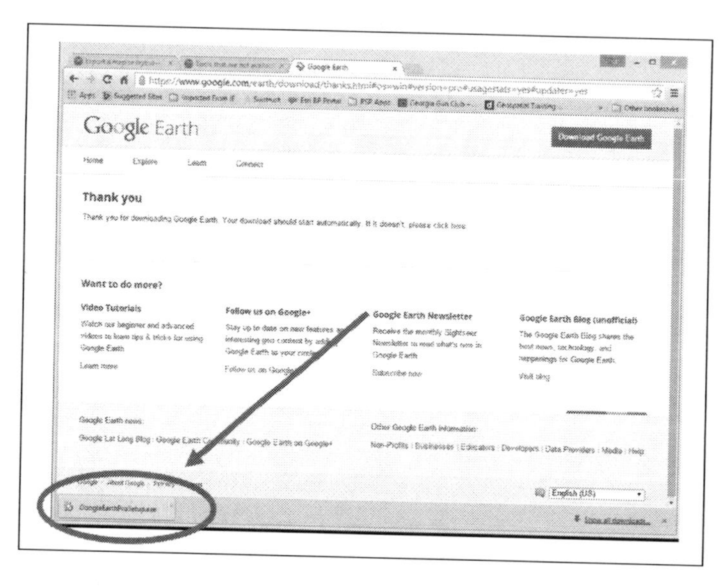

If you use Internet Explorer, just click on the **Run** button on the download bar, which appears at the bottom as follows:

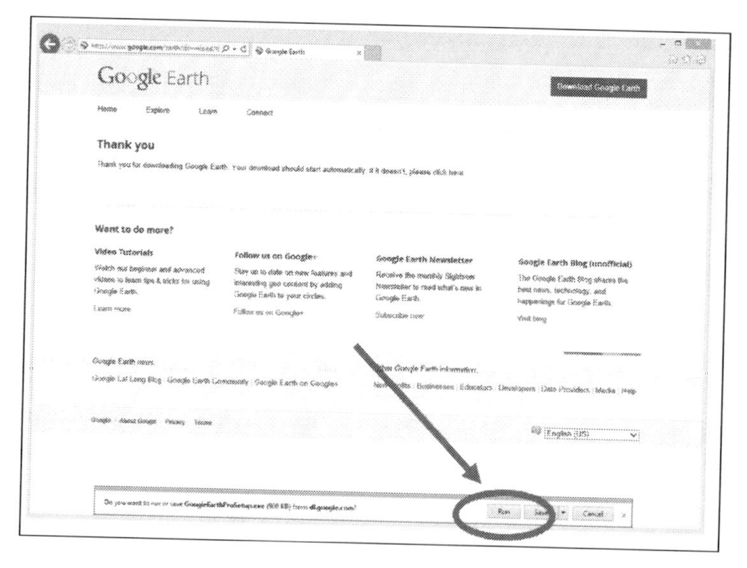

8. Follow the instructions in the install routine accepting the default values.

Step 2 – exporting to KML

In this step, you will export the zoning layer from the official zoning map in the `Ex11` project.

1. Open ArcGIS Pro and the `Ex11` project if you closed it earlier.
2. Select the **Official Zoning Map** tab at the top of the view area to make it the active map.
3. Select the **ANALYSIS** tab from the ribbon.
4. Click on the **Tools** button in the **Geoprocessing** group.
5. In the **Geoprocessing** pane on the right-hand side of the interface, click on **Toolboxes** if it is not already active.
6. Expand the **Conversion Tools** toolbox and then the **KML** toolset.
7. Select the **Layer To KML** tool.
8. Click on the small drop-down arrow located to the left of the **Layer** input cell. Select the **Zoning** layer.
9. Make the **Output File**: `C:\Student\IntroArcPro\Chapter11\Zoning_Layer.kmz` as follows:

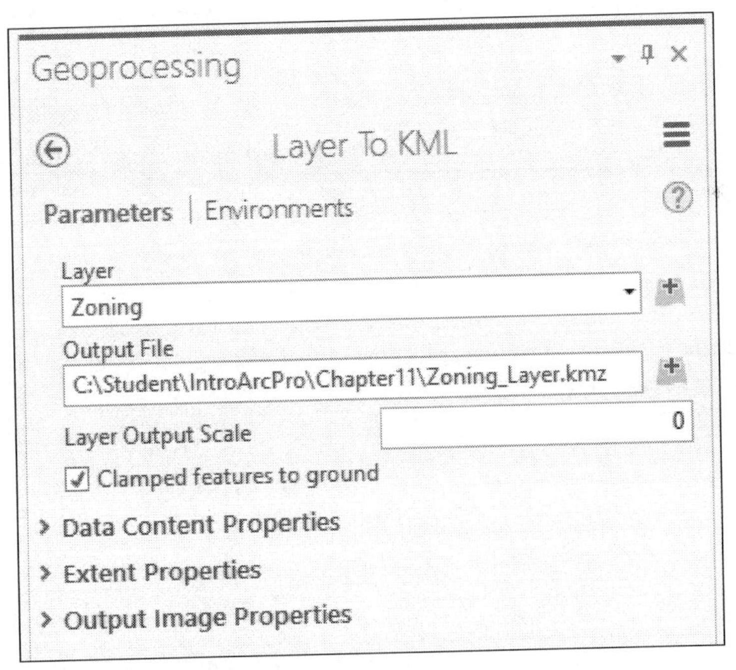

10. Once you have verified that the **Layer To KML** tool is properly configured, click on the **Run** button located at the bottom of the **Geoprocessing** pane.

This tool will produce a KMZ file. This may be confusing since it is called the Layer to KML file. So, why does it produce a KMZ instead? Well, a KMZ is a compressed KML. Compressing the KML makes it smaller and easier to upload to the Web or e-mail to someone. However, they contain the same data and will generally work interchangeably. You can open the KMZ file using an application such as WinZip or 7Zip to see the KML inside the KMZ file.

Step 3 – opening KML/KMZ in Google Earth

In this step, you will open the new KMZ file you just created in Google Earth to check whether it works. This will also allow you to understand the process if you need to provide instructions to the director.

1. Open Google Earth or Google Earth Pro by going to **Start** | **All Programs** | **Google Earth** (or **Google Earth Pro**) and selecting **Google Earth** (or **Google Earth Pro**). If you are using Windows 8.1 or 10, the process will be slightly different. Click on the **Start** icon and then the small down arrow. From there, navigate to the **Google Earth** group and select the application.
2. Feel free to explore the **Start Up** window that appears. When you are done, click on **Close**.
3. Open File Explorer or Windows Explorer depending on your operating system. You should be able to do this by clicking on the icon located in your task bar, which resembles a small file folder in a silver file holder.
4. In File Explorer, navigate to the `Chapter11` folder located in `C:\Student\IntroArcPro` using the tree on the left-hand side of the Explorer window. Once you have opened the `Chapter11` folder, you should see the `Zoning_Layer.kmz` file you created in ArcGIS Pro.
5. Move and resize the Explorer Window and Google Earth, so they are both visible at the same time similar to the following image. If you are using multiple monitors, you can place the Explorer window on one and Google Earth on the other.

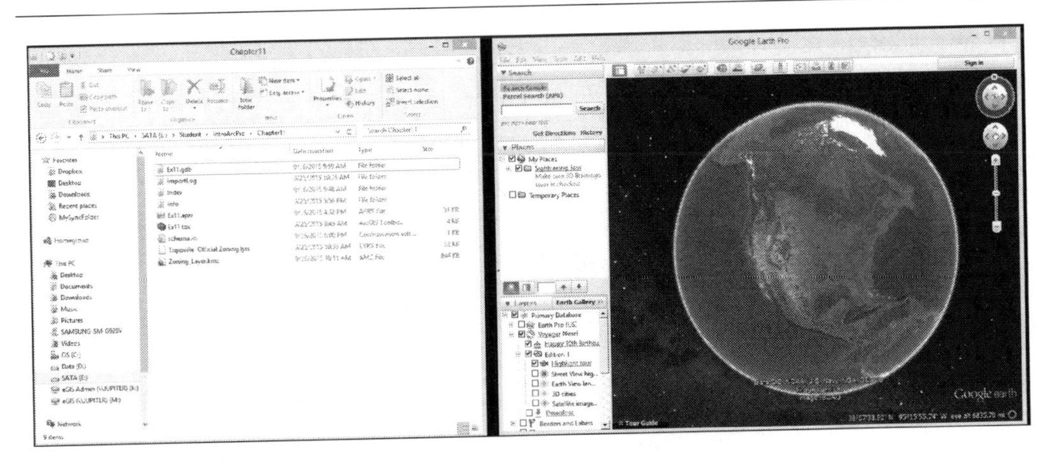

6. Select the `Zoning_Layer.KMZ` file in Explorer. Drag and drop it into Google Earth. This should add the **Zoning** layer to Google Earth.
7. Close the Explorer window.

You should see that the **Zoning** layer for the City of Trippville has now been added to Google Earth. The symbology should be very close to what you saw in ArcGIS Pro. It will not be exact because the two applications apply symbology differently.

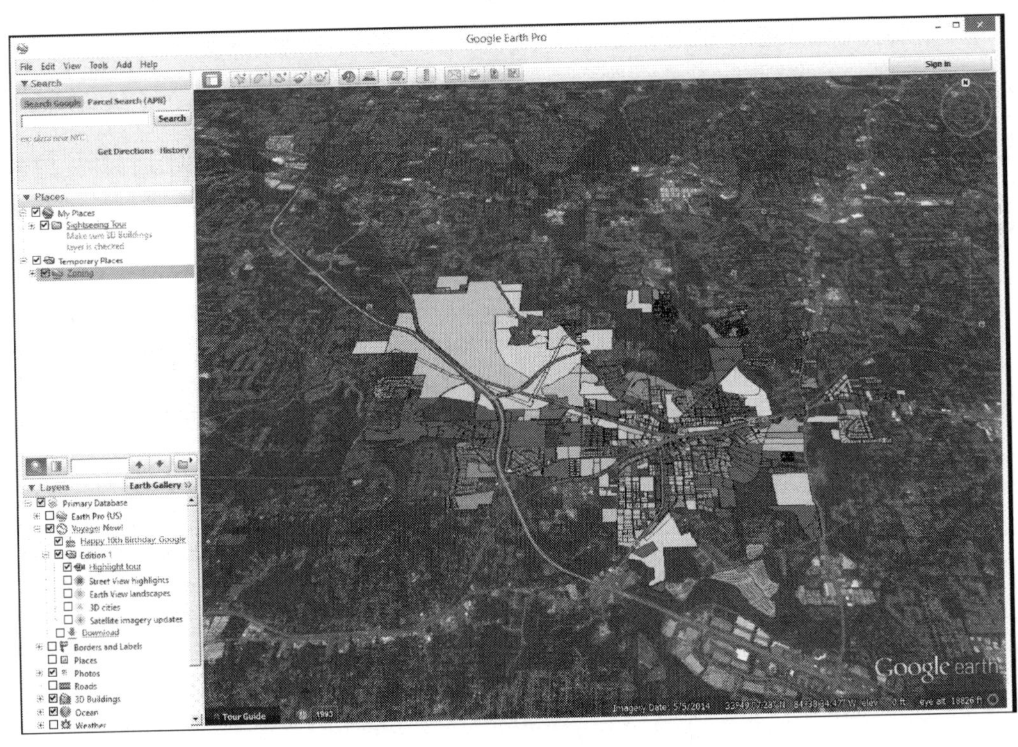

Feel free to explore Google Earth and the layer you just added. See if you can find ways to add other types of data.

8. Close Google Earth once you are done exploring. You can save or discard your changes as desired.
9. In ArcGIS Pro, save your project and close the application.

You have just successfully exported your GIS data to another non-Esri GIS format, which you were then able to open in a completely different application. Now you will explore ways that you can export GIS data to non-GIS formats, so others without GIS software or skills can use it.

If you only wish to view a single KMZ or KML file, you can just double-click on the file in File Explorer, and it will automatically open with Google Earth as long as that is the default program assigned to that file type.

Exporting to non-GIS formats

ArcGIS Pro does support exporting maps, layouts, and tables to other non-GIS formats. This can allow others that don't have ArcGIS Pro to view what you have created.

Exporting maps and layouts

You can export maps and layouts to various types of vector graphic or raster formats. This includes the following:

- BMP: Raster
- EMF: Vector graphic
- EPS: Vector graphic
- GIF: Raster
- JPG or JPEG: Raster
- PDF: Vector graphic
- PNG: Raster
- SVG: Vector graphic
- SVGZ: Vector graphic
- TGA: Raster
- TIFF: Raster

Many of these formats can then be easily opened in other applications, such as web browsers (for example, Chrome, Internet Explorer, or Firefox), free applications (Adobe Acrobat Reader and Microsoft Paint), or other GIS applications.

Most of the raster formats can also be embedded or inserted into documents, spreadsheets, and presentations. This allows them to be included in reports, studies, letters, exhibits, and more.

The PDF format actually supports some ability to create an interactive file. When you export to PDF, you have the option to include layers and/or attributes if you desire as illustrated later. This is under the **Export Options...**, as shown in the following screenshot:

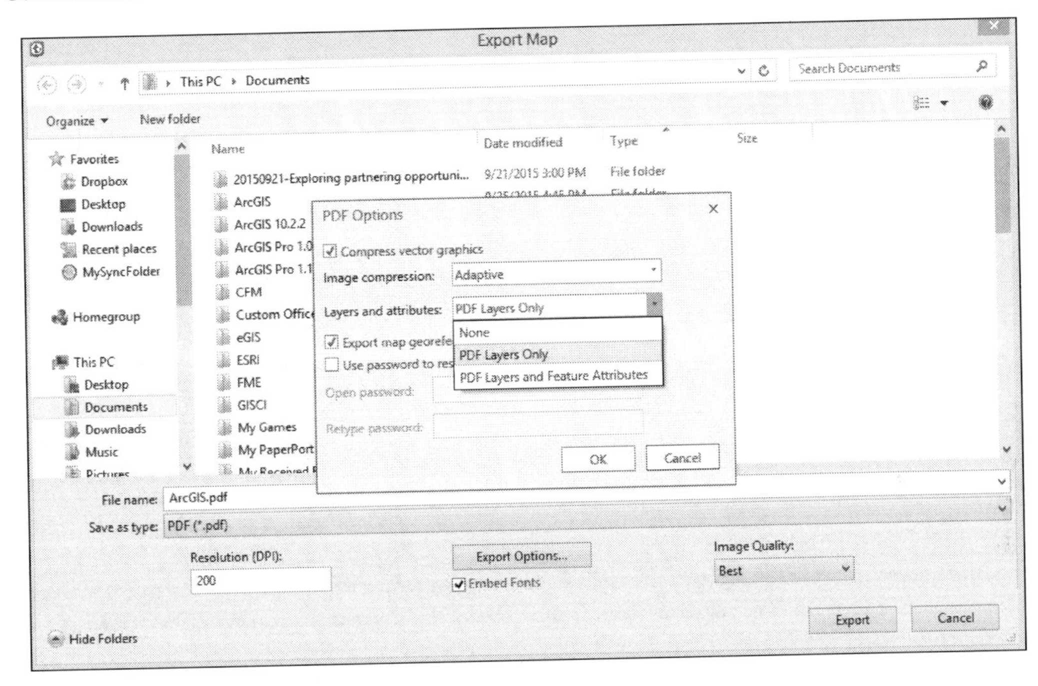

PDFs also allow you to secure the file so that it requires a password to open it.

Exporting tables

If you just need to export the attributes found in the **Attribute** table for a layer, you can export the information to various formats as well. This can be accomplished two ways. The first is from the **Table** pane, click on the options button at the upper-right corner and select Export as shown here:

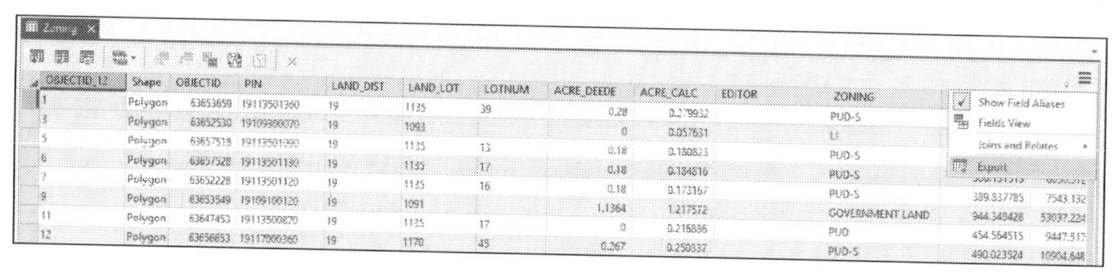

The second way is to go straight to the **Copy Rows** geoprocessing tool, which is located in the **Data Management Tools** toolbox and the **Tables** toolset. Both methods end up producing the same result.

These methods will allow you to export an entire table or a set of selected records to another table. You have the option of exporting to:

- Geodatabase standalone table
- dBase (`.dbf`)
- Comma-delimited text file (`.csv`)
- Tab-delimited text file (`.txt`)
- Info

ArcGIS Pro will also allow you to export a table to a spreadsheet. You will need to use the Table to Excel Python geoprocessing script found in the **Conversion Tools** toolbox and the Excel toolset. This will allow you to export an entire table or a set of selected records to an Excel spreadsheet for others to use.

Sharing content with those not on your network

Sharing content with those not on your network is a bit more challenging. As you have learned, maps and layouts don't actually store the GIS data they reference. Instead, they point to the original data source. The data source can be a database stored within the project folder structure, or it might be located somewhere else entirely. It is even possible for a project, map, and layout to include multiple layers, all of which reference data in completely different sources.

This is why, it is difficult to share your work with those that are not able to connect to the same resources as you. You can't just send them a map or layer file because neither of those include the underlying data. You could possibly zip the project folders together into a single file, which you could e-mail or put on an FTP site for download. But that only works if you have all the data you used within a project that is stored in the project structure.

If only you could create a file that packaged all the data you referenced in a project, map, or layer. Luckily, with ArcGIS Pro, you can do just that. It has tools that allow you to package the project, map, or layer together with its referenced data, so you can then share it with those not on your network.

Packages are also a great way to archive data. A package creates a snapshot of the data in the state it was in when you created the package. This provides you with a backup of the data at that point, which you can reference later as needed. This also means packages are disconnected from your live data. So, if you make changes to that data, you will need to recreate the package if you want it to include the changes.

Now you will explore the various packages you can create and how to create them. You will start with the smallest type of package, the layer package, and work your way up to the largest, being the project package.

A layer package

A layer package is very similar to a layer file. It too stores all the property settings for a layer so that when it is added to a map those settings are automatically applied. In addition to saving those settings a layer package also includes the data referenced by the layer. This means that a layer package can be shared with someone that does not have access to the source data. They can then add the layer package to a map and see the layer with your settings and the data in the state it was when it was packaged.

Because layer packages only include a single layer, they tend to be much smaller than other packages. This makes them more ideal for e-mailing to others. Creating a layer package is similar to creating a layer file. You select the layer in the map you wish to package. Then, select the **SHARE** tab in the ribbon. Next, you select the **Layer** button in the **Package** group. This will launch the packaging wizard.

Exercise 11C – creating a layer package

The Community Development Director has engaged the services of a consulting firm to help with the preparation of the city's long range comprehensive plan. The consulting firm needs a copy of the city's zoning layer. So, the Director has asked whether you can create a file that he can send to the consultant, which contains the standard city zoning symbology and data.

In this exercise, you will create a layer package of the **Zoning** layer. Then, you will test to make sure that it works.

Step 1 – creating the layer package

In this step, you will create the **Zoning** layer package. You will work through the package wizard. You will provide all the data and input it needs to successfully create the package.

1. Start ArcGIS Pro and open the `Ex11.aprx` project.
2. Select the **Official Zoning Map** tab at the top of the view area to activate the map.
3. Select the **Zoning** layer in the **Contents** tab.
4. Select the **SHARE** tab in the ribbon.
5. Click on the **Layer** package tool in the **Package** group.
6. Select the **Save** package to a file option at the top of the **Package Layer** pane. Note that you can automatically upload a layer package to ArcGIS Online to share with those in or outside your organization.
7. Click on the **Browse** button at the end of the cell to provide a name and a location for the layer package file.
8. Using the tree on the left-hand side, navigate to `C:\Student\IntroArcPro\My Projects` and name the package `Trippville Zoning`. Click on **Save**.
9. Type the following in for the **Summary** and **Tags**:
    - **Summary**: `Showing the zoning classifications of parcels in the City of Trippville.`
    - **Tags**: `Zoning`, `Trippville`, and `Parcels`

The layer package pane should now look like this:

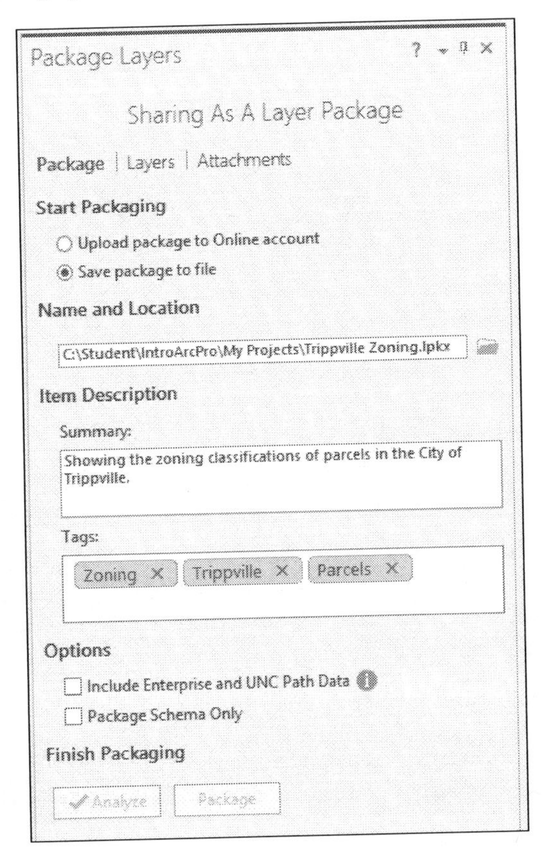

10. Click on the **Analyze** button to see if there will be any problems creating the package.

The analysis of the layer should indicate a single error. The layer description is missing and is required for packaging. Since this is an error and not just a message, it must be fixed before you can proceed with creating the package.

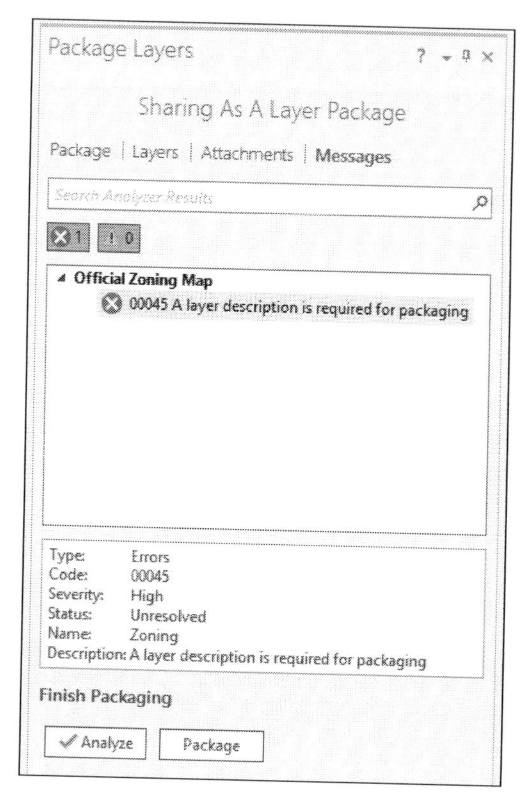

11. Right-click on the error and select **Change Layer Description** from the presented context menu. This will open the **Layer Properties** window for the **Zoning** layer.
12. Add the following description in the **Layer Propertied** window – `Zoning layer for the City of Trippville`. Feel free to update any of the other missing values as well.
13. When you are done updating the layer properties, click on **OK**. The error should now have a green check mark in place of the red and white error icon.
14. Click on the **Analyze** button once again to ensure that there are no issues, which would cause the package creation to fail.
15. If no errors are returned, click on the **Package** button.
16. When the package is successfully created, close the **Package Layers** pane.

Now you need to verify that the package is indeed created successfully.

Step 2 – testing the package

In this step, you will create a new map in your project and add the layer package you created to it.

1. In the **Project** pane, right-click on the **Maps** folder.
2. Select **New Map** from the presented context menu. A new map containing only a basemap should appear.
3. Click on the **MAP** tab in the ribbon.
4. Click on the **Add Data** button in the **Layer** group.
5. In the **Add Data** window, navigate to the `C:\Student\IntroArcPro\My Projects` folder. There, you should see the `Trippville Zoning.lpkx` file you just created.
6. Click on the `Trippville Zoning.lpkx` file and then the **Select** button.

The **Zoning** layer should now be visible in your new map. You will now verify the source of the **Zoning** layer.

7. Click on the **List by Data Source** button in the **Contents** pane.
8. If needed, resize the **Contents** pane, so you can see the entire path to the **Zoning** layer's data source.

> **Question**: What is the path to the **Zoning** layer's data source?
>
> ____
>
> ____
>
> **Question**: Is this different than the data source used by the **Zoning** layer you used to create the layer package?
>
> ____
>
> ____

9. Save your project and close ArcGIS Pro.

Congratulations, you just created and tested your first package. You now have the ability to share your layer data and settings with others. Now you will investigate a map package.

A map package

As a layer package bundles the data for a layer together with its property settings, a map package will bundle all the data associated with the layers contained in a map plus the property settings for those layers. So, a map package bundles everything you see in a map together into a single file with a `.mapx` extension.

Map packages provide an easy way to share an entire map with others that may not be able to connect to your data sources. One thing to consider though is size. Because a map package will contain all the layers and their referenced data within a single file, it is possible for them to get very large. I have personally created one for a client that was over 20 GB in size. The reason it was so large is the client had pictures stored in the database of the features. In addition, the map included local aerial photography. So, when the map package was created, it included the layers, the layer properties, the attached pictures of the features, and the aerial photographs, which were used in the map.

Of course, that is not the norm and actually represents a worst case scenario for size when creating a map package. However, it does illustrate how map packages can be very large. When creating a map package, it is first best to remove any unneeded layers from the map to keep the map package size to a minimum. Turning off their visibility will not cut it. Invisible layers will still be included in the map package. They must be physically removed from the map.

Creating a map package follows the same basic process as creating a layer package. You activate the 2D or 3D map you wish to package and then proceed to the **SHARE** tab. You then click on the **Map package** tool and work through the wizard.

The map package challenge

Use the skills you learned creating a layer package and the information earlier to create a map package for the official zoning map contained in the `Ex11.aprx` project. Save the new map package to the same location you saved the layer package created in the *Exercise 11C – creating a layer package* section. Then, compare the size of the layer package to the size of the map package you created.

Project packages

Project packages tend to be the biggest package you can create. Project packages bundle everything you include in a project into a single file. This includes maps, layers, data, toolboxes, tasks, attachments, geoprocessing history, and connections. Project packages will have a `.ppkx` file extension.

Because a project package literally includes everything that is part of a project, they tend to be large, certainly, much larger than either a map or layer package. This means that they are not typically well-suited to share via e-mail. However, it's possible to publish them to an FTP site for download, save to an external device, or even publish to ArcGIS Online. If you do publish to ArcGIS Online, you will be charged credits to store the package. The amount of credits you are charged will depend on the size. This is also true of layer and map packages, which you may publish to ArcGIS Online.

Project packages are a great way to archive a project once it is complete or as specific milestones are reached. They are a snapshot of the project at the time the package is created. Since it is all stored in a single file that makes them much easier to manage and backup.

Again, project packages are created using the same basic method as map or layer packages. You will be required to provide certain information, such as a summary, description and tags, before the package will be created successfully. Once created, a package can be successfully imported into ArcGIS Pro as a complete project, including the maps, layers, layouts, connections, and so on, that were included when the package was created. However, like the map and layer package, the newly imported project package will point to a copy of the original data sources and not the original data sources themselves.

The project package challenge

Using the skills and knowledge you have gained throughout this book, try creating a project package for one of the projects you have used as you worked through this book. Once created, compare the size of the project package to the size of the map and layer packages you created.

Summary

In this chapter, you learned many different methods, which can be used to share your GIS content with others. You have seen how you can use project, map, and layer files to share and standardize content for those on your computer network that have access to the same data sources you do.

You also learned how you can export your content, so others without ArcGIS Pro can possibly access and use your data. This chapter showed how you can export your GIS data to non-GIS formats, so those without GIS software might be able to use information from your GIS in documents, reports, presentations, and spreadsheets.

Finally, you learned how you can share data with others not on your network using packages. Packages are not limited to just sharing data but can also be used to archive or back up important data at the specific states of existence, so they can be recalled if needed.

A
ArcGIS Pro Glossary

This is a glossary of commonly used terms associated with ArcGIS Pro and GIS. This is by no means a complete compilation of all terms you may encounter.

Annotation: Annotation is one option in ArcGIS in order to store text to place on your maps. With annotation, each piece of text stores its own position, text string, and display properties. Annotation is managed individually unlike labels, which are managed as a group. This means a user can change the properties of single annotation features, including color, font, size, bold, underlines, or italic.

ArcGIS: This is Esri's GIS platform that includes Desktop, Enterprise, Data, Mobile, and Web solutions. This can be deployed singularly or as an integrated solution. The ArcGIS Platform includes ArcGIS for Desktop (ArcMap and ArcCatalog), ArcGIS Pro, ArcGIS Server, ArcGIS Online, ArcPAD, ArcGIS Mobile for Windows, ArcGIS for Android and iOS, and so on.

ArcGIS for Desktop: This is Esri's primary Desktop GIS solution. It includes two applications—ArcMap and ArcCatalog. This is used to manage, maintain, visualize, and analyze data.

ArcGIS for Server: This is Esri's 64 bit GIS Server-side enterprise solution. It includes functionality to serve GIS resources via the Web or local area network, which can be accessed using web, mobile, or desktop applications. It also allows organizations to have a multiple user GIS database and includes several components, such as Web Adaptor, **Spatial Database Engine** (**SDE**), and Portal for ArcGIS.

ArcGIS Online: This is Esri's cloud solution to share maps, data, services, and applications with GIS users, elected officials, citizens, staff members, or the general public, whether or not they have GIS software.

ArcGIS Pro: This is Esri's newest Desktop GIS application. This 64-bit application uses a modern ribbon interface to manage, maintain, and analyze data.

ArcCatalog: This is an application included within ArcGIS for Desktop. It is used primarily to manage data. It is similar in concept to File Explore found in Microsoft Windows.

ArcMap: This is an application included within ArcGIS for Desktop. It is primarily used to visualize, edit, and analyze data.

Attribute: This is a specific characteristic (value) of a feature, such as a name, date, size, or material, that can be edited, deleted, and have operations performed on. Normally, it is associated and stored within an attribute table, which is automatically linked to a feature class in GIS.

Attribute Table: This is a database table that is linked to a collection of features in a feature class. Attribute Table stores specific information associated directly with the spatial features stored in a feature class. It can normally be identified by having a `Shape`, `Shape_Length`, and/or a `Shape_Area` field depending on data format and geometry type (point, line, or polygon).

Basemap: A layer or groups of layers that provide contextual information for primary layers in a map or a scene. Basemaps often include aerial photography, roads, railroads, political boundaries, natural water features, and points of interest. Basemaps cannot be edited or queried. However, they do redraw faster than normal operational layers and therefore can increase map performance. Esri provides several pre-canned basemaps via ArcGIS Online or you can create your own.

CAD: **Computer-Aided Design** (**CAD**) is used by engineers, land surveyors, and other design professionals to create the plans and drawings for their projects. It can be **2D** and **3D**. **DXF**, **DWG**, and **DGN** are common CAD formats.

Cardinality: This is a database term that refers to how many records in one table match to records in another table. There are four types of cardinality: one to one, one to many, many to one, and many to many. In ArcGIS, this will determine when you should use a join or a relate to link two tables.

Concatenate: This is to join two or more items. In GIS, this usually refers to joining the data found in two or more attribute fields together into a single field.

Contents pane: This displays a list of all layers or tables included within a map or scene. This also allows users to control layers and access functionality associated with them. It is similar to the **Table of Contents** in ArcMap.

Contextual tab: This tab contains tools related to a specific function. These tabs appear and disappear depending on what the user has selected with ArcGIS Pro.

Coverage: This is Esri's native data format for their older product, **ArcInfo Workstation**. It uses multiple folders and files to store both the spatial (location) and attribute (descriptive) data for geographic features. A single coverage can include multiple feature classes.

Data view: This is one of the two views within ArcMap, an application included with ArcGIS for Desktop. This is similar to a Map in ArcGIS Pro. It contains layers in the **Table of Contents**, which share a related theme or map purpose.

Datum: This is part of a coordinate system. Datums are used to tie the ellipsoid back to the true earth. There are two types of datums, horizontal and vertical. Horizontal datums can be local or earth centered. Common horizontal datums used in North America include NAD 27, NAD 83, and WGS 84.

Feature: This is any item contained in your spatial data (that is, a fire hydrant, a manhole, a parcel, a waterline, a building, and so on)

Feature class: The general meaning of this refers to a collection of features that share a common geometry (point, line, or polygon), attribute table, and spatial reference (coordinate system, datum, projection). It is normally associated with Esri's Geodatabase format but can be applied to other formats as well, including Shapefiles, which store a single feature class or CAD and coverages, which store multiple feature classes. In a Geodatabase, these can be standalone or grouped in a feature dataset.

Feature dataset: This is a collection of feature classes stored together in a geodatabase that share the same spatial reference; that is, they share a coordinate system, and their features fall within a common geographic area. Feature classes with different geometry types (points, lines, and polygons) may be stored together in a single feature dataset. Feature datasets are required for the use of geodatabase topologies and geometric networks.

Fields: These are the columns in a database table and are used to store values associated with records. These must have unique names that do not contain special characters with the exception of an underscore (_). Fields must also be assigned a data type that determines what type of information can be stored within the field. Common field types include `Text`, `Date`, `Long Integer`, `Short Integer`, `Float`, and `Double`.

Geodatabase: This is the native data format for ArcGIS. There are three basic types: personal, file, and SDE. SDE Geodatabases come in three types personal, workgroup and enterprise. Workgroup and enterprise geodatabases require ArcGIS for Server. Geodatabases store various types of geographic datasets including feature classes, tables, raster datasets, network datasets, topologies, address locators, custom toolboxes, and many others.

Geoprocessing: This means the manipulation of data within ArcGIS. Usually, it is associated with the use of geoprocessing tools. The process of converting, managing, and analyzing data is considered as geoprocessing.

Geoprocessing tools: These are tools that allow users to manipulate data within the ArcGIS Platform. Tool availability is determined by licensing levels and available extensions. They can be accessed in many different ways, including toolboxes, Python, ModelBuilder, and custom applications.

Geocoding: This is the act of turning an address into a point location. It requires the following three components: address or list of addresses, reference data (GIS data with address information), and a locator, which translates between the other two components.

Integer: This is a whole number; that is, it has no fractions or decimal values.

Join: This is one of the two ways to link two tables together in ArcMap or ArcGIS Pro. This requires a cardinality of one to one or many to one. It creates a virtual link between the two tables so that within your map document or ArcGIS Pro project, the two tables appear as one. You may then use the data from both tables to query, label, and symbolize features.

Labels: Labeling is an easy way to add descriptive text to features on your map. Labels are dynamically placed, and label text strings are based on feature attributes. You can turn labels on or off as a group. They can also be locked, so their locations stay fixed as you zoom or pan on your map.

Layer: This is any spatial or tabular data displayed or accessed within a map or scene. It is located in the **Contents** pane in ArcGIS Pro or the **Table of Contents** in ArcMap. Layers have properties, including **Name**, **Symbology**, **Label Settings**, **Display settings**, and **Source Location**. Layers do not store data. They point to or reference data stored as a Geodatabase, Shapefile, CAD file, coverage, or raster format.

Layer file: This is an external file that stores layer property settings, such as symbology, data source, display settings, and label settings, so that it may be easily used in other maps and by other users that have a connection to your GIS data. This has a `.lyrx` or `.lyr` file extension and is used to standardize layer settings within an organization.

Layer package: This is a file that contains a layer's property settings (symbology, labels, definition queries, field visibility, display, and so on) along with a copy of the referenced data. This is used to share layer information with those who do not have access to your data. It has either an `.lpk` or `.lpkx` file extension.

Layout or layout view: This is a virtual page where you design your final output map product. They can include maps, tables, graphics, legend, map scale, text, and north arrow.

Map: This is a collection of related layers within an ArcGIS Pro Project. It is similar to a data frame in an ArcMap map document.

Map document: This is a collection-related data frames, layers, and a layout used by ArcGIS for Desktop's ArcMap application. It is similar to a project file used by ArcGIS Pro though more limited. It has an `.mxd` file extension.

Map package: This is a file that contains layer settings, layout, and referenced data used by the original map document or map. It is used to share maps with those that cannot access your data. This has an `.mpk` or `.mpkx` file extension.

ModelBuilder: This is a visual programming interface found in ArcGIS for Desktop and ArcGIS Pro. It allows users to create tools to automate and standardize processes without having to know how to program.

NAD 27: This is North American Datum 1927. It is a local datum, which is located in Meades Ranch, Kansas. That is the approximate center of the continental United States.

NAD 83: This is North American Datum 1983. This is an Earth-centered datum. There is more than one version of this datum. It has been updated and adjusted many times.

Pane: This is a dockable window in the ArcGIS Pro interface, which contains information and access to tools or tool parameters. Commonly used panes include **Contents, Geoprocessing**, and **Project**.

Portal: This is a connection to ArcGIS Online or Portal for ArcGIS. It is used by ArcGIS Pro to manage licenses. Also, it provides an access to shared content stored in an organization's ArcGIS Online account or Portal for ArcGIS.

Project: This is used by ArcGIS Pro. Projects contain a collection of GIS resources used for a specific project, map, or analysis. They have an `.aprx` file extension.

Project: This is the act of converting data from one coordinate system to another. This can be done on the fly by ArcGIS or done permanently using the project geoprocessing tool.

Project package: This is a file that contains all items included in a single project including the referenced data. It is used to share projects with those that do not have access to your data or to archive projects at specific points in a project cycle. Since these contain everything found in a project, they can get very large. It has a `.ppkx` file extension.

Projection: This is the representation of the Earth's curved 3D surface on a 2-dimensional surface (flat map). There are many types of projections. Each is designed to reduce distortion, which can be caused in four different properties: **Shape**, **Area**, **Distance**, and **Direction**. The projection is part of a projected coordinate system's definition.

Python: This is the primary scripting language for ArcGIS. Python is a programming language that allows you to create the script to help automate and schedule processes. Python is not only limited to ArcGIS but can also be used with many other applications.

Query: Basically, this is a question used to select features that have specific attribute values or relationships. Within ArcGIS, there are two basic types of selection queries: **Select by Attribute** and **Select by Location**. **Select By Attribute** will select features in a layer based on value criteria the user enters, such as `Parcel Owner Name = John Smith` or `Pipe Size > 6`. **Select By Location** sometimes refers to a spatial query, selects features in one or more layers based on their spatial relationship, such as all parcels within the city limits or all roads in a distance of 100 feet from a hospital.

Raster: This is a simple storage model for spatial data. It stores information using equal-sized cells. Each cell is assigned a number. This number can represent things such as counts, color, elevation, temperature, wind speed, average rainfall, population density, and so on. A raster is often associated with, but not limited to, **aerial photography**, **Digital Elevation Models** (**DEMs**), **Land Use Classification**, and **Vegetation Classification**.

Relate: This is one of the two ways to link two tables together in ArcMap or ArcGIS Pro. It can be used with any cardinality but recommended when you have a one to many or many to many cardinality. It links the two tables together so that when you select records in the primary table, the related records in the secondary table are highlighted. You cannot symbolize, query or label using data found in the secondary table when a relate is used.

Ribbon: This is the primary user interface used by ArcGIS Pro. It consists of a rectangular area located at the top of the interface, which contains various tabs. Each tab contains tools that access ArcGIS Pro functionality. Which tabs are visible will depend on the user's current actions and what is highlighted or selected within the interface.

Scene: This is a collection of related layer that can be viewed in 3D. These are stored within ArcGIS Pro Projects. They can also be created and viewed with ArcGlobe and ArcScene, which are part of the 3D Analyst extension for ArcGIS for Desktop. In this case, they are actually separate files with either a `.3dd` or `.sxd` file extension.

Shapefile: Shapefiles are the native format for Esri's ArcView 3.x and earlier. This format has become the *de facto* data transfer format for GIS. Many GIS and GPS packages have the ability to read, import, and export Shapefiles. A Shapefile stores a single feature class (that is, point, line, or polygon). Viewed as a single file by ArcGIS software, it is actually made up of multiple files (SHP, SHX, DBF, and others). Shapefiles are one of the two data formats that are editable within ArcGIS.

Spatial query: This is a query that uses the spatial relationship between features in one or more layers to select data. In ArcGIS, this is accomplished with **Select By Location**. This allows users to select features in one or more layers based on their spatial relationship, such as all parcels within the city limits or all roads at a distance of 100 feet from a hospital.

Standalone or nonspatial table: This is a database table that is used in ArcGIS but is not an attribute table. It may be joined or related to a layer to provide additional information about features. Normally, it does not contain `shape`, `shape_length`, or `shape_area` fields.

String: This is another word for text. It is used by many databases as a field type. It can also be used in various expressions in ArcGIS.

Symbology: This includes the use of color, line types, fill patterns, thickness, and styles to differentiate features on one or more layers. Symbology options will depend on whether the features are points, lines, or polygons.

Tab: This is a collection of related tools located within the ArcGIS Pro ribbon.

Table: This is a collection of data that is stored in rows and columns. Rows within a table are called records, and columns are called fields. Each field must have a unique name.

Task: This is a collection of steps needed to complete a process or workflow. It is used to standardize workflows, define best practices, and help train new users. It is stored within a Task Item in a project or can be shared as a task file with an `esriTasks` file extension.

Task Item: This is an organizational unit in an ArcGIS Pro project. This is used to store related tasks and Task Groups.

Topology: This generally means a mathematical model of how spatial features are related to one another. Do they connect? Are they next to each other? Do they overlap? In ArcGIS, it is a part of the geodatabase that defines rules on how features in one or more feature classes must relate to one another. A geodatabase topology can only be created within a feature dataset, and only the feature classes with that dataset can participate in the topology.

Vector: This is a simple storage model for spatial data. It stores information using specific coordinates as points, lines, or polygons.

WGS 84: This is both a datum and a geographic coordinate system. When used as a datum, it is an Earth-centered datum. As a geographic coordinate system, it uses degrees as its units. The United States GPS uses WGS 84 as its native coordinate system.

B
Chapter Questions and Answers

Chapter 1 – Introducing ArcGIS Pro

- What happens when you use the Explore tool within the views?

  Both views pan and zoom together.

- What happens when you click on a parcel in one of the map views?

  An informational pop-up window appears containing information about the parcel clicked on.

Chapter 2 – Using ArcGIS Pro – Navigating through the Interface

- What happens when you click on a parcel using the Explore tool?

  A pop-up window appears that contains information about the parcel clicked on.

- What happens when you click on one of the other values in the drop-down list?

  Information about that feature is now displayed in the popup.

- What happens when you roll the scroll wheel away from you?

  The map zooms in. (Unless you have adjusted your options for ArcGIS Pro.)

- What happens when you select the new symbol?

  The symbol used to visualize the building layer changes to the gray building footprint symbol, which was selected.

Chapter 3 – Creating and Working with ArcGIS Pro Projects

- What do you see listed under maps?

  A single map called Map.

- What do you see listed under database?

  ArcGIS Pro Exercise 3A.

- Does the map contain all layers which reference all the feature classes in the geodatabase?

  No it does not.

- Does the project geodatabase contain feature classes that represent the floodplains or drainage basins?

  No it does not.

- What other items are currently available in this project?

  A project toolbox, multiple styles, connection to the Chapter3 folder, and Esri World Geocoder.

- What groups do you see?

  Answers to this question will vary depending on the groups the user is a part of in ArcGIS Online or Portal for ArcGIS.

- What is available or listed under Maps?

  Scene.

- What is available or listed under toolboxes?

  %your name% 3B.

- What is available or listed under databases?

  %your name% 3B.gdb.

Chapter 4 – Creating 2D Maps

- What databases are currently connected to this project?

  %your name% Ex4A, and Map.

- Which geodatabase is the default geodatabase?

  %your name% Ex4A.

- What folders are available in this project?

  %your name% Ex4 and Chapter4.

- What is the field name or alias that contains the size of the sewer pipes?

  Pipe size or size.

Chapter 5 – Creating 3D Maps

- What is the name of the currently assigned ground surface?

  WorldElevation3D/Terrain3D.

- What is the location of the currently assigned ground surface?

  http://elevation3d.arcgis.com/arcgis/services/WorldElevation3D/Terrain3D/ImageServer.

Chapter 6 – Creating a Layout

- Purpose of Map 1:

  Highlight specific attribute values associated with features.

- Purpose of Map 2:

  Show location of features in the sewer system.

- Purpose of Map 3:

  Show spatial relationships between parcels and wetlands.

- If you know your audience might include someone who is color blind, what can you do or change so that they could successfully use your map as well?

  For line layers, you could use different patterns to represent different values or layers. For polygons, you could use different fill patterns, and for points, different marker symbols.

- You are preparing a map of your water system, which will be given to the field crews to help them locate the system in the field. The maps will be stored in their trucks and used in all kinds of weather. How might this impact your design?

  Since this map is being used in the field, it might be better to keep the size limited. This may mean needing to produce a series of maps that show different parts of the system instead of using a single map. Since these are water system workers, who will be performing maintenance and repairs, they will need a higher level of detail than others. Scale will also be important since they will be using these maps to help them locate features while in the field. Given the all-weather use of these maps, they may need to be laminated. Lamination can also cause colors to fade, so a more intense color palette might be a good idea as well.

- You are preparing the official zoning map for a city. This will be the legal zoning map as required by the city's zoning ordinance and will be hanging in City Hall for city officials and the citizens to use. What factors should you consider that might impact your design?

  Since this will be hanging in City Hall and used by multiple people, it should probably be designed as a large map. This would allow multiple people to view it at all times. Also given the number and variety of people who may view the map, it might be good to use a combination of patterns and colors to identify the zoning classifications. Being the legal official zoning map for the City, you would want to review the zoning ordinance to see if it spells out any specific requirements for the official zoning map, such as a required citation, signature lines, seals, and dates. You may also want to print this map on high-quality paper to ensure that the map is able to be printed at the best possible quality and will hold up over time.

- What is your current scale?

  The answer will vary but should be somewhere between 1:4600 and 1:5000.

Chapter 7 – Editing Spatial and Tabular Data

- What geodatabase is being referenced by the layers and where is it located?
  Trippville_GIS.gdb located in C:\Student\IntroArcPro\Databases.

- What coordinate system is the parcels layer in?
  NAD 1983 StatePlane Georgia West FIPS 1002 Feet.

- Are all your layers within the same coordinate system?
  Yes.

- What snapping position options are enabled?

  Answers will vary depending on what may have been set and used in previous sessions.

- What templates are available?

  Manhole, Parcels, RW, 8 inch Ductile Iron, 8 inch PVC, and 10 inch Clay.

- What are some of the properties associated with the template?

  Name, Description, Tags, Target Layer, Drawing Symbol, Tools, and Attributes.

- What values are assigned to those fields for this template and where do you think they came from?

  Pipe Size = 8 and Material = PVC. These values are coming from the symbology settings for that layer as shown in the Contents pane.

- What attribute fields are associated with the Street_Centerline feature class and what field types are they?

  PREFIXDIR – Text, SUFFIXDIR – Text, Shape_Length – Double, RD_Class – Text, QUAD – Text, ST_NAME – Text, Shape – Geometry, ObjectID – ObjectID.

Chapter 8 – Geoprocessing

- What version of ArcGIS Pro are you using?

  Answers will vary depending on the user's installed version. At the time the book was written, ArcGIS Pro 1.1.0 was the current version.

- What license level of ArcGIS Pro do you have?

  Answers will vary depending on what has been assigned to the user.

- What extensions if any can you use?

  Answers will vary depending on the user.

- What geoprocessing tool that you have read about in this chapter do you think you should use to create a layer that only contains the streets inside the city limits?

  The Clip tool. This will create a new layer that only contains the streets that are within the City limits.

- What field identifies what road each segment belongs to?

  ST_NAME.

- After sorting the records in the table, what do you notice about the number of segments for each road?

  There are multiple segments for each road. Many have over 10 different segments, which will make calculating the total length more difficult.

- How many records are there with the same road name?

  There is only one record for each road name.

Chapter 9 – Creating and Using Tasks

- What Task Items do you see included in this project?

  Selecting features and editing parcels.

- How many tasks are included in the Task Item you have opened and what are they?

  Three (3). Selecting features in the Map, Selecting features by attributes, and Selecting features based on location to other features.

- How many steps did this task have?

  There were two active steps, which required user interaction.

- How many steps do you see in this task?

  Four (4)

- How does this compare to the number of steps you counted when you ran the task in the last exercise?

  There are more steps.

- Why do you think the number of steps you counted in Exercise 9A differs from the actual number of steps contained in this task?

  Some of the steps in this task were set to run automatically and were hidden from the user, so they were not aware that they existed.

Chapter 10 – Automating Processes with ModelBuilder and Python

- How has the graphics for the Buffer tool and its associated variables changed?

  A drop shadow has been added to the yellow rounded square for the Buffer tool and the ovals for the variables.

- Did the model try to rerun the Buffer tool?

  No it did not.

- What tool or tools did the model run when you clicked on the Run button and why?

  The model only ran the Union tool with its associated variables because it was the only process in the ready-to-run state. The Buffer tool was in the Has been Run state, so it was not run again.

- What happens to all the processes in the model which were in the has been Run state?

  They have all been reset to the Ready-to-Run state.

- Which feature class is now in the Ex10 geodatabase and how does that compare to when you ran the model from inside ModelBuilder?

  There is only one feature class in the geodatabase: Parcels_StreamBuff_Union. When the model was run from ModelBuilder, it produced two feature classes not just one.

Chapter 11 – Sharing Your Work

- What is the path to the zoning layers data source?

  C:\Users\%Your User Name%\Documents\ArcGIS\Packages\Trippville Zoning_CCA6670E-7141-4BCA-8D83-6617F5EDB909\p11\trippville_gis.gdb. (Your actual path may be slightly different depending on your OS and what other packages you may have opened in the past.)

- Is this different than the data source used by the Zoning layer you used to create the layer package?

  Yes, the layer used to create the layer package references the Zoning feature class in the Trippville_GIS geodatabase located in the Databases folder included in the training data for the book. The Zoning layer that was added using the package.

Index

Symbols

A

B

C

D

E

F

G

I

J

K

L

Q

R

S

T

U

V

W

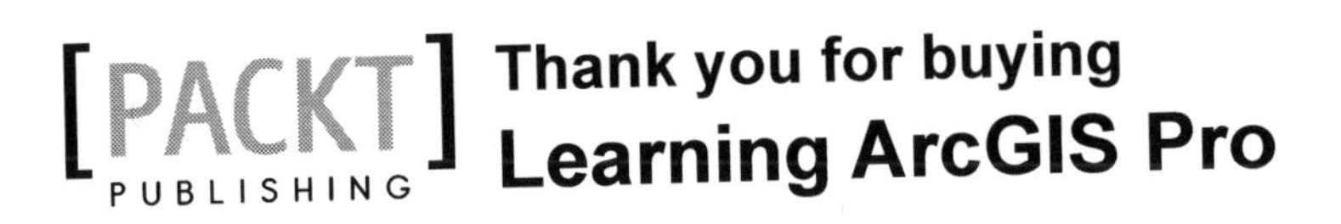

About Packt Publishing

Packt, pronounced 'packed', published its first book, *Mastering phpMyAdmin for Effective MySQL Management*, in April 2004, and subsequently continued to specialize in publishing highly focused books on specific technologies and solutions.

Our books and publications share the experiences of your fellow IT professionals in adapting and customizing today's systems, applications, and frameworks. Our solution-based books give you the knowledge and power to customize the software and technologies you're using to get the job done. Packt books are more specific and less general than the IT books you have seen in the past. Our unique business model allows us to bring you more focused information, giving you more of what you need to know, and less of what you don't.

Packt is a modern yet unique publishing company that focuses on producing quality, cutting-edge books for communities of developers, administrators, and newbies alike. For more information, please visit our website at `www.packtpub.com`.

Writing for Packt

We welcome all inquiries from people who are interested in authoring. Book proposals should be sent to `author@packtpub.com`. If your book idea is still at an early stage and you would like to discuss it first before writing a formal book proposal, then please contact us; one of our commissioning editors will get in touch with you.

We're not just looking for published authors; if you have strong technical skills but no writing experience, our experienced editors can help you develop a writing career, or simply get some additional reward for your expertise.

[PACKT] PUBLISHING

Programming ArcGIS 10.1 with Python Cookbook

ISBN: 978-1-84969-444-5 Paperback: 304 pages

Over 75 recipes to help you automate geoprocessing tasks, create soultions, and solve problems for ArcGIS with Python

1. Learn how to create geoprocessing scripts with ArcPy.
2. Customize and modify ArcGIS with Python.
3. Create time-saving tools and scripts for ArcGIS.

Mastering ArcGIS Server Development with JavaScript

ISBN: 978-1-78439-645-9 Paperback: 366 pages

Transform maps and raw data into full-fledged web mapping applications using the power of the ArcGIS JavaScript API and JavaScript libraries

1. Create and share modern map applications for desktops, tablets, and mobile browsers.
2. Present and edit geographic and related data through maps, charts, graphs, and more.
3. Learn the tools, tips, and tricks made available through the API and related libraries with examples of real-world applications.

Please check **www.PacktPub.com** for information on our titles